# FRONTIERS IN POLYMER RESEARCH

# FRONTIERS IN POLYMER RESEARCH

**ROBERT K. BREGG**
**EDITOR**

**Nova Science Publishers, Inc.**
*New York*

For permission to use material from this book please contact us:
Telephone 631-231-7269; Fax 631-231-8175
Web Site: http://www.novapublishers.com



**Library of Congress Cataloging-in-Publication Data**

Frontiers in polymer research / Robert K. Bregg (editor).
p. cm.
Includes bibliographical references and index.
ISBN 1-59454-824-2
1. Polymers--Research. 2. Polymerization--Research. I. Bregg, Robert K.
QD381.F76 2006
547'.7072--dc22 2005030493

*Published by Nova Science Publishers, Inc. ✣ New York*

# CONTENTS

# PREFACE

Polymers are substances containing a large number of structural units joined by the same type of linkage. These substances often form into a chain-like structure. Starch, cellulose, and rubber all possess polymeric properties. Today, the polymer industry has grown to be larger than the aluminum, copper and steel industries combined. Polymers already have a range of applications that far exceeds that of any other class of material available to man. Current applications extend from adhesives, coatings, foams, and packaging materials to textile and industrial fibers, elastomers, and structural plastics. Polymers are also used for most composites, electronic devices, biomedical devices, optical devices, and precursors for many newly developed high-tech ceramics. This new book presents leading-edge research in this rapidly-changing and evolving field.

As described in Chapter 1, recently, biocomposites made from natural fiber and biobased polymers are receiving more attention particularly because of their structural properties and their environmentally friendly nature. Glucose based polymers are considered as a biobased polymer with potential application for use in biocomposites. The curing characteristics and water absorption behavior of the glucose based polymer and epoxy resin are investigated for the purpose of utilization as a matrix in biocomposites. A cured matrix containing epoxy resin and 50wt% of glucose based polymer was prepared and characterized. Exothermic reactions attributed to esterification and etherification reactions between the hydroxyl and carboxyl functionalities of the glucose polymer with epoxy groups were identified as taking place during the curing process. Exothermic reactions were differentiated according to the degree of carboxyl group substituent of the glucose based polymer. The results showed that the esterification reaction occurs in the early stage of cure and then etherification followed after completion of the esterification. The cured matrix showed thermal stability up to 300°C. The average glass transition temperature and storage modulus of the matrix were as high as 95°C and 2700 MPa, respectively. The cured matrix of epoxy resin and glucose based polymer with higher carboxyl group content was found to have a lower density due to the formation of bulky groups in the crosslinks. The matrix cured at higher temperature shows compact crosslinks due to the higher concentration of ether bonds comparing to the matrix cured at the lower temperature. The effects of water absorption on the hydrophilic polymer matrices with glucose based polymer and epoxy resins were also studied as a function of curing temperature. The polymer matrices cured at different temperatures were immersed in water at room temperature for 1000 hours and the thermomechanical properties of the cured polymers were characterized using DMA and TGA. Two types of sorbed water were identified. Type I sorbed water contributed mainly to increasing the weight and to a decrease in $T_g$ due to a plasticization effect. Type II sorbed water was not removed after heating the polymer to 110°C for an hour. The strongly sorbed Type II water causes changes in the mechanical properties of the polymers cured at different temperatures depending on the crosslinking of the matrix. The matrix cured at the higher temperature has a comparatively tightly crosslinked network structure which exhibited microcracking as a degradation mechanism due to the trapping of the sorbed water within the polymer.

As presented in Chapter 2, nitrites are highly harmful compounds. They are extremely undesirable in surface and municipal water. Its permissible content in natural water is very low and should not exceed 0.01 mg $NO_2^- / L$. A redox copolymer, a macromolecular analogue of Dichloramine T (i.e. a macroporous S/DVB copolymer containing $SO_2NCl_2$ groups) was used here for removal of nitrite ions from aqueous solutions by its oxidation to the hundred folds less toxic nitrates. The resin was prepared starting from *Amberlyst 15* by a three-step transformation of the sulfonic- via chlorosulfonyl and sulfonamide- to the N,N–dichlorosulfonamide groups. The resulting copolymer contained 8.2 meq of active chlorine/g and showed strong oxidizing properties. It was employed in batch and flow processes for treatment of $NaNO_2$ solutions containing 115, 230 or 460 mg $NO_2^- / L$. The effects of various parameters on the reaction course have been studied (mole ratio of reagents, pH of the reaction media, flow rate in the column processes). The solid phase oxidation carried out in a dynamic regime provided to drive the reaction to completion. Thus, nitrite free effluents (< 3.0 μg/L) were obtained in the column processes. The reaction of nitrite oxidation by means of this heterogeneous oxidant was fast and therefore the permissible flow rate was very satisfactory - close to 20-25 bed volumes/h. Under the examined reaction conditions, 1 mol of $NO_2^-$ ions was oxidized by 1 mol of active chlorine, so the oxidation capacity of the resin was nearly 200 mg $NO_2^- / g$ of the copolymer. The N,N–dichlorosulfonamide copolymer is very useful for purification of neutral or medium acidic solutions from nitrites. In the case of alkaline solutions the nitrite oxidation reaction proceeds slowly, and what more o blocking of a part of the active chlorine atoms in the copolymer takes place. The intermediate $SO_2NClNa$ groups do not oxidize the nitrite ions. The here determined copolymers redox potentials and carried out complex redox titration measurements proved why the macromolecular oxidant shows various reactivity in dependence of the solution pH.

Aromatic polyimide(PI)s are heterocyclic polymers and have excellent thermal stability, good chemical resistance, electrical and mechanical properties. Most of these PIs are insoluble in common organic solvents. Fluorination of PIs is one of the many approaches to overcome the difficulty in the processing of these materials. Owing to the easy availability of 2,2-bis(3,4-dicarboxyphenyl) hexafluoropropane dianhydride (6FDA) as a commercial sample and the good properties that it imparted to the resulting polymer such as good solubility, decreased dielectric constant, increased thermal and hydrolytic stability, 6FDA based fluorinated PIs are extensively studied and used in various high technology applications.

Chapter 3 reviews the work done on 6FDA based fluorinated PIs with respect to its synthesis and various copolymers, polymerisation methods, poly(ether-imide), photosensitive polyimide, hyperbranched polyimide, addition polyimide, poly(amide-imide), poly(urethane-imide), poly(epoxy-imide), poly(ester-imide), poly(siloxane-imide), nanocomposites and nonlinear optical polyimides. Finally, its application in electronics and use as a material for gas separation and corrosion protection are discussed.

Chemical separation with functional polymers is a virtually important part of technologies producing different chemical and biochemical substances by cultivation of yeast, bacteria, or fungus. Each cultivation mixture is extremely complex and the product extraction/purification could be the most costly step. Chromatographic techniques (low-pressure liquid chromatography) can be considered as the main option to fulfil the extraction/purification task. The sorbents used in such processes differ from the materials used in analytical chromatography due to demands on the product quantity/purity rather than quality of the analytical signal. Functional polymers are highly advantageous for such separations. Even more, a careful selection of the polymer and operating conditions could allow replacing the costly chromatographic separation by more economically and environment-friendly processes based on selective sorption and stripping interactions.

Chapter 4 describes applications of functional polymers for separations of bio-cultivated substances combining primary data with review of previously published works. An attention is paid to relationships between properties of the selected polymer, target product(s), and contaminants. Exploitation of these relationships for benefits of the separation efficiency is described. Specific phenomena and interactions taking place in phase of the polymer are discussed as well as effect of these phenomena on the separation processes. This includes specific interactions with functional groups and three-dimensional polymeric networks, transformations of substances in the polymer phase, dimerization, ion exchange isothermal supersaturation, etc.

A special section discusses changes taking place in phase of the functional polymers at continuous industrial use. This includes phenomena of semi-reversible and irreversible sorption, chemical and physical deterioration, aging, etc.

The systems selected to serve as major examples include amino acids and different ion exchange resins. This selection was done due to representativity of amino acids as an example of biochemical substances (labile charge, ability to form zwitterions that is almost specific for bio-products, tendency to form associates, etc.) while structures of these materials are well-known. Functional polymers and, particularly, ion exchange resins are materials of first choice for sorption-based separation of such substances.

Cationic photopolymerization of vinyl ether, epoxy and oxetane systems has been reported. A structure-properties relationship by varying different additives in the cationic photocurable formulations has been investigated. The presence of hydroxyl containing compounds in the photocurable formulation induced an increase on rate of polymerization and final conversion. The use of fluorine containing monomers as additives to UV curable systems allowed the modification of the surface properties of the UV-cured films obtaining a high hydrophobic surface while the bulk properties remained unchanged. A wide range of hyperbranched polymers (HBP) was also investigated in Chapter 5, as additives in cationic photopolymerization of epoxy systems. The HBP were inserted into the polymeric network either by a copolymerization or through a chain transfer reaction involving the hydroxyl groups. Notwithstanding the low commercial availability of photocurable monomers, it has been shown that is possible to modulate and tailor the final properties of the UV-cured films by varying properly the additive in the system controlling the mechanism of the curing process.

Wines, particularly red wines contain numerous biologically active compounds, the most important of which are polyphenols, whose nutritional importance is attributed to their antioxidant power.

The research in Chapter 6 was carried out to evaluate the phenolic capacity of two red wines produced in Turkey. Analysis was performed by using enzyme electrodes constructed by the immobilization of tyrosinase in conducting copolymers.

Immobilization matrices were synthesized by copolymerization of terephthalic acid bis-(2-thiophen-3-yl ethyl) ester (TATE) with pyrrole. Immobilization of enzyme was performed via entrapment in conducting copolymers using electrochemical polymerization of pyrrole.

Measurements were performed by using Besthorn's Hydrazone method which includes spectrophotometric analysis of quinones produced by enzyme. Enzyme electrodes were characterized in terms of maximum reaction rate ($V_{max}$) and Michaelis-Menten constant ($K_m$). In addition to kinetic parameters, stability of enzyme electrodes towards environmental conditions such as pH and temperature was investigated. Usage stability and shelf-life analysis were also examined.

It is known from previous studies that free enzyme could not be used in phenolic determination studies in wines because of inhibitory effects of various substances naturally found in red wines. To understand the behavior of immobilized tyrosinase toward the inhibition, benzoic acid was used as the inhibitor and inhibition constant $K_i$ was determined.

Due to their technological importance, polymer blends have attracted considerable attention during the past decade. For thermodynamic reasons, most polymer pairs are immiscible and their degree of compatibility is of underlying importance to the microphase structure and consequently, to the mechanical properties of the blend. The Flory-Huggins $\chi$ interaction parameter for the polymer pair plays a dominant role in explaining critical phase behavior of a compatible pair and in estimating interfacial tension and interfacial thickness for semicompatible or incompatible pairs. Direct measurement of this parameter is not always possible, thus the obtained information, in conjunction with suitable theoretical models of polymer solutions may lead to an assessment of the interaction parameters for the actual polymeric case.

In Chapter 7 we present a theoretical discussion regarding this interaction parameter for 10 polymer-polymer-solvent systems, 4 copolymer-solvent systems along with their corresponding polymer pairs. Our polymer blends are real mixtures of 5 homopolymers consist of poly(N,N-dimethyl methacrylamide) (PDMAA), poly(2-dimethyl aminoethyl methacrylate) (PDMAEMA), poly(acrylic acid) (PAA), a typical membrane of commercial soft-contact lens i.e. poly(2-hydroxyethyl methacrylate) (PHEMA), and poly(N-vinyl-2-pyrrolidone) (PVP) all with water solvent. Copolymers studied are poly(acrylonitrile-co-butadiene) in acetonitrile, poly(styrene-co-acrylonitrile) in 1,2-dichloroethane, poly(acrylonitrile-co-butadiene) in hexane and poly(acrylonitrile-co-butadiene) in pentane.

For ternary systems, the results are expressed in terms of $\chi_{1,23}$ which reduces to the classical Flory-Huggins $\chi_{12}$ interaction parameter for the case of binary mixtures. The data on $\chi_{1,23}$ may be used for an approximate estimation of the $\chi'_{23}$ interaction parameter for the limiting case of zero solvent concentration. For this purpose, at the end of each subsection of the tables, the limiting value of $\chi^{\infty}_{1,23}$ is given. The limiting values of $\varphi^{\infty}_{2}$, $\varphi^{\infty}_{3}$ and $\chi'_{23}$ are also appeared at the end of each table. It should be noted that these values are obtained by graphical extrapolation of the data to the zero concentration of solvent.

As outlined in Chapter 8, vinylidene chloride polymers find important application in the barrier plastics packaging industry. These materials display low permeability rates for both oxygen (and other small molecules) and for food aroma and taste constituents. On the one hand, they function to prevent spoilage of packaged food items and, on the other, to prevent the loss of flavor agents that make these items palatable. While these materials have excellent barrier properties they may be processed only with difficulty owing to the propensity to undergo thermally-induced degradative dehydrochlorination. In fact, the homopolymer cannot be processed. Incorporating simple acrylate comonomers into the polymer structure lowers the melt temperature and improves processibility. However, it is insufficient to prevent significant degradation at process temperatures. Incorporation into vinylidene chloride polymers of a series of comonomers which result in the formation of polymer pendant groups with the potential 1.) to react with hydrogen chloride as it is formed (and thus prevent its interaction with the walls of process equipment to form Lewis acids, principally iron(III) chloride, which accelerate the dehydrochlorination reaction) and 2.) to expose phenolic units (which may scavenge chlorine atoms and other radical species) on reaction with hydrogen chloride has been examined as a means of stabilizing these materials.

As presented in Chapter 9, the polyolefins have an established position in packaging and in agriculture as a result of their technological properties, which include water and microbe resistance. However, it is generally accepted that commercial polyolefins for durable goods do not biodegrade rapidly enough in the environment where they accumulate as litter. Their behaviour in the environment can be compared with that of natural rubber (*cis*-poly(*iso*prene)), which, when fabricated to an automobile tyre, is resistant to biodegradation for decades, although it is oxo-biodegradable as produced by nature. This results in both cases from the addition of antioxidants during manufacture. There is convincing evidence to show that the rate-determining step in the biodegradation of hydrocarbon polymers is the rate of peroxidation. This process is accelerated by transition metal ions both thermally and by light so that abiotic peroxidation and oxo-biodegradation lead synergistically to the bioassimilation of polymers in the outdoor environment. Special antioxidants inhibit the formation of low molar mass highly biodegradable oxidation products and hence inhibit bio-degradation during use.

Contaminated mixed plastics wastes from domestic sources present a difficult challenge to traditional recycling techniques. On the other hand, the hydrocarbon portion of mixed domestic wastes can be made oxo-biodegradable by the incorporation of transition metal ions that accelerate both perooxidation and biodegradation. Oxo-biodegradable plastics thus make a realistic contribution to the recovery of value from waste packaging as fertilisers and soil-improvers for agriculture and horticulture.

Standards are essential to ensure the environmental safety of compost. Hydrocarbon plastics do not biodegrade rapidly in compost or in soil and it must be demonstrated that, like of nature's wastes, they do not accumulate in the soil. Standards for biodegradability and compostability of plastics must therefore address, not only the question of non-accumulation of any long-lasting plastics residues in the soil substances but also the safety of any non-degradable residues. These aspects will be discussed in the light of recent scientific studies.

In: Frontiers in Polymer Research
Editor: Robert K. Bregg, pp. 1-26
ISBN: 1-59454-824-2


*Chapter 1*

# CURING CHARACTERISTICS AND WATER ABSORPTION BEHAVIOR OF GLUCOSE BASED POLYMERS

***Seong Ok Han*[1] *and Lawrence T. Drzal*[2]**
[1] Functional Materials Research Center, Korea Institute of Energy Research, 71-2, Jang-dong, Yuseong-gu, 305-343, Daejeon, Korea
[2]Composite Materials and Structures Center, Michigan State University, 2100 Engineering Building, East Lansing, MI 48824-1226, USA

## Abstract

Recently, biocomposites made from natural fiber and biobased polymers are receiving more attention particularly because of their structural properties and their environmentally friendly nature. Glucose based polymers are considered as a biobased polymer with potential application for use in biocomposites. The curing characteristics and water absorption behavior of the glucose based polymer and epoxy resin are investigated for the purpose of utilization as a matrix in biocomposites. A cured matrix containing epoxy resin and 50wt% of glucose based polymer was prepared and characterized. Exothermic reactions attributed to esterification and etherification reactions between the hydroxyl and carboxyl functionalities of the glucose polymer with epoxy groups were identified as taking place during the curing process. Exothermic reactions were differentiated according to the degree of carboxyl group substituent of the glucose based polymer. The results showed that the esterification reaction occurs in the early stage of cure and then etherification followed after completion of the esterification. The cured matrix showed thermal stability up to 300°C. The average glass transition temperature and storage modulus of the matrix were as high as 95°C and 2700 MPa, respectively. The cured matrix of epoxy resin and glucose based polymer with higher carboxyl group content was found to have a lower density due to the formation of bulky groups in the crosslinks. The matrix cured at higher temperature shows compact crosslinks due to the higher concentration of ether bonds comparing to the matrix cured at the lower temperature. The effects of water absorption on the hydrophilic polymer matrices with glucose based polymer and epoxy resins were also studied as a function of curing temperature. The polymer matrices cured at different temperatures were immersed in water at room temperature for 1000 hours and the thermomechanical properties of the cured polymers were characterized using DMA and TGA. Two types of sorbed water were identified. Type I sorbed water contributed mainly to increasing the weight and to a decrease in $T_g$ due to a plasticization effect. Type II sorbed water was not removed after heating the polymer to 110°C for an hour. The strongly sorbed Type II water causes changes in the mechanical properties of the polymers cured at different temperatures depending on the crosslinking of the matrix. The matrix cured at the higher temperature has a comparatively tightly crosslinked network structure which exhibited microcracking as a degradation mechanism due to the trapping of the sorbed water within the polymer.

**Key Words**: Epoxy resin, Carboxyl functionalized glucose copolymer, Curing characteristics, Water absorption behavior

# Introduction

The increasing global environmental awareness, high rate of depletion of petroleum resources, drive for sustainable technology, and new environmental regulations have together triggered the search for new products and processes that are compatible with the environment. Research on biocomposites consisting of a natural fiber reinforcement and a polymer matrix, especially a biodegradable polymer matrix, are the basis for environmentally friendly polymer composites. Biocomposites can be used as a replacement for glass fiber reinforced polymer composites in applications ranging from automobiles to building materials because of their comparable strength and stiffness [1, 2, 3].

A large number of interesting applications are emerging for biocomposites. In Europe the emphasis is on the automotive industry which is seriously looking into the use of plant fiber based composites as a way to reduce environmental pressures and at the same time save weight (and therefore fuel) and cost. In the US, Canada and Australia, wood fiber based composite building materials have been developing for some time. In India and South America jute and sugar cane fibers are used in low cost housing and in China there has been recent development of rice husk based composites for planking [4].

Classical polymers such as polyethylene, polypropylene, and polyester, and epoxy resin have been combined with biofibers to produce partially biobased composites. However, once these materials are discarded, they persist in the environment without being degraded thus giving rise to ecological and environmental concerns. Recently, the use of a biodegradable polymers itself or blending a classical polymer with a biodegradable polymer as the biocomposite matrix is an alternative approach for the development of an environmentally friendly biocomposite matrix [5, 6].

Epoxy resins are widely used for many important applications such as coatings, adhesives, reinforced plastics, and matrix resins for advanced composites, due to their high thermal resistance, high tensile strength and modulus [7]. Epoxy resins have also been investigated as a polymer matrix for biocomposites with natural fibers in order to apply them in automobile and construction industries [8, 9]. The wide range of applications of epoxy resin can be attributed to the fact that epoxy resin can be crosslinked with a variety of functionalized compounds that contain hydroxyl, carboxyl and amine groups [10]. On the other hand, epoxy resin as a matrix shows relatively poor fracture resistance and a great deal of effort has been made to improve it. For example, epoxy resins have been successfully toughened by various means including chemical modification of epoxy molecular structure

[11], addition of rubber particles [12, 13], addition of thermoplastic phases [14, 15], and the addition of inorganic fillers [16].

Glucose maleic acid ester vinyl copolymer (GMAEVC) as a carboxyl functionalized glucose copolymer has been developed as a biodegradable adhesive for the paper and packaging industry [17]. Since GMAEVC contains reactive carboxyl and hydroxyl functional groups in its structure, GMAEVC can be crosslinked with epoxy resins and used as a part of polymer matrix for biocomposites [18]. The incorporation of biodegradable materials into an epoxy resin has advantages for the development of environmentally friendly biocomposites [19]. Also, the toughening effect of epoxy resin can be expected by the addition of a thermoplastic phase of GMAEVC.

The relative reactivity of hydroxyl and carboxyl groups with epoxy has been of great interest to researchers [20, 21]. Shechter et al. has studied the reactivity of different alcohols and carboxylic acids towards different epoxide groups such as styrene oxide, phenyl glycidyl ether and benzyl oxide [22]. Wu et al. has investigated the crosslinking reactions for hydroxyl and carboxyl functionalized acrylic copolymer with cycloaliphatic epoxy resin [10]. However, the abundant hydrophilic functional groups of the cured matrix of GMAEVC and epoxy resin react to the water molecules much easier in wet environments than the cured epoxy matrix itself. Therefore, the sorbed water in the cured matrix of GMAEVC and epoxy resin may change the properties of the cured matrix significantly.

Wet environments compromise the mechanical stability of advanced composites due to water sorption of the matrix material. The hydrophilic groups of epoxy resins absorb water molecules that, consequently, alter the physical and mechanical properties of the matrix. The main effects of water absorption in epoxy matrices are plasticization, changes of physical properties and hygrothermal degradation. Plasticization occurs through several mechanisms depending on the interaction of sorbed water molecules with the matrix. Changes in physical properties include a decrease of the mechanical properties, microcracks, chain scission, degradation of fiber and matrix interface bonding [23].

The water absorption effects in epoxies have been investigated using various techniques. Zhou et al. [24, 25] has investigated the hygrothermal effects of epoxy resin using DSC, TMA and NMR. They reported that the water molecule bonds with epoxy resins via hydrogen bonding. Furthermore, two types of bound water exist depending on the difference in the bond complex and activation energy. Pethrick et al. [26] studied the water absorption of epoxy resin using dielectric and gravimetric measurements and explained that changes in the final cure temperature affect both the extent and distribution of the types of water molecules present in the matrix. Mikols et al. [27] reported that water exists in the polymer matrix in two distinct forms: free water that fills the microcavities of the network and water strongly bound to polar groups in the epoxy resin. Xiao et al. [28] investigated the reversibility of water uptake and swelling of epoxy resin during hygrothermal ageing. The results show that the rate of the swelling of the polymer is less than that attributed to the mass of water absorbed initially and the swelling process is not fully reversible. Apicella et al. [29] proposed three different modes for water sorption in an epoxy resin: (a) bulk dissociation of water in the polymer network; (b) moisture sorption onto the surface of holes that define the excess free volume of the glassy structure; and (c) hydrogen bonding between hydrophilic groups of the polymer and water. They also reported that the chemical structure of the polymer matrix constituents, as well as the processing conditions, influence the resulting polymer networks and hence the properties of the crosslinked polymer in a wet environment.

The objectives of this research are to investigate the curing mechanisms between epoxy resin and GMAEVC that have both hydroxyl and carboxyl groups in its structure and to determine if this matrix can be used in biocomposites. Also, the properties of the hydrophilic cured matrix of GMAEVC and epoxy resin are investigated by examining the effects of the water absorption. The sorbed water effects on the cured polymer matrix are compared as a function of curing conditions.

# Experimental

## 2.1 Materials

Epoxy resin, Tactix 123, based on diglycidyl ether of bisphenol A (DGEBA) was obtained from Ciba Chemical Co. The viscosity of the resin is 5000 cps at 25°C and the epoxy equivalent weight is 172-176 g/mol. The structure of the monomer is shown in Structure 1.

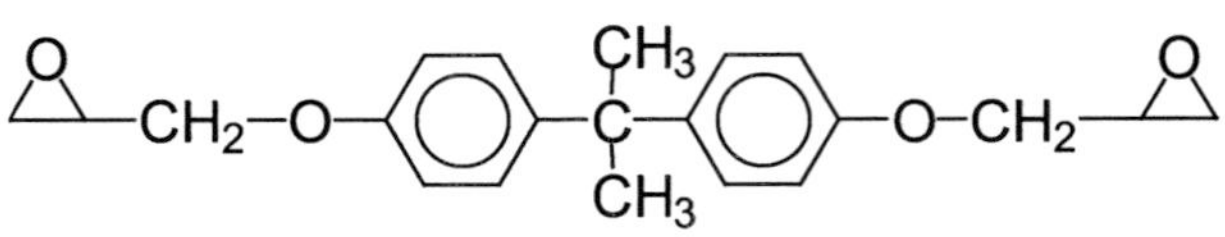

Structure 1: The DGEBA molecule

Glucose maleic acid ester vinyl copolymer was obtained from EcoSynthetix Inc. and used as received. The structure of GMAEVC is shown in Structure 2. The average degree of polymerization of glucose (DP) and substitution of the maleic ester group (DS) of the samples used for this research are summarized in Table 1. Sample denoted as GMAEVC-0 is hexyl-α-D-glycoside that does not have any carboxyl group in its structure. The average degree of polymerization of glucose (DP: n) and substitution of the maleate ester group (DS) of the samples used for water absorption are 1.2 and 1.4, respectively.

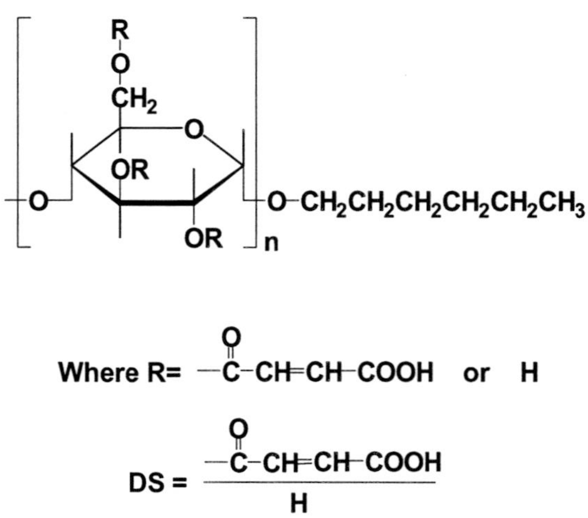


Structure 2: Glucose maleic acid ester vinyl copolymer molecule

GMAEVC starts to decompose at 165°C and thermally activated at temperatures over 110°C. Comparison of the FTIR spectra between original GMAEVC and the activated GMAEVC shows that the vinyl group peak and the -C-O-C- peak characteristic of the glucose disappear for the thermally activated GMAEVC. A new peak, due to the presence of a free carboxyl group appears as the GMAEVC thermally activated. It was also found that the broad peak for the hydroxyl group changed to a sharp peak when the glucose ring opened and formed free hydroxyl groups. These carboxyl and hydroxyl groups can now easily react with the epoxy group of the resin [18].

Table 1: DP and DS of GMAEVC samples

| Samples | DP (n) | DS | Main functional groups of GMAEVC for expected reaction with epoxy group |
|---|---|---|---|
| GMAEVC-0 | 1.2 | 0 | -OH |
| GMAEVC-1.4 | 1.2 | 1.4 | -OH/-COOH |
| GMAEVC-2.0 | 1.2 | 2.0 | -COOH |

## 2.2 Sample Preparation

### 2.2.1 Samples for Curing Behavior of Epoxy Resin and GMAEVC

GMAEVC and epoxy resin were heated to 90°C separately and mixed by a melt-blending process. The formulation of 50 wt% epoxy resin and GMAEVC with different DS of 0, 1.4, and 2.0 was used for experiments, respectively. Once a homogeneous solution was obtained by mixing two solutions for 10 seconds, the solution was degassed at 90°C for 10 min. This mixture was used to study the curing behavior of epoxy resin and GMAEVC by differential scanning calorimetry (DSC) and FTIR in real time. Any curing reactions during the blending and degassing were not detected by FTIR examinations.

For preparation of the cured matrix, the degassed mixture was poured into a silicone mold (1.2 cm × 7.5 cm × 0.3 cm) and cured in an air-circulating oven at a heating rate of 5°C/min. The curing conditions were completed by heating the mixture at 175°C for 2 hours and 200°C for 2 hours, consecutively, allowing it to gradually cool down to room temperature. Thermomechanical properties of the cured samples of epoxy resin and GMAEVC with different degrees of carboxyl group substitution were analyzed and compared.

### 2.2.2 Samples for Water Absorption Test

The effects of water absorption on the hydrophilic polymer matrices with GMAEVC and epoxy resin were studied as a function of curing temperature. The formulation of 50 wt% GMAEVC (DP=1.2, DS=1.4) was used for all experiments. The matrix of epoxy resin and GMAEVC cured at 175°C for 2 hours and 200°C for 2 hours, consecutively, was prepared. The matrix cured at a higher temperature of 175°C for 2 hours and 260°C for 2 hours was also prepared for comparison. The curing conditions of 175°C(2 hrs), 200°C(2 hrs) and 175°C(2 hrs), 260°C(2 hrs) are expressed as lower and higher curing temperature for water absorption test, respectively.

The cured matrices were dried in an oven at 110°C for one hour. Immediately upon cooling, the specimens were weighed. The specimens were immersed in distilled water at ambient temperature and weighed at predetermined times. Every procedure was performed by following ASTM D570-98: Standard test methods for water absorption of plastics [30]. The specimen size was 1.2 cm × 7.5 cm × 0. 3 cm and the water gain percentage, M%, was determined from the equation:

$$M\% = (W - W_d) / W_d \times 100$$

W is the weight of the water absorbed specimen and $W_d$ is the initial weight of the dry specimen. To ensure the removal of excessive surface water, specimens were gently wiped dry using clean, lint-free tissue paper and allowed to stand free at ambient environment for 2 minutes. To examine the reaction between water and the matrix, the specimen that was immersed in the water for 1000 hours was dried in an air-circulating oven at 110°C for an hour. The weight gain of this specimen was compared to the weights of both dried and water absorbed for 1000 hours specimen.

Weight gain of the epoxy matrix cured with a catalyst, SarCat CD1012 (CD1012) cationic initiator from Sartomer was examined to compare with the hydrophilic cured matrix of GMAEVC and epoxy resin. The active molecule of the catalyst is the diaryl iodonium hexafluoroantimonate that decomposes under thermal curing (140°C) [31].

## 2.3 Instrumental Analysis

### 2.3.1 DSC Monitoring of the Reaction Between Epoxy Resin and GMAEVC

A differential scanning calorimetry (DSC) study was performed under a nitrogen atmosphere using a DSC2920 modulated differential scanning calorimeter from TA instruments. High purity indium was used to calibrate the calorimeter. The sample (6±0.2 mg) was taken in the DSC aluminum pan at room temperature. Real time monitoring of the curing of epoxy resin and GMAEVC was performed in an aluminum pan in the 30°C to 320°C temperatures ranges with heating rate of 5°C/min. The sample weight did not change before and after the DSC measurement.

### 2.3.2 FT-IR Monitoring of the Reaction between Epoxy Resin and GMAEVC

Curing of epoxy resin with GMAEVC was quantitatively analyzed by transmission FTIR spectroscopy using a Perkin Elmer FTIR system 2000 model, equipped with a conventional TGS detector. Samples were prepared by casting a thin film of resin onto a sodium chloride plate and placed in a heating cell in the spectrometer to carry out the reaction from 100°C to 170°C at a heating rate of 2°C/min. The temperature of the heated cell was monitored with a DigiSense temperature controller from the Cole Parmer Co. The FTIR spectra were collected at different temperatures and compared to the FTIR spectra of fully cured samples prepared in an oven to confirm the presence of polymerization products.

The conversion of epoxy and hydroxyl functions in the formulation based on epoxy resin and GMAEVC were calculated from the FTIR spectra. The 1509 $cm^{-1}$ band was unchanged upon curing, and subsequently, was used as an internal standard [31]. The decrease of the band at 912 $cm^{-1}$ assigned to the epoxy function permits accurate measurement of the monomer conversion via the following relation, where Π is the functional conversion and T is temperature. The epoxy function expresses the opening of the epoxy ring in the mixture. For the conversion of the hydroxyl function the maximum point of the hydroxyl peak in the region of 3666-3113 $cm^{-1}$ was measured at each temperature. The hydroxyl function expresses the generation of the hydroxyl group in the mixture.

$$\Pi(\text{epoxy}) = 1 - \frac{A_{912(T)} / A_{1509(T)}}{A_{912(T=25)} / A_{1509(T=25)}}$$

$$\Pi(\text{hydroxyl}) = 1 - \frac{A_{max(T)} / A_{1509(T)}}{A_{max(T=25)} / A_{1509(T=25)}}$$

Scheme 1

### 2.3.3 Surface Analysis

X-ray Photoelectron Spectroscopy examination was used to determine the functional groups on the surface of the cured matrix. A Perkin Elmer Physical Electronics PHI 5400 ESCA spectrometer equipped with standard magnesium x-ray source operated at 300 W (15 kV and 20 mA) was used for surface analysis.

### 2.3.4 Density Measurement

Density of the cured matrix of epoxy resin and GMAEVC with different degrees of carboxyl group substitution was measured using a NaBr solution method in a density measurement column of Techne Inc. [32].

### 2.3.5 Thermomechanical Analysis

The glass transition temperature and the modulus of the cured matrix were measured by dynamic mechanical analysis (DMA) in the single cantilever mode, at a frequency of 1Hz. DMA runs were recorded with a DMA 2980 Dynamic Mechanical Analyzer from TA instruments. The glass transition temperature ($T_g$) was measured at the maximum of the Tan delta (δ) curve deduced from DMA experiments. Storage modulus of the matrix was determined at 40°C.

### 2.3.6 Thermal Stability Analysis

Thermal stability of the cured matrix was analyzed under a nitrogen atmosphere using a TGA2950 thermal gravimetric analyzer from TA instruments.

### 2.3.7 Phase Morphological Analysis

The phase morphology of room temperature fractured specimens were characterized using a model 2020 ElectroScan environmental scanning electron microscopy (ESEM) operating with a beam energy of 20 kV and a water vapor pressure between 2 and 3 Torr.

# Result and Discussion

## 3.1 Curing Behavior of Epoxy Resin and GMAEVC

The curing behavior of epoxy resin and GMAEVC was investigated using GMAEVC with different DS of 0, 1.4, and 2.0. Since GMAEVC contains reactive carboxyl and hydroxyl functional groups in its structure, GMAEVC can be crosslinked with epoxy resins and shows different curing behavior depending on the degree of DS, namely, degrees of carboxyl group substitution.

### 3.1.1 DSC Investigation of Epoxy Resin and GMAEVC Mixture

Figure 1 shows DSC scans for the mixture of epoxy resin and GMAEVC with DS equal to 0, 1.4, 2.0. No exothermic reactions are observed in the DS=0 for the mixture of epoxy resin and GMAEVC in the region of 100-300°C. This is because there are no carboxyl groups in the mixture. For the reaction of epoxy resin and GMAEVC-1.4 three exothermal peaks are observed in the regions of 100-170°C(A), 200-240°C(B) and 260-300°C(C), respectively. The exothermic peaks in the regions of 100-170°C(A) and 260-300°C(C) increase for the DS=2.0, however, the exothermic peak in the region of 180-240°C(B) disappears. From this result it is postulated that the exothermic reaction observed in the regions of 100-170°C is attributed to the exothermic reaction of epoxy with carboxyl groups [33]. This is coincident with the results of Nakamura S. et al. [34] that shows an exothermic reaction between epoxy resin and carboxyl group of 1,3,5-triacetoxybenzene in the region of 100-180°C.

Oh et al. [35] has studied the reaction between epoxy resin and hydroxyl groups of hyperbranched polymers. The research shows that the reaction between the epoxy group and the hydroxyl group of hyperbranched polymers has an exothermic reaction in the range of 250-350°C. From these results the peak in the region of 260-300°C can be explained as a exothermic reaction peak resulting from the reaction of an epoxy group of the epoxy resin and hydroxyl group of GMAEVC.

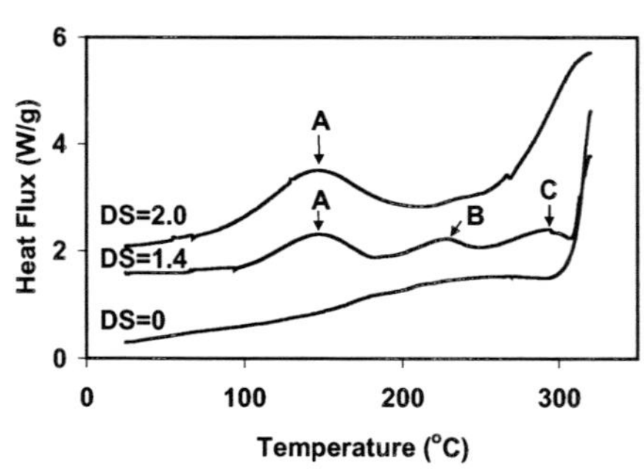


Figure 1: DSC scans for mixture of epoxy resin and GMAEVC with different degree of carboxyl group

### 3.1.2 FTIR Spectra of Epoxy Resin and GMAEVC Curing

The curing behavior of epoxy resin and GMAEVC with different degrees of carboxyl group substitution was monitored in real-time while heating the mixtures from room temperature to 170°C by FT-IR. FTIR spectra were obtained at 25°C, 100°C and 170°C for each mixture of epoxy resin and GMAEVC. Spectra from samples with different degrees of carboxyl group substitution are compared in Figure 2.

The mixture of epoxy resin and GMAEVC, shown in Figure 2, indicates how the hydroxyl peak changes from a broad to a sharp peak and shows the shift to higher frequencies with increasing temperature. A change from a broad to a sharp peak results from the opening of glucose ring and that produces free hydroxyl groups in the GMAEVC structure. Peak sharpening can also be explained by changing the intermolecular hydrogen bonding between hydroxyl groups of glucose ring to intramolecular hydrogen bonding. Peak shifting to higher frequencies indicates that new ester or ether bonds are formed near hydroxyl group due to crosslinks between epoxy resin and GMAEVC [36].

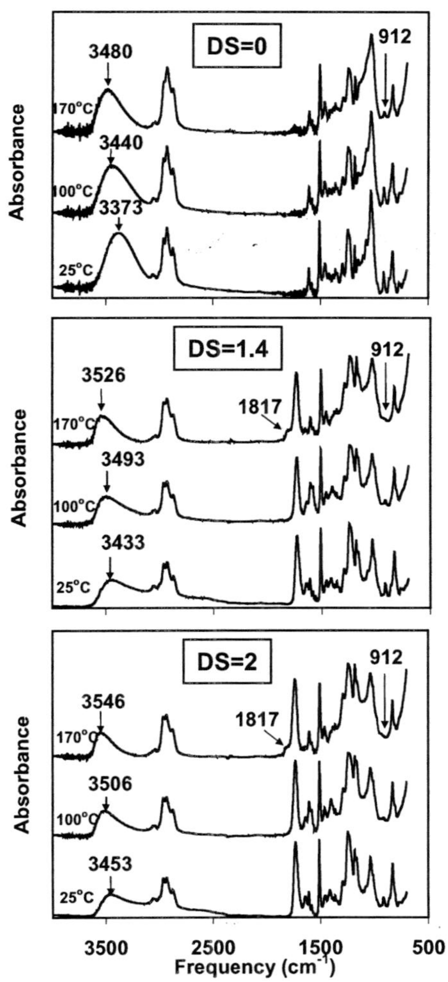


Figure 2: FTIR Spectra of epoxy resin and GMAEVC mixtures with different degree of carboxyl group substitution

FTIR spectra for the mixture of epoxy resin and GMAEVC-1.4 or GMAEVC-2 shows a new peak formation in the region of 1780-1856 $cm^{-1}$ when the mixture is heated above 160°C. This peak can be explained as the formation of saturated carboxylic acid anhydride between carboxyl groups of GMAEVC or saturated aliphatic acid peroxide during the curing of epoxy resin and GMAEVC [35]. This peak appears and increases faster for the mixture of epoxy resin and GMAEVC-1.4 comparing to that of epoxy resin and GMAEVC-2.0. This indicates that the peak usually results from the formation of aliphatic acid peroxide. The reaction between the carboxyl group and hydroxyl group can proceed faster than the reaction between two carboxyl groups in the mixture. This peak was also observed in the cured matrix at temperatures higher than 170°C. The increase in the peak near 1200 $cm^{-1}$ is due to the stretching of –C-O-C- groups generated from crosslinks between epoxy resin and GMAEVC [37].

The characteristic peak for the epoxy ring is also identified at 912 $cm^{-1}$ in Figure 2. As the cure reaction proceeded, the intensity of this peak decreases due to the opening of epoxy ring. This peak disappeared faster in the mixture of epoxy resin and GMAEVC with a higher degree of carboxyl group substitution (DS=2.0) resulting from the easy opening of epoxy ring under acidic conditions. However, this peak can be observed in the mixture of DGEBA and GMAEVC-0 when the mixture is heated to 170°C.

Figure 3 shows the shift of maximum hydroxyl peak of the mixture with increasing temperature. The maximum peak shifts to the higher frequencies with the higher DS in the mixture of epoxy resin and GMAEVC. The difference of frequencies between DS=0 and DS=2.0 is almost 100 $cm^{-1}$ resulting from proximity of the carboxylic groups to the hydroxyl groups. Peak shifting to higher frequencies of a mixture of epoxy resin and GMAEVC with increasing temperature indicates that new ester or ether bonds are formed near hydroxyl groups due to the formation of crosslinks during the curing process.

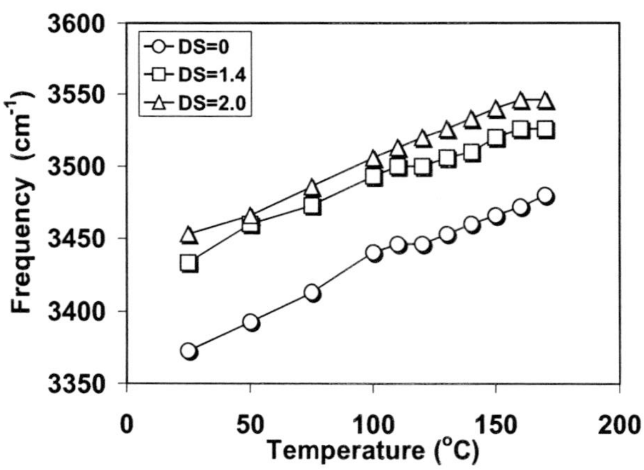


Figure 3: The maximum point of hydroxyl peak for the mixture of epoxy resin and GMAEVC with increasing temperature

### 3.1.3 Epoxy Conversion of Epoxy Resin and GMAEVC Mixture

Figure 4 shows the epoxy conversion in the mixture of epoxy resin and GMAEVC when the mixture is heated from room temperature to 170°C. The epoxy conversion increases continuously with increasing temperature and reaches 100% at 170°C for the mixtures of epoxy resin and GMAEVC-1.4 and GMAEVC-2.0. For the mixture of epoxy resin and GMAEVC-0, however, the epoxy conversion reaches 60% at 170°C. This result shows that the epoxy ring opens very easily in acidic conditions resulting from the reaction with the hydroxyl group of GMAEVC-0 .

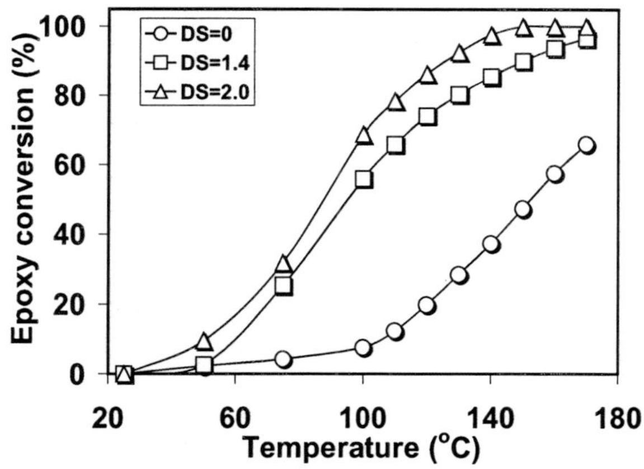


Figure 4: Epoxy conversion in the mixture of epoxy resin and GMAEVC with temperature

### 3.1.4 Hydroxyl Conversion of Epoxy Resin and GMAEVC Mixture

Figure 5 shows the hydroxyl conversion in the mixture of epoxy resin and GMAEVC when the mixture is heated from room temperature to 170°C. Because there is no carboxyl group in the mixture of epoxy resin and GMAEVC-0, the epoxy ring opening is mainly caused by the reaction with the hydroxyl group of GMAEVC-0. During the reaction, one hydroxyl group is generated with the opening of each epoxy ring as a result of the consumption of one hydroxyl group of GMAEVC. Therefore, the change of the hydroxyl group concentration in the mixture is not substantial. In the early stages of the curing reactions between the epoxy resin and GMAEVC-0, the amount of the hydroxyl groups generated from the epoxy rings is small and, hence, the hydroxyl groups of the GMAEVC are mainly responsible for initiating the cure process. This decreases the hydroxyl group concentration below 100°C. Above 100°C the generation and consumption of hydroxyl groups reaches a steady state due to epoxy ring opening; therefore, the hydroxyl conversion remains constant. This result is consistent with the results of Oh et al. [35].

The data in Figure 5 shows that curing of DGEBA and GMAEVC at DS equal to 1.4 increases the hydroxyl conversion from room temperature to 100°C. Again this is a result of the easier epoxy ring opening in acidic condition. The esterification reaction between the carboxyl group and the hydroxyl group decreases the hydroxyl function in the range of 100-140°C. Above 140°C the esterification reaction between the carboxyl group and the epoxy group produces an increase of hydroxyl conversion. This results from the opening of epoxy ring and generation of hydroxyl group.

This pattern is more distinct in the mixture of epoxy resin and GMAEVC-2.0. The hydroxyl conversion increases dramatically below 100°C since the opening of epoxy ring and production of the hydroxyl group in the acidic condition are easier. The higher the hydroxyl group concentration is, the easier the esterification reaction between the carboxylic group and the hydroxyl group occurs. This produces a sharp decrease in hydroxyl conversion in the temperature range of 100-140°C. The hydroxyl conversion increases again due to the epoxy ring opening that results from the esterification of carboxylic groups and epoxy resin. DSC and FTIR results show that the esterification reaction mainly occurs in the early stage of cure and then etherification proceeded after completion of the esterification. These results are comparable to the results of Park et al. [37].

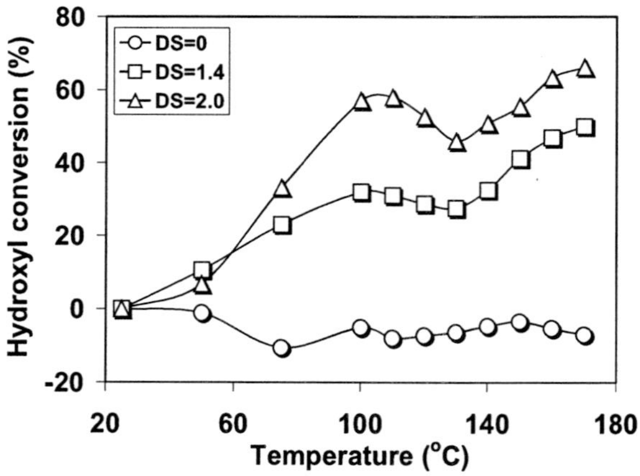


Figure 5: Hydroxyl conversion of epoxy resin and GMAEVC mixture

### 3.1.5 Curing Mechanisms of the Epoxy Resin and GMAEVC Mixture

The reaction mechanisms between the epoxy group and the carboxyl and hydroxyl groups have been studied and three different reactions - two esterifications and one etherification - are identified [21]. The results of DSC, FTIR and reaction mechanisms indicate that the curing process between the epoxy resin and GMAEVC can be explained by condensation esterification (carboxyl-hydroxyl reaction), addition esterification (carboxyl-epoxy reaction) and etherification (hydroxyl-epoxy reaction) from lower temperature to higher temperature.

## 1. Condensation esterification(carboxy-hydroxyl reaction)

## 2. Addition esterification(carboxy-epoxide reaction)

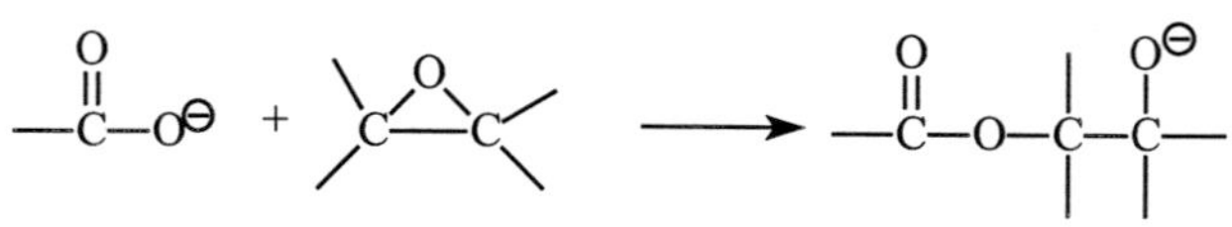

## 3. Etherification(hydroxyl-epoxide reaction)

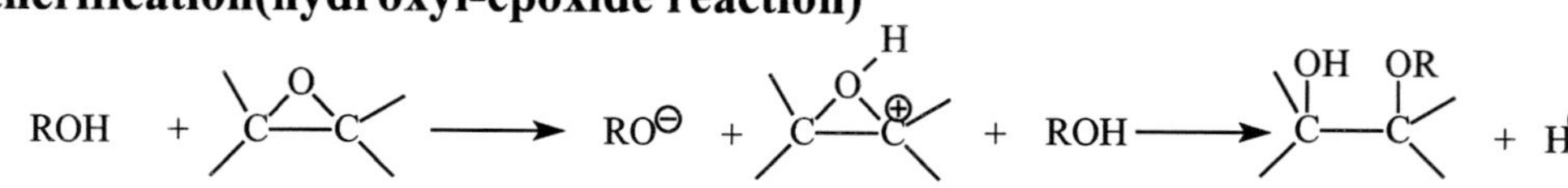


Scheme 2

## 3.2 Characterization of the Cured Matrices of Epoxy Resin and GMAEVC with Different DS

The mixture of epoxy resin and GMAEVC was cured at 175°C for 2 hours and 200°C for 2 hours, consecutively. The properties for the cured matrix of epoxy resin and GMAEVC with different degree of carboxyl group substitution were characterized and compared. The mixture of DGEBA and GMAEVC-0 was not fully cured and was very brittle and sticky when the mixture was heated at 175°C for 2 hours and 200°C for 2 hours. Therefore, the mechanical properties of the mixture of epoxy resin and GMAEVC-0 could not be investigated.

### 3.2.1 Thermal Stability of the Cured Matrix of Epoxy Resin and GMAEVC

Figure 6 compares the thermal stability of the cured matrix of epoxy and GMAEVC with different degrees of carboxyl group substitution. The cured matrices of epoxy resin and GMAEVC-1.4 or GMAEVC-2 show thermal stability up to 320°C. However, the cured matrix of DGEBA and GMAEVC-0 starts to decomposes around 250°C and shows a fast decomposition at 370°C. The residues for the cured matrix of epoxy resin and DMAEVC-0, DMAEVC-1.4, DMAEVC-2 after thermal decomposition over 600°C are 15.4%, 23.3% and 23.8%, respectively. This means that the mixture of epoxy resin and GMAEVC-0 could not be fully cured under the curing condition of 175°C for 2 hours and 200°C for 2 hours.

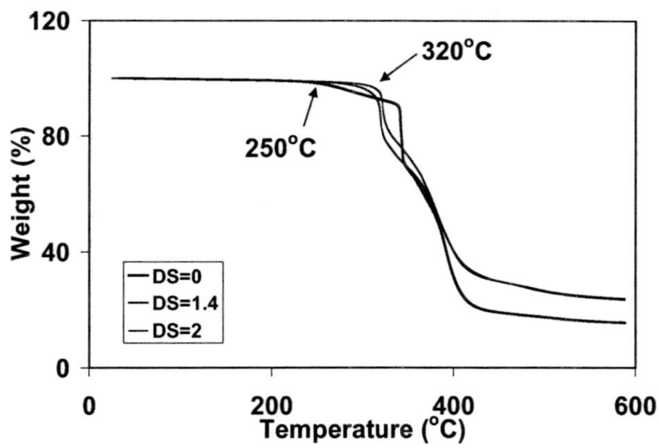


Figure 6: TGA curves of the cured matrix for epoxy resin and GMAEVC mixture

### 3.2.2 Surface Analysis of Cured Matrix of Epoxy Resin and GMAEVC

Surface analysis of the cured matrix of epoxy resin and GMAEVC with different degrees of carboxyl group substitution is compared in Table 2. The atomic ratio of oxygen to carbon increased for the cured matrix of epoxy resin and GMAEVC with the higher degree of carboxyl group substitution. Table 3 shows the changes of the carbon 1s and oxygen 1s of the cured matrix. The carbon 1s spectra is deconvoluted to four peaks that are assigned to -C-C*-C- (284.6 eV), -C*-O-H, -C*-O-C- (286.1 eV), -O–C*-O-, -C*(=O)- (287.6 eV) and -O-C*(=O)- (289.0 eV), respectively. The oxygen curve is deconvoluted to two peaks that are assigned -O-C-O*- (530.0 eV) and –C-O*H (532.0 eV).

Generally, C1s and O1s concentrations are similar in the cured matrix of epoxy resin and GMAEVC-1.4 or GMAEVC-2. Table 3 shows that C1s concentration of the ether bonds increase with the cured matrix of epoxy resin and GMAEVC-1.4 compared to the cured matrix of epoxy resin and GMAEVC-2. On the other hand, the free hydroxyl groups are more abundant in the cured matrix of epoxy resin and GMAEVC-2. From this result the curing of epoxy resin and GMAEVC-1.4 proceeds to the formation of ether bonds resulting in compact crosslinks comparing to the formation of bulky crosslinks in the cured matrix of epoxy resin and GMAEVC-2.

Table 2: Atomic ratio of the cured matrix

| Samples | [C] % | [O] % | [O]/[C] |
|---|---|---|---|
| GMAEVC-1.4 | 75.7 | 24.3 | 0.322 |
| GMAEVC-2 | 74.1 | 25.9 | 0.349 |

Table 3: Carbon 1s and oxygen 1s of the cured matrix

| Samples | C1s concentration (%) | | | | O1s concentration (%) | |
|---|---|---|---|---|---|---|
| | -C-C*-C- | -C*-O-H, -C-C*-O- | -O–C*-O- -C*(=O)- | -O-C*(=O)-C- | -C-O*-C- | –C-O*H |
| GMAEVC-1.4 | 38.0 | 24.2 | 9.6 | 3.8 | 11.0 | 13.7 |
| GMAEVC-2 | 38.4 | 25.8 | 6.5 | 3.4 | 10.0 | 15.8 |

### 3.2.3 Density of Cured Matrix of Epoxy Resin and GMAEVC

Figure 7 shows the density of the cured matrix of epoxy resin and GMAEVC-1.4 or GMAEVC-2 with different curing conditions. The cured matrix of epoxy resin and GMAEVC-1.4 shows a higher density compared to that of the epoxy resin and GMAEVC-2 under all curing conditions. This indicates that the curing process of epoxy resin and GMAEVC with a higher degree of carboxyl group produces bulky crosslinks due to the formation of crosslinks between carboxyl groups or carboxylic and hydroxyl groups. However, the curing between epoxy resin and GMAEVC-1.4 produces compact crosslinks like ether bonds in the network of the cured matrix.

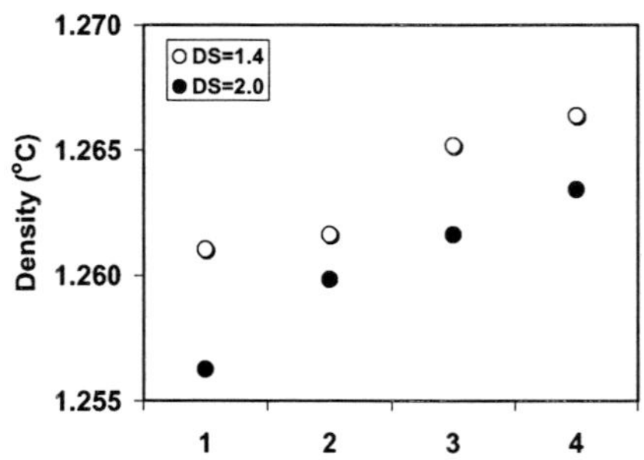


Figure 7: Density of the cured matrix for DGEBA and GMAEVC mixture with different curing conditions

1: 175°C for 2 hours and 200°C for 2 hours, 2: 175°C for 2 hours and 220°C for 2 hours
3: 175°C for 2 hours and 240°C for 2 hours, 4: 175°C for 2 hours and 260°C for 2 hours

The density of the cured matrix increases with increases in the curing temperature. This can be explained since the compact crosslinks such as ether bond forms easily with increasing temperature.

### 3.2.4 Glass Transition Temperature and Storage Modulus of Cured Matrix of Epoxy Resin and GMAEVC

The glass transition temperature ($T_g$) and the storage modulus of the cured matrix of epoxy resin and GMAEVC-1.4 or GMAEVC-2 are compared in Figure 8. The storage modulus of the cured matrix is not different much between GMAEVC-1.4 and GMAEVC-2; however, the $T_g$ of the cured matrix of epoxy resin and GMAEVC-2 decreased 10°C. The decrease in $T_g$ can be explained by considering that the cured matrix of epoxy resin and GMAEVC-1.4 contains a higher concentration of dense bonds such as ether bonds in the crosslinks. The bulky crosslinks of the epoxy resin and GMAEVC-2.0 compared to the tight crosslinks of epoxy resin and GMAEVC-1.4 gives flexibility to the chains, resulting in the $T_g$ decrease.

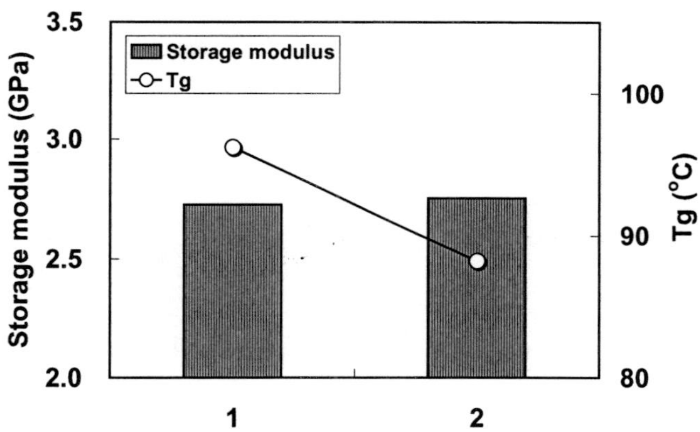


Figure 8: Glass transition temperature of the cured matrix for epoxy resin and GMAEVC mixture at 175°C for 2 hours and 200°C for 2 hours

1: epoxy resin and GMAEVC-1.4, 2: epoxy resin and GMAEVC-2

## 3.3 Characterization of the Cured Matrix of Epoxy Resin and GMAEVC with Different Curing Profile

For the water absorption test, the cured matrices of epoxy resin and GMAEVC (DP=1.2, DS=1,4) with different curing profiles were prepared. The matrix of epoxy resin and GMAEVC cured at 175°C for 2 hours and 200°C for 2 hours, consecutively. The matrix cured at a higher temperature of 175°C for 2 hours and 260°C for 2 hours was also prepared for comparison. The curing conditions of 175°C(2 hrs), 200°C(2 hrs) and 175°C(2 hrs), 260°C(2 hrs) are expressed as lower and higher curing temperature for water absorption test, respectively.

### 3.3.1 Surface Analysis of Matrix Cured at Different Temperature

Surface analysis of the cured matrix of epoxy resin and GMAEVC at different curing temperatures is compared in Table 4. The atomic ratio of oxygen to carbon increased for the higher curing temperature. Table 5 shows the changes of the carbon 1s and oxygen 1s of the cured matrix. The carbon 1s spectra is deconvoluted to four peaks that are assigned to -C-C*-C- (284.6 eV), -C*-O-H, -C*-O-C- (286.1 eV), -O–C*-O-, -C*(=O)- (287.6 eV) and -O-C*(=O)- (289.0 eV), respectively. The oxygen curve is deconvoluted to two peaks that are assigned -O-C-O*- (530.0 eV) and –C-O*H (532.0 eV).

Table 5 also shows that C1s concentration of the ether bonds increases for materials cured at higher temperature, however, the surface of the matrix cured at lower temperature contains the ester bond and the free hydroxyl groups more abundantly. This indicates curing at higher temperature proceeds the formation of ether bonds in the polymer network resulting in compact crosslinks of the matrix compared to the bulky crosslinks associated with ester bonds found in the matrix cured at the lower temperature.

This result is consistent with the study of Barral et al. that showed the relationship between the structure and the mechanical properties of the epoxy matrix as a function of the degree of cure [38]. It explained that etherification reactions are important in the highest temperature treatment of epoxy resin and influenced the mechanical properties of the matrix. The higher activation energy for a polymer matrix cured at high temperature is attributed to the higher crosslink density of the network, which diminishes with the availability of molecular sized holes in the polymer structure.

Table 4: Atomic ratio of the cured matrix at different curing conditions

| **Curing condition** | **[C] %** | **[O] %** | **[O]/[C]** |
|---|---|---|---|
| 175°C(2hrs), 200°C(2hrs) | 75.7 | 24.3 | 0.322 |
| 175°C(2hrs), 260°C(2hrs) | 74.1 | 25.9 | 0.350 |

Table 5: Carbon 1s and oxygen 1s of the cured matrix at different curing conditions

| **Samples** | **C1s concentration (%)** | | | | **O1s concentration (%)** | |
|---|---|---|---|---|---|---|
| | **-C-C*-C-** | **-C*-O-H, -C-C*-O-** | **-O–C*-O- -C*(=O)-** | **-O-C*(=O)-C-** | **-C-O*-C-** | **–C-O*H** |
| 175°C(2hrs), 200°C(2hrs) | 38.0 | 24.2 | 9.6 | 3.8 | 11.0 | 13.7 |
| 175°C(2hrs), 260°C(2hrs) | 36.0 | 25.8 | 9.0 | 3.3 | 14.0 | 11.8 |

### 3.3.2 Density of the Matrix Cured at Different Temperature

Figure 9 shows the density of the matrix of GMAEVC and epoxy resin cured at different curing conditions. The density of the matrix increased with increasing curing temperature. This result is consistent with the XPS results that show the cured at higher temperature contains the more abundant tight crosslinks such as ether bonds in the network of the cured matrix rather than bulky ester bonds in the network of the matrix cured at lower temperature.

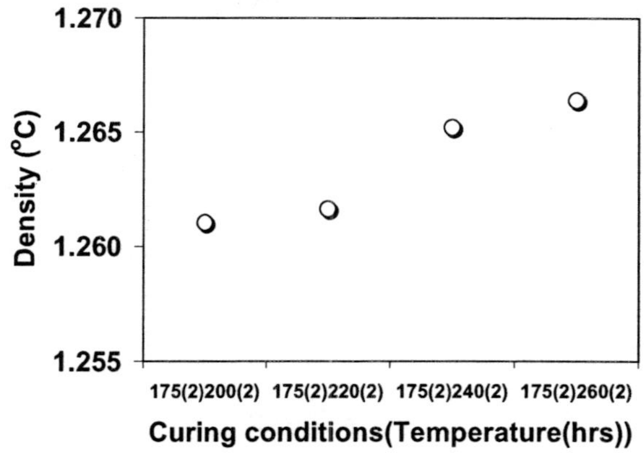


Figure 9: Density of the cured matrix for GMAEVC and epoxy resin at different curing conditions

### 3.3.3 Thermal Stability and Themomechanical Properties of Cured Matrix of GMAEVC and Epoxy Resin at Different Temperature

Figure 10 compares the thermal stability of the polymer matrix of GMAEVC and epoxy resin cured at different temperatures. The cured matrix shows thermal stability up to 300°C and three decomposition products between 300°C-400°C. The decomposition pattern of the matrix cured at higher temperature shows a decrease of the peak at 316°C and an increase of the peak at 381°C. This result indicates that curing at higher temperature produces high molecular weight products more abundantly than the matrix cured at lower temperature.

The $T_g$ and the storage modulus of the matrix of GMAEVC and epoxy resin cured at the lower temperature are as high as 95°C and 2700Mpa, respectively. The matrix cured at the higher temperature shows a comparatively higher $T_g$ and storage modulus (101°C and 3100Mpa), respectively.

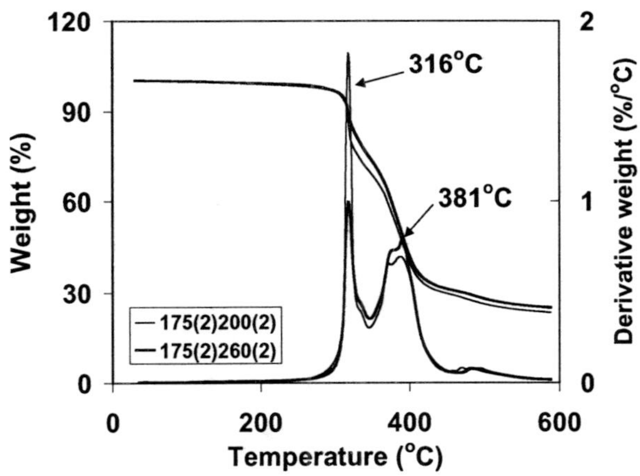


Figure 10. TGA comparison of the matrices cured at different curing conditions

## 3.4 Water Absorption Test on the Cured Matrix

The cured matrix of GMAEVC and epoxy resin contains hydrophilic functionality and easily absorbs water molecules. The absorbed water leads to dimensional variations in composites and also affects the mechanical properties of the composites. Water absorption tests were performed on the matrices prepared at different curing conditions and the thermomechanical performance of matrices was compared to the dry, original samples.

### 3.4.1 Water Absorption Profile of Cured Matrix

The weight increases of the matrices of GMAEVC and epoxy resin cured at different temperatures are compared to the epoxy matrix itself in Figure 11. The matrix of GMAEVC and epoxy resin shows a faster weight increase than the epoxy matrix due to the abundant hydrophilic groups on the surface of the matrix. The weight gain of the cured matrix of GMAEVC and epoxy resin increased four times more than the epoxy matrix after immersion in water for 1000 hours.

During the early stages of the test, the matrix cured at lower temperatures shows a faster weight increase comparing to the matrix cured at higher temperature. In the latter stage of the test, the weight increase for both matrices is 4% after immersion for 1000 hours. When these matrices are heated at 110°C in an oven for one hour, the weight gains decrease by 2.1%.

This phenomenon indicates that the matrix cured at lower temperature contains a higher concentration of hydroxyl group so it can bind with the water molecules easily through hydrogen bonding. The bound water molecules on the surface of the matrix can also be removed easily from the matrix when heated above 100°C. In contrast, the sorbed water in the network of the matrix is hard to remove and remains after heating for one hour at 110°C. The sorbed water molecules also change the mechanical properties of the cured matrix depending on the crosslink structure.

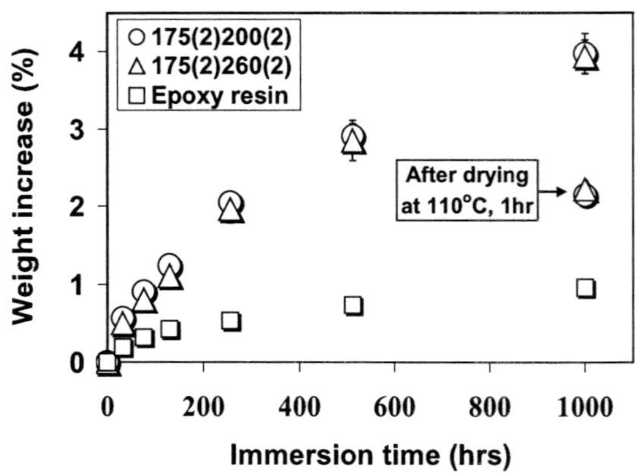


Figure 11: Comparison of weight increase for the matrices with immersion time

### 3.4.2 TGA Curves of the Water Absorbed Matrix

Figure 12 compares the TGA results between the matrix cured at lower temperature and the matrix immersed in water for 1000 hours. The TGA curves do not show any obvious changes except for the slight weight decrease for the immersed matrix in the region of 100 °C-200°C.

This is a result of water evaporation. For the matrix cured at higher temperature TGA curves show a difference between the cured matrix and the immersed matrix. In Figure 13 the immersed matrix shows a new peak for the derivative of the weight change that decomposes at 361°C. This peak can be interpreted as the formation of new products with smaller chains due to the chain scission that results from the reaction of the water molecule in the tight crosslink of the polymer network.

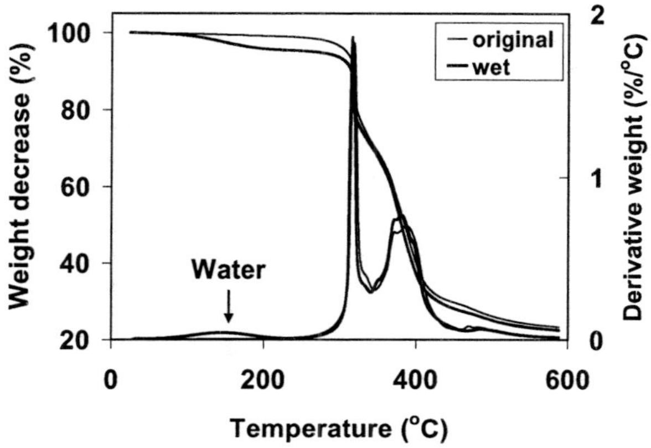


Figure 12: TGA comparison of the original matrix and the immersed matrix in water for 1000 hours (curing condition : 175°C(2 hrs)200°C(2 hrs))

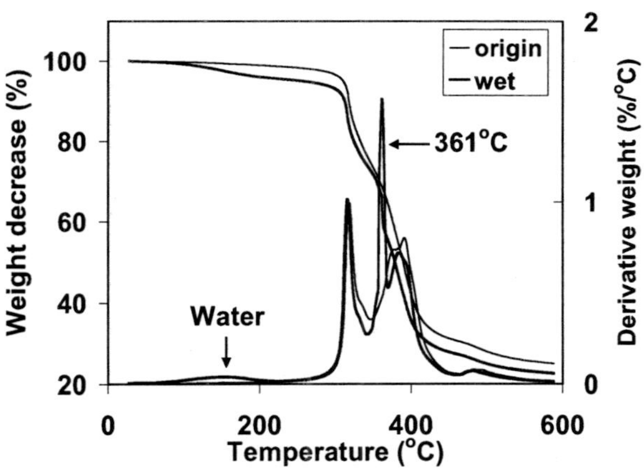


Figure 13: TGA comparison of the original matrix and the immersed matrix in water for 1000 hours (curing condition : 175°C(2 hrs)260°C(2 hrs))

Figure 14 shows the TGA water evaporation patterns of the matrix immersed in water for 1000 hours as a function of curing conditions. The matrix cured at lower the temperature shows a higher evaporation of water content in the region of 100°C-200 °C compared to the matrix cured at the higher temperature. This suggests that matrix cured at the lower temperature has a greater abndance of hydroxyl groups in the matrix so the water molecules bound through hydrogen bonding can be easily removed when heated above 100 °C. The matrix cured at the higher temperature contains higher concentrations of sorbed water that evaporate at temperatures over 200 °C. This result shows that the matrix cured at the higher temperature contains the sorbed water molecule in the tight crosslinks of the network. This sorbed water trapped in the tight crosslinks can not be removed from the matrix in the temperature region of 100°C-200 °C and can contribute to the changing the mechanical performance of the matrix.

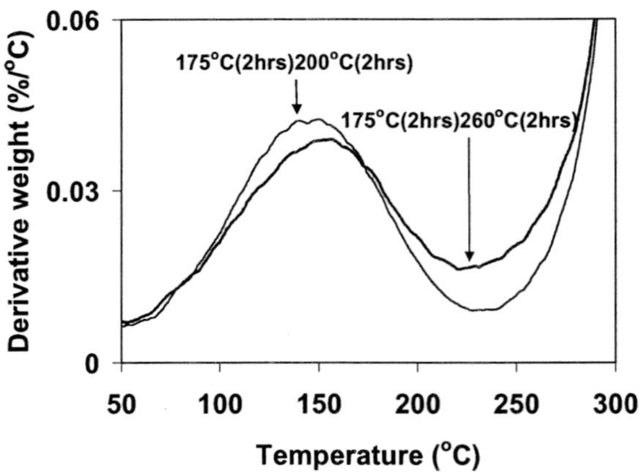


Figure 14: Comparisons of the derivative weight of immersed matrices in water for 1000 hours as a function of curing conditions

### 3.4.3 Water Absorption Effects on Mechanical Performances of the Cured Matrix

Figure 15 shows changes in storage modulus and tan delta (δ) for the matrix cured at the lower temperature as a function of the immersed time. The maximum peak at 0 hour of tan delta (δ) - that is related to $T_g$ - changes to a convoluted peak after immersion in water for 1000 hours. The overall reductions in tan delta values and the shift of $T_g$ to lower temperatures with increasing immersion time are identified. The $T_g$ of the cured matrix changes from 95°C to 82°C and 98°C for the matrix immersed in water for 1000 hours. This indicates that the sorbed water acts as a plasticizer by increasing the mobility of the chains and decreases $T_g$.

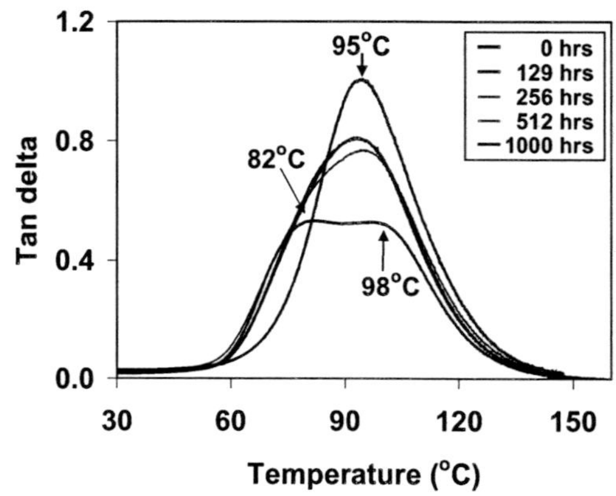


Figure 15: Changes of storage modulus and $T_g$ of the cured matrix with immersion time in water (curing condition: 175°C(2 hours), 200°C(2 hours))

Increasing values of the tan delta (δ) height at temperatures below the $T_g$ of the dry sample were related to the changes in the effective crosslink density caused by water molecules and polymer interactions. The reductions in the effective crosslink density increased the distribution of molecular weights participating in the glass relaxation and caused the observed broadening of the tan delta peaks in the glass transition region to lower temperatures, that is, longer relaxation times [25, 39]. The storage modulus of the matrix does not change significantly with increasing immersion time.

Figure 16 shows the recovery pattern of the storage modulus and $T_g$ for the immersed matrix cured at the lower temperature. The convoluted peak of the matrix immersed in water for 1000 hours changes to a single peak similar to the original matrix when the water uptake matrix is conditioned using the DMA cycling test. Heating the specimen to 150°C and cooling it down to room temperature is one DMA cycle. This $T_g$ shows a constant value of 105°C after the second DMA cycle

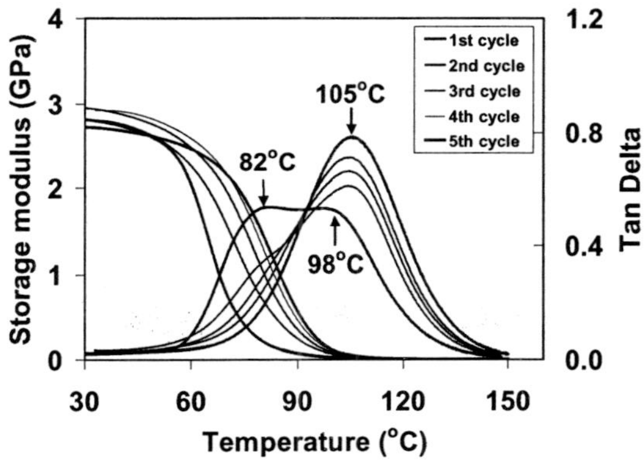


Figure 16: The recovery pattern of the storage modulus and $T_g$ of the immersed matrix in water for 1000 hours (curing condition: 175°C(2 hours), 200°C(2 hours))

This reversion behavior can be interpreted on the basis of studies of hygrothermal effects of epoxy resin. Because the cured matrix of GMAEVC and epoxy resin has hydrophilic functional groups on the surface, water easily binds to the surface. The absorbed water exists as two different types: bound water and free water. Bound water is characterized by strong interactions with hydrophilic groups of the matrix and free water is present in capillaries and microvoids within the matrix [26]. The bound water molecules are divided into two types. Type I bonding corresponds to a water molecule that forms a single hydrogen bond within the matrix network. This water molecule possesses lower activation energy and is easy to remove. Type II bonding is the result of a water molecule forming multiple hydrogen bonds within the matrix network. This water molecule possesses higher activation energy and is correspondingly harder to remove. Type I bound water acts as a plasticizer and decreases $T_g$. In contrast, Type II bound water contributes to an increase of $T_g$ by forming secondary crosslinked networks [24, 40].

Figure 17 shows the changes of storage modulus and tan delta (δ) of the matrix cured at the higher temperature as a function of immersion time. These results are very different from the results for the matrix cured at the lower temperature. The storage modulus and tan delta (δ) increases and the storage modulus decreases with increasing immersion time.

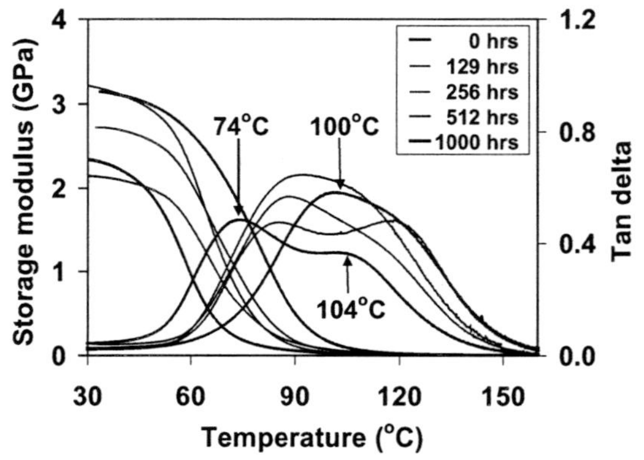


Figure 17: Changes of storage modulus and $T_g$ of the cured matrix with immersion time in water (curing condition: 175°C(2 hrs), 260°C(2 hrs))

Figure 18 shows the recovery pattern of the storage modulus and $T_g$ of the immersed matrix cured at the higher temperature. The convoluted peak of $T_g$ for the immersed matrix shows a higher peak at the lower temperature, however, this peak changes to a convoluted peak that has the higher peak in the higher temperature with increasing of the cycling. The storage modulus shows different value from each cycles and does not be fully recovered after the fifth cycle.

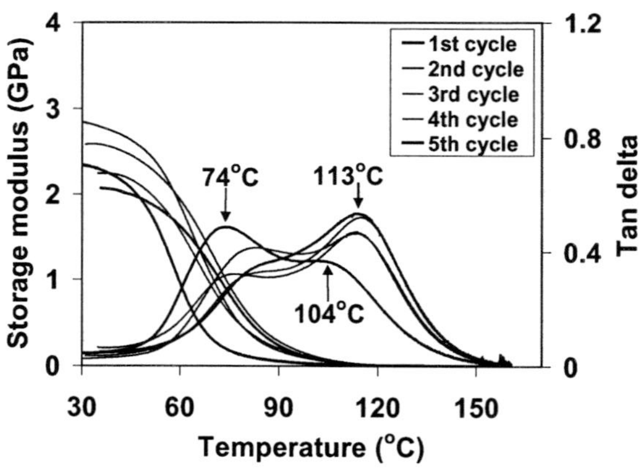


Figure 18: The recovery pattern of the storage modulus and $T_g$ of the immersed matrix in water for 1000 hours (curing condition: 175°C(2 hours), 260°C(2 hours))

The decrease in storage modulus and the changes in $T_g$ observed for increasing immersion time for the matrix cured at the higher temperature can be interpreted as s reaction between the crosslink structure and the sorbed water. The matrix cured at the higher temperature has a tighter crosslink structure than the matrix cured at the lower temperature. The sorbed water in the tight network of the matrix cannot be easily removed. Therefore, the sorbed water is trapped into the tight network of the matrix and causes degradation processes such as chain scission. The degradation of polymer network can be deduced from the reaction of the water molecule with polymer chain or the vibration of the water molecule resulting in the production of shorter chain. This result is consistent with the TGA result of the immersed matrix cured at the higher temperature that showed a new peak at 361°C. This results from the new products of shorter chain.

### 3.4.4 Water Absorption Effects on Phase Morphology of the Cured Matrix

Though most of the fracture surface of the matrices similar to a brittle epoxy specimen, the matrix cured at a lower temperature has more ductile regions in the shape of globules compared to the matrix cured at a higher temperature. Figure 19 compares the phase morphology of the immersed matrices cured at different temperatures. For the immersed matrices in water for 1000 hours the fracture surface indicated the presence of large globules of plasticized material and very large stringers of plastic regions for the matrix cured at a lower temperature; however, the specimen cured at a higher temperature shows typical brittle fracture, but the small hackle structure and the thin stringers indicate localized ductility. Some large globular structures are still present. Therefore, immersing the matrix in water for extended periods of time seems to preferentially plasticize regions of the sample. The water molecules trapped in the tight crosslinks of the matrix cured at higher temperature have a significant effect on the typical brittle fracture.

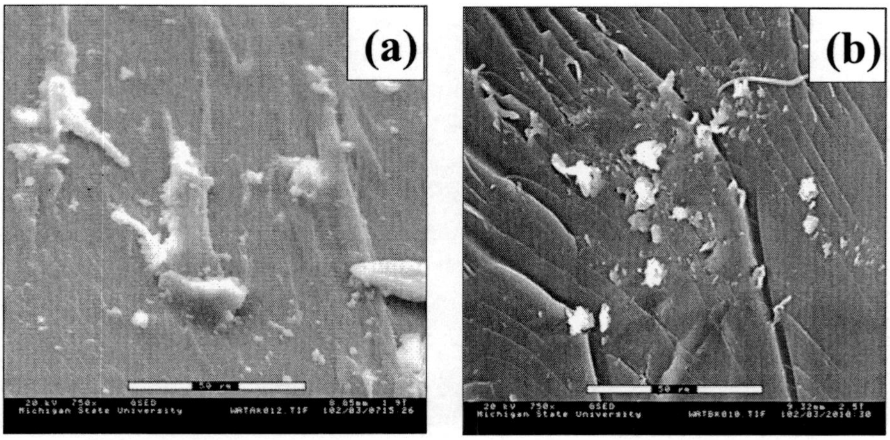


Figure 19: Phase morphology of the immersed matrices cured at different temperatures (X750)

(a) After 1000 hrs immersion of the matrix cured at 175 °C(2hrs)200°C(2hrs)
(b) After 1000 hrs immersion of the matrix cured at 175 °C(2hrs)200°C(2hrs)

The relationships between the water absorption and the crosslink structure for the polymer matrix have been previously studied. Barrel has investigated the water absorption of an epoxy matrix and showed that water absorption depends on the availability of molecular sized holes in the polymer structure. The availability of these holes depends on the polymer structure, morphology, and crosslink density [41]. Apicella et al. has reported that absorbed water in an epoxy can lower $T_g$ and storage modulus. Furthermore, sorption of water may act both as a plasticizer and a crazing agent for the epoxy matrix. Therefore, the presence of water in a polymer can lead to marked changes in chemical and physical state of the polymer [29]. Neve et al. has reported that changes in the final cure temperature have the effect of changing both the extent and distribution of the types of water molecules present in the polymer matrix resulting from the structure and the crosslink density of the matrix and that the water molecule can induce physical and chemical modifications of the matrix [42].

From comparisons of the sorbed water effects on the polymer cured under different conditions, it can be concluded that the sorbed water in the polymer contributes to the changes of the thermomechanical properties differently depending on the crosslink structure. It is concluded that water absorption leads both to plasticization effects and chemical modification of the hydrophilic polymer matrix of GMAEVC and epoxy resin depending on the structure and crosslinks of the matrix.

## Conclusion

An ecofriendly polymer matrix of epoxy resin and GMAEVC has been investigated as a candidate matrix for biocomposites. Curing characteristics of epoxy resin and carboxyl functionalized glucose copolymer (glucose maleic acid ester vinyl copolymer: GMAEVC) have been studied by DSC and FTIR methods. The curing mechanism of the epoxy resin and GMAEVC is identified as esterification and etherification reactions of the hydroxyl and carboxyl functionalities of GMAEVC with the epoxy groups of the epoxy resin. The results showed that esterification reaction occurs in the early stage of cure and then etherification proceeded after completion of the esterification.

The cured matrix containing 50wt% of GMAEVC with different degrees of carboxylic group substitution was prepared and characterized. The cured matrix with a higher degree of carboxyl group shows an increase in the atomic ratio of oxygen to carbon of the cured matrix and a decrease of density resulting from the bulky crosslinks. The cured matrix of epoxy resin and GMAEVC showed a remarkable thermal stability up to 300°C. The average glass transition temperature and storage modulus of the matrix with epoxy resin and GMAEVC were as high as 95°C and 2700 MPa, respectively. This study illustrated the potential for development of environmentally friendly polymer matrix based on epoxy resin and GMAEVC for biocomposites. Biocomposites made from this matrix material could be used in durable goods applications such as those found in the automobile and construction industries.

The effects of water absorption on hydrophilic polymer matrices based on glucose maleic acid ester vinyl resin (GMAEVC) and epoxy resin were also studied as a function of curing temperature. The matrix cured at higher temperature shows the compact crosslinks due to the etherification in curing process comparing to the matrix cured at lower temperature. The matrix cured at different temperature was immersed in water for 1000 hours and the sorbed water effects on properties of the matrix were characterized. The hydrophilic property of the cured matrix of GMAEVC and epoxy resin shows the faster weight increase comparing to the epoxy matrix itself.

Two types of sorbed water were identified from the immersed matrix in the water for 1000 hours. Type I of sorbed water contributed mainly on the weight increases and $T_g$ decrease of the matrix due to the plasticizer effect. Decrease of $T_g$ of the matrix was recovered after heating the matrix up to 150°C. However, Type II of sorbed water was bound in the crosslinks network of the matrix and did not removed from the matrix after heating the matrix to 110°C for an hour. The water molecule trapped in the crosslinks network of the matrix contributed to the thermomechanical properties of the matrix differently depending on the curing conditions.

The cured matrix at higher temperature has the comparatively tight crosslinks in the network structure and the sorbed water molecule disturbed the polymer network resulting in the degradation of the matrix such as chain scission. The thermomechanical properties of the matrix cured at higher temperature were changed with the immersion time in water and did not recovered after heating the matrix to 150°C. Perturbation of crosslinks due to the sorbed water degraded the polymer. Brittle fracture was observed in the polymer matrix cured at higher temperatures; however, the comparatively bulky polymer matrix cured at lower temperatures showed the better stability in wet environment.

This study illustrated the potential for development of environmentally friendly polymer matrix based on epoxy resin and GMAEVC when the curing condition is considered properly for the application environments of the polymer matrix. Biocomposites made from this thermally stable polymer matrix could be used in high and wet environments such as those found in the automobile industries.

## Acknowledgement

The authors of this paper would like to thank EcoSynthetix Inc. for supplying the glucose maleic acid ester vinyl copolymer. The authors would like to thank to Dr. P. A. Askeland for the XPS analysis and Dr. Richard Schalek for ESEM work and the review of this paper.

## References

[1] Nickel, J.; Riedel, U. *Materials Today*. 2003, 44-48.
[2] George, M. *Materials Today*. 2003, 36-43.
[3] Mohanty, A. K.; Misra, M.; Drzal, L. T. *J Polym Environ*. 2002, vol. 10, no.1/2, 19-26.
[4] Caroline, B. *Composites Science and Technology*. 2003, vol. 63, 1223-1224.
[5] Mohanty, A. K.; Misra, M.; Hinrichsen, G. *Macromol. Mater. Eng*. 2000, vol. 276/277, 1-24.
[6] Van de Velde, K.; Kiekens, P. *Polymer Testing* 2002, vol. 21, 433-442.
[7] Rong, M.; Zeng, H. *Polymer* 1996, vol. 37, no. 12, 2525-2531.
[8] Bledzki A.K.; Gassan, *J. Prog Polym Sci*. 1999, vol.24, 221-274.
[9] Saheb, D. N.; Jog, J. P. *Advances in Polymer Technology* 1999, vol.18, no. 4, 351-363.
[10] Wu, S.; Soucek, M. D. *Polymer*. 1998, vol. 39, no. 23, 5727-5759.
[11] Sue, H. J.; Puckett, P. M.; Bertram, J. L.; Walker, L. L. In: Pearson, R. A.; Sue, H. J.; Yee, A. F. editors. *Toughening of plastics. ACS Symposium Series* **759**, Washington, DC: American Chemical Society; 2000, 171.
[12] Kinloch, A. J.; Shaw, S. J.; Tod, D. A.; Hunston, D. L. *Polymer* 1983, vol. 24, 1341-1354.
[13] Pearson, R. A.; Yee, A. F. *J Mater Sci* 1986, vol. 21, 2475-2488.
[14] Bucknall, C. B.; Gilbert, A. H. *Polymer* 1989, vol. 30, 213-217.
[15] Pearson, R. A.; Yee, A. F. *Polymer* 1993, vol. 34, 3653-3657.

[16] Azimi, H. R.; Pearson, R. A.; Hertzberg, R. W. *J Appl Polym Sci* 1995, vol.58, 449-463.
[17] U.S. Pat. 5,872,199 (Feb. 16, 1999) Steven B, Ian JM, Ramani N; *U.S. Pat.* **6**, 242, 593 (June. 5, 2001) Steven B, Ian JM, Ramani N.
[18] Han, S. O.; Defoort, B.; Drzal, L. T. 222nd American Chemical Society Meeting *Polymer Priprint.* 2001, vol. 42, no. 2, 218-219.
[19] Mohanty, A.K.; Misra, M.; Hinrichsen, G. *Macromol Mater Eng.* 2000, vol. 276, 1-24.
[20] May, C.A.; Tanaka, Y. *Epoxy resins chemistry and technology*. Marcel Dekker: New York, 1988. pp. 60-66.
[21] Lee, H.; Nevelle, K. *Handbook of epoxy resins*, McGraw-Hill Book Company, 1982 (chapter 5, 6).
[22] Shechter, L.; Wynstra, *J. Ind. Eng. Chem.* 1956, vol. 48, no. 1, 86-93.
[23] Cotugno, S.; Larobina, D.; Mensitieri, G.; Musto, P.; Ragosta, G. *Polymer* 2001, vol.42, 6431-6438.
[24] Zhou, J.; Lucas, J. P. *Polymer* 1999, vol.40, 5505-5512.
[25] Zhou, J.; Lucas, J. P. *Polymer* 1999, vol.40, 5513-5522.
[26] Pethrick, R. A.; Hollins, E. A.; Mc Ewan, I.; Pollock, E. A.; Hayward, D. *Polym Int* 1996, vol. 39, 275-288.
[27] Mikols, W. J.; Seferis, J. C.; Appicella, A.; Nicolais, L. *Polym Comps* 1982, vol. 3, 118-124.
[28] Xiao, G. Z.; Shanahan, M. E. R. *Polymer* 1998, vol. 39, no. 14, 3253-3260.
[29] Apicella, A.; Nicolais, L.; Cataldis, C. *Adv Polym* 1985, vol. 66, 189-207.
[30] ASTM D570-98; *Standard test methods for water absorption of plastics.*
[31] Defoort, B.; Drzal, L.T. *SAMPE International Symposium* 2001, 2550-2561.
[32] Density measurement instruction manual, *Techne Inc.*
[33] Han, S. O.; Defoort, B.; Askeland, P. A.; Drzal, L. T. *33rd International SAMPE Technical Conference*: Advancing affordable materials technology, 2001, 1466-1478.
[34] Nakamura, S.; Saegusa, Y.; Yanagisawa, H.; Touse, M.; Shirai, T.; Nishikubo, T. *Thermochimica Acta* 1991, vol. 183, 269-277.
[35] Oh, J. H.; Jang, J.; Lee, S. *Polymer* 2001, vol. 42, 8339-8347.
[36] Socrates, G. *Infrared Characteristic Group Frequencies*, John Wiley & Sons, Ltd., 1980, p. 46, pp. 67-71.
[37] Park, W. H.; LEE, J. K. *J Appl Polym Sci* 1998, vol. 67, 1101-1108.
[38] Barral. L.; Cano, J.; López, J.; López-Bueno, I.; Nogueira, P.; Abad, M. J.; Torres, A.; Ramírez, C. *J. Applied Polymer Science* 2000, vol.77, 2305-2313.
[39] Nogueira, P.; Ramirez, A.; Torres, M.; ABAD, J.; Cano, J.; Lopez, J.; Lopez-Bueno, I.; Barral, L. *Applied Polymer Science* 2001, vol. 80, 71-80.
[40] Ping, Z. H.; Nguyen, Q. T.; Chen, S. M.; Zhou, J. Q.; Ding, Y. D. Polymer 2001, vol. 42, 8461-8467.
[41] Barral, L. *Journal of Thermal Analysis* 1996, vol. 47, no. 3, 791-797.
[42] Neve, B. D.; Shanahan, M. E. R. *J. of applied Polymer Science* 1993, vol. 34, 5099-5105.

In: Frontiers in Polymer Research
Editor: Robert K. Bregg, pp. 27-42
ISBN: 1-59454-824-2


***Chapter 2***

# A MACROMOLECULAR OXIDANT, THE N,N–DICHLOROSULFONAMIDE FOR REMOVAL OF RESIDUAL NITRITES FROM AQUEOUS MEDIA

***Romuald Bogoczek, Elżbieta Kociołek-Balawejder and Ewa Stanisławska***
Chair of Industrial Chemistry, Wroclaw University of Economics
ul. Komandorska 118/120, 53-345Wrocław, Poland

## Abstract

Nitrites are highly harmful compounds. They are extremely undesirable in surface and municipal water. Its permissible content in natural water is very low and should not exceed 0.01 mg $NO_2^- / L$. A redox copolymer, a macromolecular analogue of Dichloramine T (i.e. a macroporous S/DVB copolymer containing $SO_2NCl_2$ groups) was used here for removal of nitrite ions from aqueous solutions by its oxidation to the hundred folds less toxic nitrates. The resin was prepared starting from *Amberlyst 15* by a three-step transformation of the sulfonic- via chlorosulfonyl and sulfonamide- to the N,N–dichlorosulfonamide groups. The resulting copolymer contained 8.2 meq of active chlorine/g and showed strong oxidizing properties. It was employed in batch and flow processes for treatment of $NaNO_2$ solutions containing 115, 230 or 460 mg $NO_2^- / L$. The effects of various parameters on the reaction course have been studied (mole ratio of reagents, pH of the reaction media, flow rate in the column processes). The solid phase oxidation carried out in a dynamic regime provided to drive the reaction to completion. Thus, nitrite free effluents (< 3.0 μg/L) were obtained in the column processes. The reaction of nitrite oxidation by means of this heterogeneous oxidant was fast and therefore the permissible flow rate was very satisfactory - close to 20-25 bed volumes/h. Under the examined reaction conditions, 1 mol of $NO_2^-$ ions was oxidized by 1 mol of active chlorine, so the oxidation capacity of the resin was nearly 200 mg $NO_2^- / g$ of the copolymer. The N,N–dichlorosulfonamide copolymer is very useful for purification of neutral or medium acidic solutions from nitrites. In the case of alkaline solutions the nitrite oxidation reaction proceeds slowly, and what more o blocking of a part of the active chlorine atoms in the copolymer takes place. The intermediate $SO_2NClNa$ groups do not oxidize the nitrite ions. The here determined copolymers redox potentials and carried out complex redox titration measurements proved why the macromolecular oxidant shows various reactivity in dependence of the solution pH.

# Introduction

## The Nature and Use of the Reactive Polymers

Reactive (co)polymers being chemical mixtures of synthetic high - molecular compounds, insoluble in water and organic solvents, built as spatial crosslinked porous resins, include reactive functional groups. In the most widespread products - *Ion exchange resins* these groups have the ability of cations- and anions-exchange. Ion exchange resins are used primarily for industrial water treatment (65-70% of the world-production). They are ideally suited to demineralise or soften water when the most stringent treated water specifications are required. The recipients of so prepared water are the following industries: power, electronics, automobiles, oil refinery, kraftpulp mills, steel mills, chemicals, petrochemicals, pharmaceuticals, textile, synthetic fiber, brewing or sugar industry. Particular application of ionites in numerous processes increase with time because they enable efficient and elegant methods for solving many problems in chemical engineering applied in industry.

Ion-exchange resins are i. a. installed for [1]:

- treatment of potable water (e. g. for removal of nitrates or perchlorates),
- for food processing - hydrolysis of oligo- and polysaccharides; demineralization of glucose, fructose, liquid sugar, polyols, whey, gelatin, fruit juices, citric acid, lactic acid; recovery of amino-acids, nucleic acids, glutamic acid, ascorbic acid; and for decalcification in beet processing,
- as bulk pharmaceutical chemicals for finished dosage forms - taste masking agent, modified release, tablet disintegrant, drug stabilizing agent, potassium reduction, cholesterol reduction, bile acid sequestrant,
- for chemical processing - brine softening in membrane chlor-alkali plants; boron removal from concentrated $MgCl_2$ solution; removal of heavy metals from aqueous and non aqueous streams such as hydrocarbons; purification of aqueous and organic solutions (phenol, glycerin, hydrogen peroxide); hydrometallurgy (gold recovery from cyanide leach liquors, uranium recovery from ore),
- as solid polymeric catalysts for the production of: oxygenates such as MTBE, ETBE, TAME, Bisphenol A, alkylphenols, tertiary butyl alcohol, fatty alcohols, 1,4-butanediol, pentaerythritol, methyl methacrylate,
- for regenerable condensate polishing applications in nuclear and fossil power plants,
- for non-regenerable applications in nuclear power plants,
- for production of ultra pure water in semiconductor industry,
- as fertilizing agents in hydroponics the plants.

Next to the rich offer of reactive polymers designed for classic processes of water-treatment it happens that in the market place appear more and more new products having unique properties for special destination. They are produced according the needs of biochemistry, organic syntheses and for chemical analysis. These are nucleophilic and electrophilic scavenger resins, polymer-supported bases, polymer-bound coupling reagents, polymeric reagents for introduction of protecting groups to amines. Chromatographic resins were developed to solve the problems associated with large scale biomolecule purification. They are designed for laboratory and process scale purification of proteins, peptides, nucleic acids, antibiotics, and small molecular weight pharmaceuticals.

Last but by no means least, the Nobel Prize winner's Merrifield nucleic acid synthesis is accomplished thanks to reactive polymers.

To the class of the reactive polymers rank also redox polymers it is those when to the high - molecular skeleton moieties having oxidative and/or reductive properties are incorporated. According to literature, reactive polymers are divided into three groups, so far [2-5]:

1. *Electron-exchange polymers,* these are (co)polymers which include covalently fixed molecular formations with a reversible redox character and which do not possess ion-exchange properties. (these are unites like hydroquinone or catechol).
2. *Electron and ion-exchange polymers*, these are (co)polymers which include covalently bound, to the macromolecular matrix, electron-exchange as well as ion-exchange groups (e. g. thiolic cationites like [P]–$CH_2SH$) or [P]–SH).
3. *Oxidative and reductive ion-exchangers* produced from traditional market cation-exchangers or anion-exchangers or complex forming resins on which, by use of the ion-exchange or a complex-forming process or physical adsorption, a substances with redox properties was placed (e. g. on cation-exchangers a cation of a metals with variable valence, on anion-exchangers the anions being oxidants or reducers).

We propose to distinguish a forth class of reactive polymers, the so called *macromolecular oxidants* or *macromolecular reductans*. These are (co)polymers with covalently bound oxidative or reductive molecular moieties. To this class belongs the used in this paper macromolecular N,N–dichlorosulfonamide. This is an oxidant but it is by no means a reductant. Even not after its oxidation process has been completed. These kind of macromolecular oxidants have been investigated by a number of authors.

Commercially available are redox polymers both with oxidative and reductive properties, although their offer is very modest. But one can observe a development in this matter. The process of removing oxygen dissolved in water is proposed by big industrial Companies producers of this kind of reactive polymers [Bayer AG and Rohm and Haas Co.]. This method involves the use of metallic palladium-doped ion exchange resins (an anionite in its chloride form is used) combined with the injection into the oxygen containing water of stoichiometric amount of hydrogen. The catalytic reaction between oxygen and hydrogen takes place at the inner and outer surface of an ion exchange resin. The product of the reaction is water only. The residual oxygen level is usually below 10 μg/L, whereas the raw deoxidized water usually contains 5-15 mg $O_2$/L. This method has become increasingly popular, it is 70% cheaper to operate than correspondingly dimensioned vacuum degassing units. It allows also to get rid of the use of chemicals like hydrazine.

*Novabiochem* offers Redox Polymers in reagent quantities for solution phase organic synthesis. These reagents can be used to simplify reaction work-up and product isolation. Polymer-supported oxidants comprise anion-exchange resins in forms of perruthenate-, metaperiodate, oxoammonium moieties, whereas polymer-supported reductants are anionites in form of cyanoborohydride or borohydride.

An important group of the intensely studied macromolecular oxidants is a high - molecular copolymer which incorporates to a nitrogen atom covalently bound active halogen atoms being on the + 1 degree of oxidation. Attention deserve the numerous works on this subject coming from the teams of D.W. Emerson et al. [6-10] and S. D. Worley et al.[11, 12], who studied the possibility of a bead-form biocide product utilization in water disinfection. Different authors describe reactive (co)polymers used in organic syntheses as oxidizing and chlorinating factors [13-15].

## The Environmental Impact of Nitrites and Their Elimination

In our investigations we deal with the synthesis and application of copolymers which include active halogen in functional groups comprising the N–monohalogenosulfonamide and/or N,N–dihalogenosulfonamide of the S/DVB copolymer and we look forward to their application of ameliorating very diluted solutions of harmful residual matter. We showed as yet, that they are effective in removing some toxic admixtures from waters, which as a result of oxidation transform into environmental safe substances. Applied in a column process heterogeneous oxidants are especially effective for removing of residual contaminants from aqueous media. Our recent investigations refer to the removal of nitrites from waters. The nitrites are dangerous for the living biosphere. It is the property of nitrites to react with the amino groups of albumens with the result of nitrosamines formation which show mutagenic and carcinogenic activity. This is why the content of nitrites in natural waters one of the main decisive factors of its cleanliness class is. The toxicity of nitrites for human beings results also from the enzymatic reaction of the $Fe^{2+}$ ion present in the hemoglobin molecule. After this reaction the hemoglobin loses their binding ability of oxygen what can lead so to anoxaemia of the organism, especially threatening young children (methaemoglobinaemia). Therefore the content of nitrites in the natural water is one of the main decisive factors indicating the class of cleanliness. The high content of nitrites in water is always undesirable and testifies the pollution of the aqueous environment, because they form in the waters as result of the natural processes - the nitrification and the denitrification processes. The permissible content of the nitrites in the natural waters the first class of the purity, is very low and should not exceed 0.01 mg $N_{NO_2^-}/L$. This is comparable to the permissible content of the cyanides, sulfides, formaldehyde as well as cations of heavy metals, which are generally well-known poisons.

Nitrites can be formed in water-supply systems as a result of denitrification bacteria activity, and also in water treatment - e.g. during iron removal, because nitrates irrespective of oxygen, can be the oxidants of $Fe^{2+}$ ions. The increasing content of nitrates in the environment is imputing to the irrational fertilization of soils with nitrogen species, mineral as well as natural, which are the subject of a multidirectional transformation in the soil and as a consequence of their penetration via sewage into the water system. Numerous manufacturing processes using nitrites as raw materials are also well known - the production of dyes, the synthesis of nitrogen-containing organic derivatives, the galvanic metal treatment, meat pickling and auxiliary chemicals used in corrosion protection in water circuits and cooling circuits.

The removal of nitrites from natural waters and industrial solutions can be carried out by chemical methods, using oxidation to nitrates. In this case the use of a strong oxidizing agent is required. Chlorine or sodium hypochlorite, chlorine dioxide, ozone, hydrogen peroxide can be used [16]. The deep removal of the residual undesirable substances demands extensive use of a homogeneous oxidant with the result that the conditioned solution is polluted again, although with other kind of substance.

The permissible content of nitrates in waters is 5.0 mg $N_{NO_3^-}/L$. It is several hundred times higher than the permissible content of nitrites. The removing of nitrates from waters does not create any problem. Resins with a high affinity to nitrates are produced since long (Imac HP 555, produced by Rohm and Haas Co.). They were especially developed for selective nitrate removal from potable waters. They are even more selective for nitrates than for sulfates what is an important feature over conventional strongly basic resins because of the high proportion of sulfates present normally in the purified potable water.

In a previous paper we described the results of our investigation on the oxidation of nitrites using an S/DVB copolymer with N–monochlorosulfonamide functional groups [17]. Bringing a diluted $NaNO_2$ aq solution with the sodium monochloro derivative of the sulfonamide copolymer into contact, no oxidation reaction of the nitrite could be observed, either in the batch- or in the column-process. We discovered that the oxidation of the nitrite ions to the hundred folds less toxic nitrate ion can be performed by the copolymeric hydrogen form of the N–monochlorosulfonamide. As we showed, this copolymer is especially suitable for the removal of nitrites from their very diluted solutions, i.e. in the concentration of tens of mg $NO_2^-/L$. Working in more concentrated solutions, say several hundreds of mg $NO_2^-/L$, in addition to this reaction of oxidation a number of disadvantageous consecutive reactions occur, as a result of the reaction medium excessive acidification. In this case an evolution of gaseous by-products create difficulties in the column process.

## Here Disclosed Method

In the method disclosed in the now presented paper, the oxidation of nitrites by active-chlorine contained in resin has been extended to a copolymeric N,N–dichlorosulfonamide which contains extremely large amount of active chlorine. The concentration of active chlorine in this well swollen in water copolymer, attains ca. 2.5 M. Its strong oxidizing activities for cyanides, thiocyanates and sulfides we have shown in previous contributions [18-20].

The purpose of this investigation is the examination of the possibility of nitrites removal from aqueous solutions, where their primary concentration is tens to several hundreds of milligrams/L and their final concentration is less than 0.1 mg/L. The following reaction course has been studied:

$$[P]\text{–}SO_2NCl_2 + 2\,NO_2^- + 2H_2O \longrightarrow [P]\text{–}SO_2NH_2 + 2\,NO_3^- + 2HCl \qquad (1)$$

[P] stands for the copolymer styrene/divinylbenzene, macroporous structure.

The N,N–dichlorosulfonamide copolymer used here is a high molecular equivalent of the known low molecular equivalent of oxidants, such as the aromatic Dichloramine T. In literature we found the information, that monochloramine T is used by a redoxymetric method as reagents for the quantitative, titrimetric, $NO_2^-$ ions determination in analytical chemistry [21-23].

# Experimental

## Reagents

The copolymer that had N,N-dichlorosulfonamide groups (DCSR - stands for the styrene/divinylbenzene dichlorosufonamide resin) was prepared by the method previously described [24, 25]. The following procedure was used:

$$[P]-SO_3H \longrightarrow [P]-SO_2Cl \longrightarrow [P]-SO_2NH_2 \longrightarrow [P]-SO_2NCl_2$$

As a starting material Amberlyst 15 (produced by Rohm and Haas Co.), a commercially available sulfonate cation exchanger was used. This is a macroporous poly(S/20%DVB) resin which, in the air dried state, contained 4.7 mmol $SO_3H$/g, surface the area 45 $m^2$/g, average pore diameter 25 nm. Its initial functional groups we transformed to the chlorosulfonyl, and then to sulfonamide groups, which joined the active chlorine atoms as a result of the sodium hypohlorite in acetic acid medium reaction. The product contained 2.05 mmol $SO_2NCl_2$/g (i.e. 4.10 mmol of active chlorine/g or 8.20 miliequivalent active chlorine/g) and a small amount of sulfonic groups (0.60 mmol $SO_3H$/g).

Analytical grade sodium nitrite served for the preparation of the aqueous solutions containing $NaNO_2$ alone or in a mixture with sodium hydroxide or acetic acid in different proportions. Solutions used in the batch regime experiments were: 0.01 M $NaNO_2$ (i.e. 460 mg $NO_2^-/L$) in: (a) water, (b) 0.01 or 0.05 M $CH_3COOH$, (c) 0.01, 0.02 or 0.1 M NaOH. The solutions carried out in a dynamic regime were: 0.005 M $NaNO_2$ (230 mg $NO_2^-/L$) in: (a) water, (b) 0.01 M $CH_3COOH$ and 0.0025 M $NaNO_2$ (115 mg $NO_2^-/L$) in 0.005 M NaOH.

## Analytical Methods

The nitrite and nitrate ions concentrations were determined by colorimetric methods (Spekol 1200, Analytic Jena, Germany). The nitrites concentration was determined by a modified Griess-Ilsovay method. The reaction of a violet diazo dye formation of sulfanilic acid and dihydrogenchloride of N-(1-naphthyl)etylenodiamine was used. The absorbance measurement was taken at 545 nm wavelength. Nitrates were determined with sodium salicylate and the formation of the yellow nitrosalicylic acid was applied. The absorbance was determined at 410 nm wavelength [26]. Chloride ions were estimated by argentometric titration using 0.01

M $AgNO_3$ and the Ag/AgCl/calomel electrodes system. Hypochlorites in solutions and the active chlorine content in the resin were determined by the iodometry.

The redoxymetric titration of the DCSR copolymer was performed by the use of 0.01 M $NaNO_2$ in different media: in (a) water, (b) 0.01 M $CH_3COOH$, (c) 0.01 M NaOH. Into eleven separate samples of the resin (ca. 0.24 g, i.e. ~2 miliequivalent active chlorine in each sample) the following increasing solution volume rations of $NaNO_2$ were introduced. Respectively: (1) 0, (2) 12.5 mL, (3) 25 mL, (4) 37.5 mL, (5) 50 mL, (6) 62.5 mL, (7) 75 mL, (8) 87,5 mL, (9) 100 mL, (10) 125 mL and (11) 150 mL of 0.01 M $NaNO_2$. To the first (1) sample of the copolymer the distilled water was added only. Increasing solution volumes of $NaNO_2$ were needed to bring about the reduction, for example of: (1) 0%, (5) 50%, (9) 100% of the functional group active chlorine. However, the last two samples contained (10) 125% and (11) 150% of the nitrite ion relative to stoichiometry. These samples, in closed vessels, were shaken at constant temperature (20° C). After 24 hours the electric potentials of the reaction media were measured by means of the platinum/calomel electrode pair and the pH of the solutions by a glass/calomel couple were tested [27, 28]. At the end the contents of nitrites in the post reaction solution were determined.

### Nitrite Solution Treatment

In all studies in batch regime, at room temperature, a measured amount of the resin (ca. 0.25 g) placed in a flask was shaken with 0.01 M $NaNO_2$ solution in different media:

(a) 150 mL, a 150% excess of nitrites relative to the stoichiometry of Eq. (1).
(b) 75 mL, 75% relative to the stoichiometry of Eq. (1).

Time-dependant measurements of the residual $NO_2^-$ contents in solution were made. After the reaction, the copolymeric reagent was separated from the reaction medium by filtration and was analysed for its active chlorine content.

In the dynamic regime of active chlorine content) was packed into a glass column (inner diameter ~1.15 cm; height of package ~17.5 cm). $NaNO_2$ solutions of various alkalinity were passed through the column bed of DCSR. The observed flow rates were 5 to 30 bed volumes per hour. 250 mL fractions were collected to estimate their composition in terms of pH, the nitrites, nitrates, chlorides and hypochlorites. The exhausted copolymer was removed from the column, and then it was water-washed and air-dried. The exhausted copolymer was subjected to analysis of active chlorine contents.

## Result and Discussion

Two facts were taken under consideration while planning the investigation of nitrites removal from aqueous solutions, using DCSR as the macromolecular oxidant.

(a) Nitrites are unstable in acid environment - decomposition of the nitrite ion takes place accompanied by emission of gaseous nitrogen oxides under the action of strong acids. This is why the investigation should be limited to neutral, alkaline and weak medium acidic solution.

(b) Concerning the stability of DCSR in aqueous media earlier experiments showed its highest stability in neutral and acidic media. In strong alkaline media the resin loses a part of its activity by dechlorination (one of the two active chlorine atoms leaves easily the resin phase). Therefore our investigation was limited to medium alkalinity only.

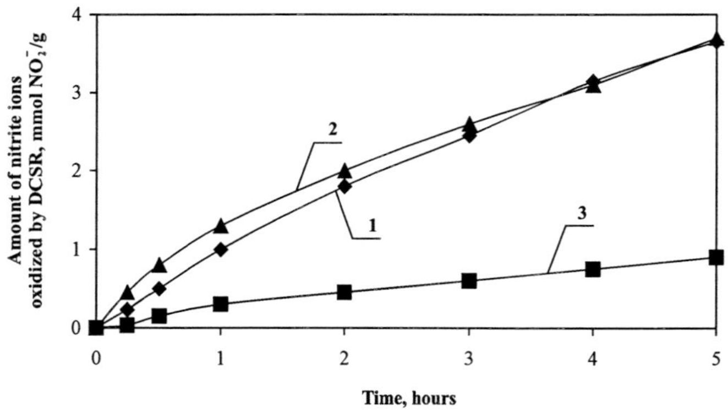


Fig. 1 Decrease of nitrates from the aqueous solution in batchwise reaction:

0.25 g DCSR + 150 mL 0.01 M $NaNO_2$ , pH=6.28
0.25 g DCSR + 150 mL 0.01 M $NaNO_2$ in 0.01 M $CH_3COOH$, pH=3.72
0.25 g DCSR + 150 mL 0.01 M $NaNO_2$ in 0.01 M NaOH, pH=12.62

In the first stage of our work, in the batch regime, the reactive copolymer samples were treated with an excess of the given 0.01 M $NaNO_2$ in various media (Fig.1). The progress of the reaction was studied by tracking the decrease of nitrites in the tested solution. The reaction proceeded well in case of the two tested solutions - $NaNO_2$ alone in solution (curve 1.) and $NaNO_2$ with acetic acid (curve 2.) After 5 hrs the concentration of the nitrites reduced markedly from 460 down to 180 mg $NO_2^-$ / L. That meant that the descent of nitrites was nearly the theoretical value, which is 4.10 mmol $NO_2$ /g of the DCSR. The reaction speed was higher if acetic acid was present in the reaction medium. When the reactions finished both solutions were acidic, pH 2.65; the chloride contents in both solutions were 0.9 mmol. In the post-reaction state the copolymers still showed a small content of active chlorine, i. e. ca. 5% relative to its initial value.

A considerably lower speed of reaction was observed in the case of alkaline solution (curve 3.). After 5 hours a small fall of nitrite concentration i. e. from 460 mg/L to 385 mg/L was observed only. So was the pH value. This dropped from 12,62 to 12.13. This denotes less than 25% of the oxidizing ability of the copolymer being in use. The copolymer itself after reaction, contained ca. 50% of its primarily active chlorine only, this means less than resulted from the decrease of nitrites in solution. The oxidation of nitrites by DCSR in alkaline media proceeds less effectively.

More experiments were carried out by the use of an excess of DCSR in relation to stoichiometry. The reason why an excess of DCSR was chosen is to investigate the efficiency of the copolymer with respect to very small concentration of nitrites and the wish to drive the reaction to completion. In the case of two solutions - $NaNO_2$ alone and $NaNO_2$ in medium acidic solution, after 24 hours the concentration of nitrites felt almost 100 times, i.e. to ca. 5.0 mg/L, as this can be seen in Table 1. The decreasing of the nitrites concentration was quicker in the acidic sample than in the neutral one. The chloride amount found in each of the both post-reaction solutions was 0.7 mmol (and this was equal to the anticipated amount) whereas the post reaction copolymers contained still ca. 25% each of their initial active chlorine contents.

In the 0.02 M NaOH solution the nitrites oxidation proceeded slowly. Their concentration decreased by half after 24 h. In the 0.1 M NaOH medium the reaction rate was the lowest, after 24 hours the nitrite concentration fell ca. 15% only.

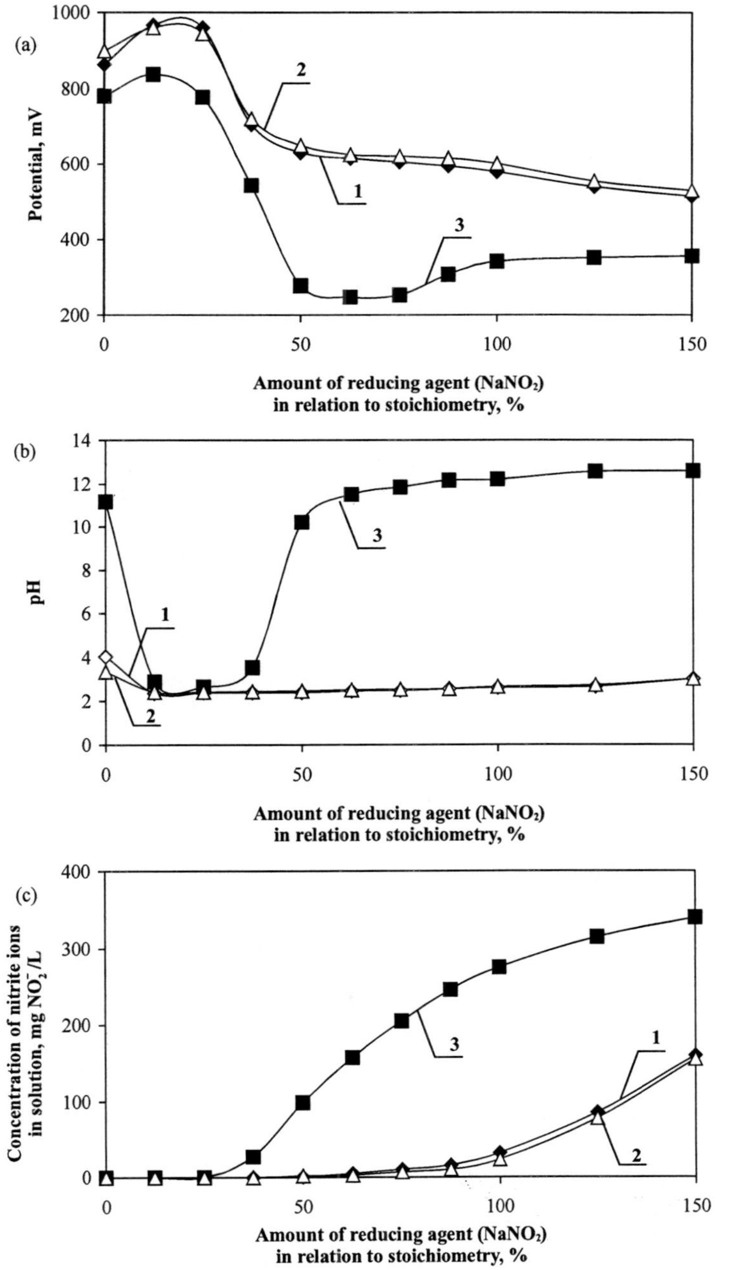


Fig. 2 (a) The redox titration curve of DCSR (0.24 g) by 0.01 M $NaNO_2$ in: (1) water, (2) 0.01 M $CH_3COOH$ and (3) 0.01 M NaOH: (b) pH value (c) the concentration of nitrites in solution

To find out why the N,N–dichlorosulfonamide copolymer, oxidizes the nitrite ion in neutral and medium acidic solution smoothly, whereas in the alkaline solution with difficulty only, the copolymer was subjected to a redox titration by means of 0.01 M $NaNO_2$ in three different media: (a) distilled water, (b) 0.01 M acetic acid, (c) 0.01 M NaOH (Fig.2a).

According to the redox titration theory [27, 28], from the centre of the titration curve on a suitable E (mV) level we can infer about the oxidizing potential and so of the oxidizing or reductive power of the redox copolymer in relation to the given low molecular substance, which is to be reduced or oxidized. The central point of the titration curve, meaning the 50% /50% of the oxidation/reduction state of the functional groups was measured by titration with 0.01 M $NaNO_2$ solution by use of a platinum electrode and a calomel reference electrode coupled to an electrometer. It was the apparent redox potential. To mark the important points at the potentiometric curve (Fig.2a), separate samples of DCSR and $NaNO_2$ were used. That method was used because of the low reaction velocity due to the fact that the reagents performed different phases.

Fig. 2a curve 1 show the redoxmetric titration of the DCSR with 0.01 M $NaNO_2$. The curve has an untypical run. One can distinguish two different levels of reaction characterized by two different redox potentials. The first stage proceeded at a very high level up to +1000 mV (up to 50% $NaNO_2$). We do understand that the DCSR redox reaction follows two different detailed reactions. The first of them is the following:

$$[P]–SO_2NCl_2 + NO_2^- + H_2O \longrightarrow [P]–SO_2NClH + NO_3^- + HCl \qquad (2)$$

According to the reaction (Eq. 2) HCl is evolved and so a decrease of pH was stated in the post-reaction mixture (Fig.2b, curve 1). This conversion proceeds quantitatively. The solution samples (2)–(5) are free of nitrites after reaction (Eq. 2) (Fig. 2c, curve 1).

The second stage of the redox reaction (addition 50-100% $NaNO_2$) as we found, proceeds at a still high but lower potential +650 mV. The nitrites oxidation is still succeeding (Eq. 3) but by means of a second chlorine atom with a lower oxidation power:

$$[P]–SO_2NClH + NO_2^- + H_2O \longrightarrow [P]–SO_2NH_2 + NO_3^- + HCl \qquad (3)$$

The course of the redoxymetric titration of a DCSR sample with $NaNO_2$ in the medium of 0.01 M $CH_3COOH$ (Figs. 2a-c, curves 2.) is similar to the previously discussed titration with an alone $NaNO_2$ solution. The numeric data for the curves were taken 24 h after the reaction start i.e. they do not show the favorable impact of the acidic medium at the beginning of the reaction run.

The course of the redoxymetric titration in 0.01 M NaOH has a different progress in relation to the both media previously discussed over. In this case two of the reaction stages are particularly clearly visible on each of the three graphs (Fig. 2a-c, curves 3).

The first reaction stage (up to 50% $NaNO_2$) represents the quantitative decrease of nitrites in the solution, lowering of the pH and a high redox potential of the reaction medium (up to + 850 mV) is the accompanying result of this. The pH felt to less than 3.0 although the amount of the NaOH in the titrant was sufficient to neutralize the HCl evolved by the oxidation reaction is remarkable.

In the second stage of titration (addition of 50-100% of the stoichiometrically foreseen $NaNO_2$) there does not occur any oxidation of the nitrites. A high concentration of the $NO_2^-$ ions and a new increase of pH were observed in the reaction medium. If oxidation does not take place, HCl does not evolve and so it does not neutralise the NaOH present in the titrant. The reason for the inhibition of the oxidation reaction was the low redox potential of the reaction medium (ca. + 250 mV). The reason for this low potential is the transformation of the macromolecular oxidant functional groups from the hydrogen-form to the sodium-form:

$$[P]\text{–}SO_2NClH + NaOH \longrightarrow [P]\text{–}SO_2NClNa + HCl \quad (4)$$

The sodium-form N-chlorosulfonamide copolymer is to weak for nitrite oxidation, as we have shown earlier [17]. Nitrites, belonging to the so called redox amfoterics show rather weak reductive proprieties; this means they are relatively difficult to oxidize. As a rule the oxidation process is rather strong pH dependent. The redox potentials are considerably higher in acid environment. In the case under examination the basic environment additionally inhibited the reaction run.

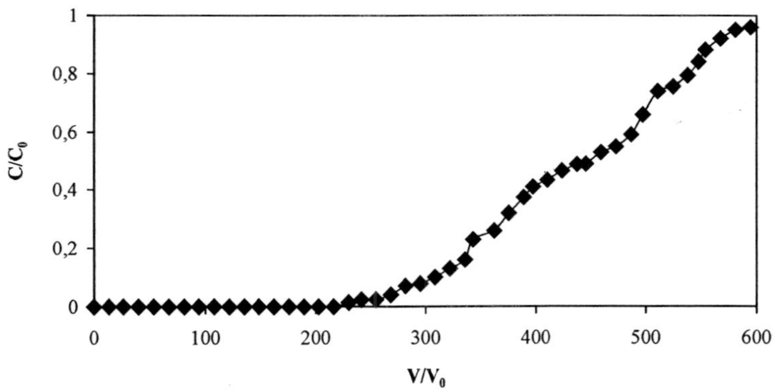


Fig. 3 Nitrites breakthrough curve for DCSR in the column process; influx 0.005 M $NaNO_2$, flow rate 15 bed volumes/hour.

The column investigation began with passing of 0.005 M $NaNO_2$ solution through a resin bed of DCSR (Fig. 3). First it was necessary to establish the flow intensity of solution through the column when the leakage of nitrites should be lowered sufficiently. 20 bed volumes/h was a right intensity and the nitrites concentration was practically undetectable (the nitrite concentration decreased to a level below 0.003 mg/L, when nitrates concentration increased to ca. 300 mg/L). A quantitative oxidation of $NO_2^-$ to $NO_3^-$ took place. The concentration of chlorides in the leakage was ca. 0.005 M accordingly to anticipation. The oxidation reaction was accompanied by acidification of the product solution - this was pH 2.65-2.80, (relatively to the influent of pH 6.28). Up to 250 bed volumes the utilized solution was nitrites free. The quantity of disappeared nitrites from the processed solution was 7800 mL x 0.005 M $NaNO_2$/L = 39.4 mmol. That means that up to the relative concentration of $C/C_o$ = 0.5 the amount of the utilized solution was V/Vo = 450 bed volumes. The quantity of oxidized nitrites in the column process responded to the contents of active chlorine in the column bed. That process came to its end when ca. 11 L of the nitrite solution passed through the column to reach the point $C/C_o = 0.98$. Thereafter the column was emptied and the spent copolymer was dried and weighed, its mass was 8.0 g. It still contained less than 0.5 meq active chlorine/g of the copolymer.

Even more efficient ran the N, N–dichlorosulfonamide copolymer column process when the influent was 0.005 M $NaNO_2$ in 0.005 M acetic acid. The solution flow intensity could be even higher, i. e. 25 bed volumes/h.

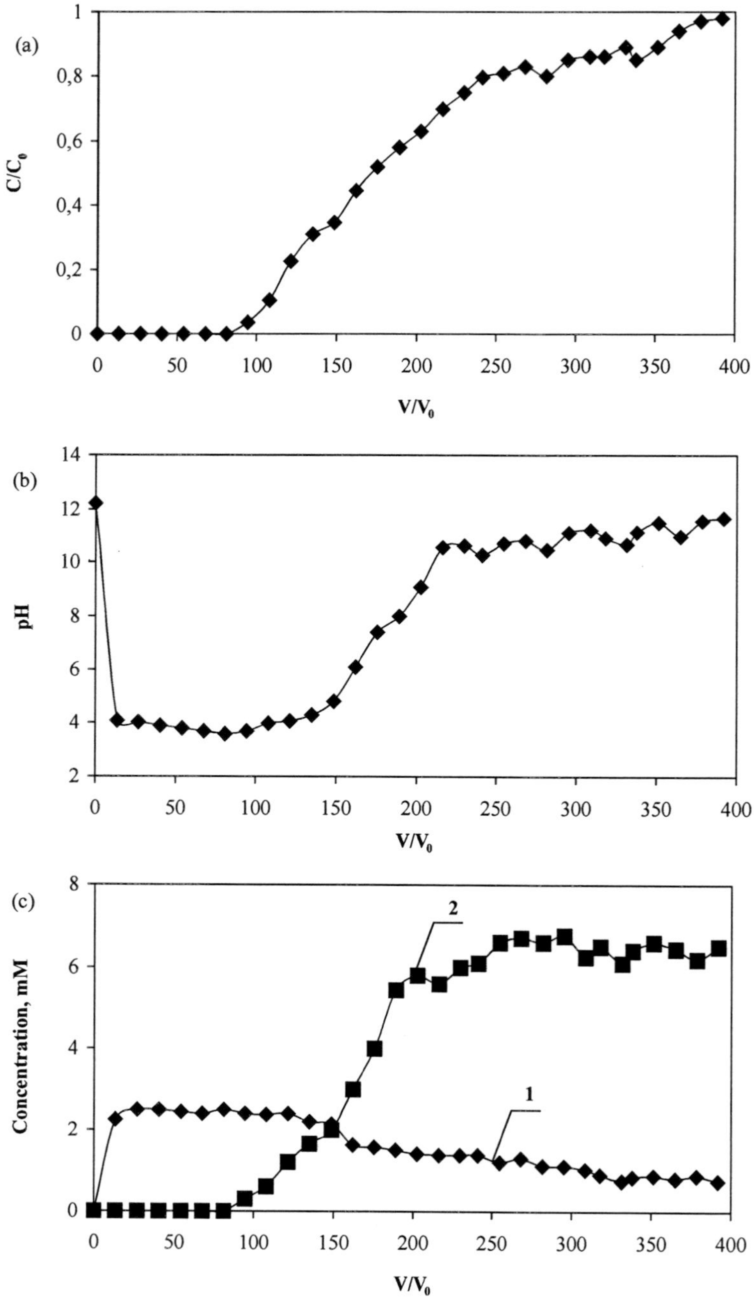


Fig. 4 (a) Nitrites breakthrough curve for DCSR in the column process; influx 0.0025 M $NaNO_2$ in 0.005 NaOH, flow rate 15 bed volumes/hour; (b) pH value of effluent.

The carried out column test gave important results. It could be stated that the N, N–dichlorosulfonamide copolymer is a very effective and efficient nitrite ion oxidant. It was shown that its superiority over the competitive N–monochlorosulfonamide copolymer results not only from its higher active-chlorine content i. e. from its higher oxidation capacity and by the higher redox potential i.e. oxidative power. The N, N–dichlorosulfonamide copolymer placed in a column removed nitrites from the processed solution efficiently, without any mechanical or chemical disturbance.

In the case of the competitive N–monochlorosulfonamide an unfavorable gas evolution was observed and gas locks made it difficult to conduct the process in a controlled way unless the nitrite concentration was lower than tens mg /L.

In the next stage of our column process investigation we checked how the oxidation would proceed if the nitrite influent would be very little alkaline. As the influent a 0.0025 M $NaNO_2$ in 0.005 M NaOH was applied (Fig. 4a). The intensity flow in that process was 15 bed volumes/h. First, up to $V/V_o$ = 100, the reaction ran normally good. The fractions were free of nitrites and the pH of the effluent was ca. 4.0 (Fig.4b). The effluent pH decreased despite of the alkaline reaction of the influent. The NaOH was used up not only for the neutralization of HCl formed in the first stage of the process, but also for the sodium-form origination of the intermediately nascent hydrogen-form N-monochlorosulfonamide copolymer (Eq. 4). Unfortunately the $NO_2^-$ ions appeared in the effluent just after ca. 2 L of the effluent was passed and the nitrite concentration grew quickly as did so the pH value in the column leakage. It turned out, that not only chlorides are present in effluent (Fig. 4c, curve 1), but hypochlorites (Fig. 4c, curve 2) too. So long as the reaction medium was acidic or neutral ($V/V_o$ up to = 150) the oxidation of nitrites followed and the concentration of chlorides in the leakage was according to our anticipation, i.e. ca. 0.0025 M. When pH grew up nitrites were oxidized in a small degree only. As the result of hydrolysis in alkaline medium active chlorine appeared in the leakage:

$$[P]\text{–}SO_2NCl_2 + 2NaOH \longrightarrow [P]\text{–}SO_2NClNa + NaOCl + H_2O \qquad (5)$$

So, the effluent fractions coming from the second part of the process showed the presence of hypochlorite ions despite of the simultaneous presence of the nitrite ions. The simultaneous presence in solution of NaOCl and $NaNO_2$ is possible if their concentration is not high enough.

One can work out from the break-through curve (Fig. 4a), that the nitrites quantity removed from the solution processed (up to $C/C_o$= 0.5 and $V/V_o$ = 180) was 3150 mL x 0.0025 M $NaNO_2$/L = 7.9 mmol only. The quantity of oxidized nitrites in this column process is considerably smaller in relation to the active chlorine content present in the column copolymer. This process was finished after letting through the column 7 L of influent solution, when the value of $C/C_o$ was close to 1.0. The mass of the used and dried copolymer from the column was ca. 9.0 g. Its analysis showed that it contained a considerable quantity of active chlorine (almost 4.0 meq/g).

On the basis of the active chlorine content in the spent copolymer and the chloride ions present in the effluent can be concluded that

a) ca. 50% of the copolymer active chlorine remained in the copolymer in the inactive form towards nitrite ions – the sodium form of the monochlorosulfonamide ([P]$SO_2$NClNa) copolymer.

b) ca. 25% of the copolymer active chlorine did oxidize the nitrite ions (especially in the first stage of the process – when the pH was sufficient low,

c) ca. 25% the rest of the copolymer active chlorine – disconnected from the copolymer in a less productive way.

## Conclusions

The copoly(N,N–dichlorosulfonamide S/DVB) oxidant in bead form used in a column process is a very efficient oxidant, of the poisonous nitrite ions even if they are in very diluted aqueous media. The nitrite ions are oxidized to the hundred fold less toxic nitrate ions. The reaction of nitrite oxidation by means of this water-insoluble heterogeneous oxidant, runs at favorable speed, as can be seen from the large flow intensities in the column experiments. The oxidative ability of the examined copolymer is high; it amounts to 0.20 g $NO_2^- / g$. It was possible to obtain effluents free of nitrites (< 10 µg $NO_2^- / L$) when the influent nitrite concentration was ca. 0.25 g $NO_2^- / L$. This macromolecular oxidant is a good choice in the case if the nitrite solution is not alkaline. In alkaline medium the copolymer is much less effective and resembles the competitive monochloro derivative of the copolymeric sulfonamide, we published elsewhere.

This macromolecular reagent is also a good choice against the oxidation of nitrites when using homogenous oxidant. In the reactive copolymer column process the diluted nitrite ion gets in contact with high concentration of oxidant (4.10 mmol of active chlorine/g of copolymer) and that is why the oxidation reaction runs fast to completeness. The post-reaction solution may be used in a mineral fertilizer production or it can be applied directly as a nutritive spray for farmlands. Neutralization with ammonia or lime-water or calcium carbonate suspension is recommended.

## References

[1] Product Data Sheet, Engineering Data Sheet, Technical Information Brochure - Rohm and Haas Company, Bayer AG; *Novabiochem* 2004/5 Catalog.

[2] Cassidy, H.G.; Kun, K.A. *Oxidation-Reduction Polymers (Redox Polymers)* , Wiley-Interscience, New York, 1965.

[3] Sansoni, B. *Neue chemische Arbeitsmethoden durch heterogene Reactionen: Redoxaustauscher und numerometrische* Titration. Als Habilitationsarbeit der Universitaet Marburg/Lahn, gedruckt in Muenchen 1968.

[4] Kozhevnikov, A. V. *Election Ion Exchangers (Elektronoionoobmenniki).* Khimiya, Leningrad, 1972.

[5] Ergozhin, E.E., Mukhitdinova, B.A. *Redox resins (Redoksionity),* Nauka, Alma-Ata, 1983.

[6] Emerson, D.W.; Shea, D.T.; Sorensen, E.M. Functionally Modified Poly(styrene-diwinylbenzene). Preparation, Characterization, and Bactericidal Action. *Ind. Eng. Chem. Prod. Res. Dev*. 1978, 17, 269.

[7] Emerson, D.W. Polymer-bound Active Chlorine: Disinfection of Water in a Flow System. Polymer Supported Reagents. 5. *Ind. Eng. Chem. Res*. 1990, 29, 448.

[8] Emerson, D.W. Slow Release of Active Chlorine and Bromine from Styrene-Divinylbenzene Copolymers Bearing N,N-Dichlorosulfonamide, N-Chloro-N-alkylsulfonamide, and N-bromo-N-alkylsulfonamide Functional Groups. Polymer Supported Reagents. 6. *Ind. Eng. Chem. Res*. 1991, 30, 2426.

[9] Emerson, D.W. Chlorine Dioxide Generated by Reaction of Sodium Chlorite with N-halosulfonamide or N-alkyl-N-halosulfonamide Groups on Styrene-Divinylbenzene Copolymers. *Ind. Eng. Chem. Res*. 1993, 32, 1228.

[10] Zhang, Y.; Emerson, D.W.; Steinberg, S.M. Destruction of Cyanide in Water Using N-chlorinated Secondary Sulfonamide Substituted Macroporous Poly(styrene-co-divinylbenzene). *Ind. Eng. Chem. Res*. 2003, 42, 5959.

[11] Chen, Y.; Worley, S.D.; Kim, J.; Wei, C.-I.; Chen, T.Y.; Santiago, J.I.; Williams, J.F. Sun, G. Biocidal Poly(styrenehydantoin) Beads for Disinfection of Water. *Ind. Eng. Chem. Res*. 2003, 42, 280.

[12] Chen, Y.; Worley, S.D.; Kim, J.; Wei, C.-I.; Chen, T.Y.; Suess, J.; Kawai, H.; Williams, J.F. Biocidal Polystyrenehydantoin Beads. 2. Control of Chlorine Loading. *Ind. Eng. Chem. Res*. 2003, 42, 5715.

[13] Salunkhe, M.M.; Mane, R.B.; Kanade, A.S. Polymer-supported analogues of halogeno-sulphpnamides preparation and applications in synthetic organic chemistry. *Eur. Polym. J*. 1991, 27, 461.

[14] Kawasoe, S.; Kobayashi, K.; Ikeda, K.; Ito, T.; Kwon, T.S.; Kondo, S.; Kunisada, H.; Yuki, Y. Preparation of polymeric analogs of N,N-dichloro-p-toluenesulfonamide and their use for oxidation of alcohols, oxidative lactonization of diols, and chlorination of carbonyl compounds. *J.M.S.-Pure Appl. Chem*. 1997, A34, 1429.

[15] Zhong, J.; Li, C.; He, B.; Wu, Z. Synthesis of Poly(styrene-co-divinylbenzene)-supported Dichloro Cyanuric Acid. *Chem. Res. Chin. Univ*. 1997, 13, 406.

[16] White, G.C. *Handbook of Chlorination and Alternative Disinfectants*; Wiley, New York 1999.

[17] Bogoczek, R.; Kociołek-Balawejder, E.; Stanisławska E. A macromolecular N–chloro-sulfonamide as oxidant for residual nitrites in aqueous media, *Ind. Eng. Chem. Res*. 2005, 44, (in press, Special Issue in Honor of D.C. Sherrington).

[18] Kociołek-Balawejder, E. A macromolecular N,N-dichlorosulfonamide as oxidant for cyanides. *Eur. Polym. J*. 2000, 36, 295.

[19] Kociołek-Balawejder, E. A macromolecular N,N-dichlorosulfonamide as oxidant for thiocyanates. *Eur. Polym. J*. 2000, 36, 1137.

[20] Kociołek-Balawejder, E. A macromolecular N,N–dichlorosulfonamide as oxidant for residual sulfides. *Eur. Polym. J*. 2002, 38, 953.

[21] Deshmuth, G.S.; Murthy, S.V.S.S. Amperometric determination of nitrite with chloramines T. *Indian J. Chem*. 1963, 1, 316.

[22] Agterdenbos, J. The volumetric determination of nitrite with chloramine T. *Talanta* 1970, 17, 238.

[23] Agraval, A.; Nahar, S.; Hussain, Z.; Sharma, P.D. Kinetics and mechanism of chloride ion catalysed oxidation of nitrite with N–chlorotoluene-p-sulphonamide (chloramines T) in aqueous acid perchlorate medium. *Oxid. Commun.* 1993, 16, 80.

[24] Bogoczek, R.; Kociołek-Balawejder, E. N-Monohalogeno- and N,N-dihalogeno (styrene-co-divinylbenzene)sulfonamide. *Polym. Commun.* 1986, 27, 286.

[25] Bogoczek, R.; Kociołek-Balawejder, E. Studies on a Macromolecular Dichloroamine - the N,N-Dichloro Poly(Styrene-co-Divinylbenzene)sulfonamide. *Angew. Makromol. Chem.* 1989, 2774, 119.

[26] Williams, W.J. *Handbook of Anion Determination*; Butterworth, London 1979.

[27] Ullmanns Encyklopaedie der Technischen Chemie 3 Ed. 1961, Vol.2/I, 591-601.

[28] Jucker, H.; Oehme, F. Das Redoxpotential und seine Anwendungen. *Chemiker-Ztg. Chem. Apparatur* 1963, 87, 381.

In: Frontiers in Polymer Research
Editor: Robert K. Bregg, pp. 43-105
ISBN: 1-59454-824-2


*Chapter 3*

# 6FDA Based Fluorinated Polyimides

***P. Santhana Gopala Krishnan*** [1]
Molecular and Performance Materials Cluster, Institute of Materials Research and Engineering, 3 Research Link, Singapore 117602

## Abstract

Aromatic polyimide(PI)s are heterocyclic polymers and have excellent thermal stability, good chemical resistance, electrical and mechanical properties. Most of these PIs are insoluble in common organic solvents. Fluorination of PIs is one of the many approaches to overcome the difficulty in the processing of these materials. Owing to the easy availability of 2,2-bis(3,4-dicarboxyphenyl) hexafluoropropane dianhydride (6FDA) as a commercial sample and the good properties that it imparted to the resulting polymer such as good solubility, decreased dielectric constant, increased thermal and hydrolytic stability, 6FDA based fluorinated PIs are extensively studied and used in various high technology applications.

The proposed chapter reviews the work done on 6FDA based fluorinated PIs with respect to its synthesis and various copolymers, polymerisation methods, poly(ether-imide), photosensitive polyimide, hyperbranched polyimide, addition polyimide, poly(amide-imide), poly(urethane-imide), poly(epoxy-imide), poly(ester-imide), poly(siloxane-imide), nanocomposites and non-linear optical polyimides. Finally, its application in electronics and use as a material for gas separation and corrosion protection are discussed.

## Introduction

Aromatic polyimide(PI)s have found widespread use in microelectronics, optoelectronics, gas separation and aerospace applications because of their excellent electrical properties, chemical resistance, high mechanical strength, high modulus and thermoxidative stability. Most aromatic PIs are intractable due to the poor solubility in organic solvents and the insolubility is attributed to the extremely strong interchain interactions of imide rings [1] or donor-acceptor interactions [2] in combination with non-specific Vander Waals interactions. They are usually processed in their soluble precursor form namely polyamic acid (PAA), its

[1] E-mail address: psgkrishnan@hotmail.com; krishnan_santhana@hyflux.com, Present Address: Hydrochem (S) Pte Ltd, Hyflux Building, 202, Kallang Bahru, Singapore 339339.

ionic salt (PAS) or polyamic ester (PAE) (Figure 1) and cured thermally, chemically, photochemically or a combination of both ultraviolet and heat. These precursors are unstable and can undergo many reactions such as hydrolysis of amide bonds or terminal anhydride groups and cyclization with the elimination of amine or water depending upon the synthesis and storage conditions [3-5]. These reactions not only affect the properties of precursors but also resulting PIs. The insolubility of most of the PIs and storage instability of their precursors, prompted the researchers all over the World to synthesize organo-soluble PIs. Further, soluble PIs are more advantageous than their precursors because they require no thermal curing for imidization, which induces excessive volume contraction due to the removal of water, alcohol or photosensitive groups.

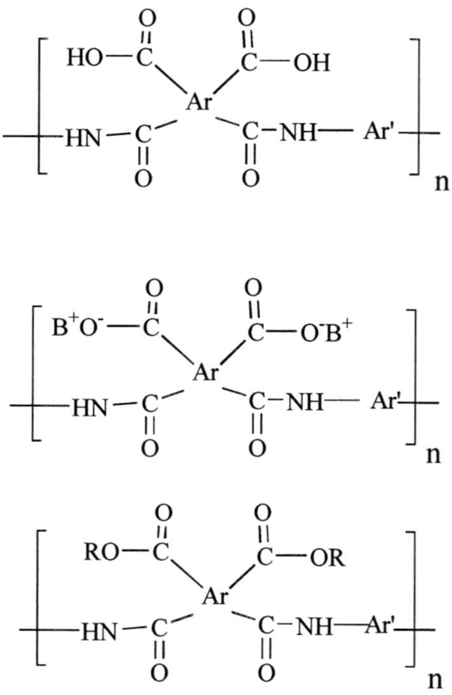

PAA

PAS

PAE

Where $B^+$ =

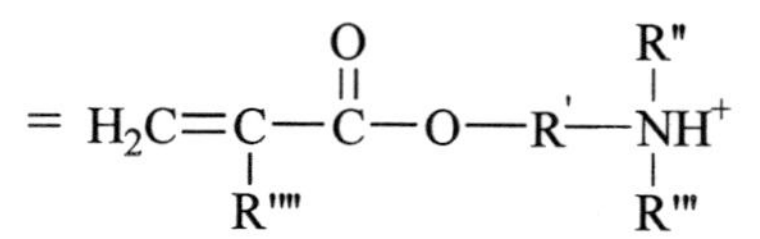

R, R', R'', R''', R'''' = alkyl

Ar and Ar' = aromatic

Fig. 1: Chemical structures of PI precursors

The solubility of aromatic PIs could be enhanced by the introduction of kinks such as ether, bulky side groups, alicyclic groups [6], long flexible chains [7] meta- or ortho-oriented phenylene rings [8], cardo rings [9] and inclusion complex compounds [10]. Fluorination of

PIs is one of the ways to impart solubility. Further, the introduction of fluorine atoms in PI structure provides the following unique properties: (i) reduction in dielectric constant and refractive indices (ii) increase in transparency in visible and near infra-red region, (iii) increase in free volume and permeability of gases, (iv) decrease in water absorption, (v) increase in thermal stability and coefficient of thermal expansion (CTE) and (vi) lowering of glass transition temperature [11-12]. These properties make fluorinated PIs suitable for electronics, aerospace, optical waveguide and gas separation applications. However, the fluorinated PIs may have low adhesion strength, low tear resisistance, high CTE and high solvent sensitivity.

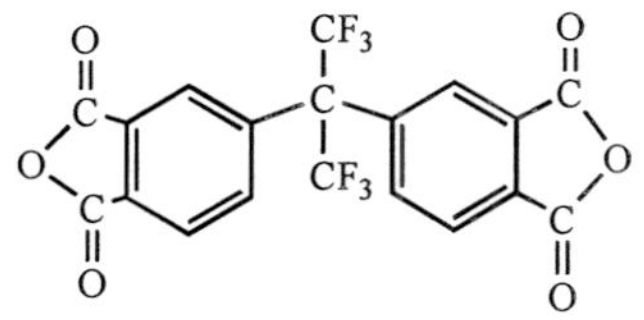

6FDA

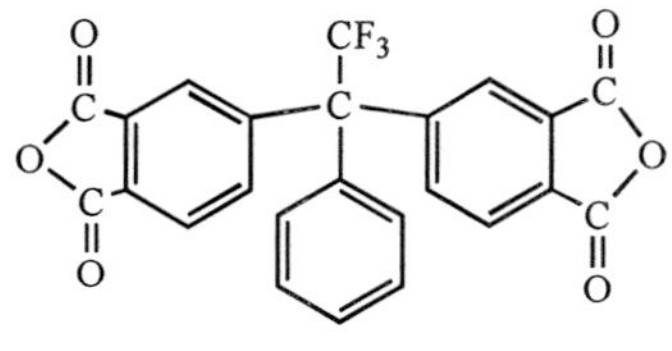

3FDA

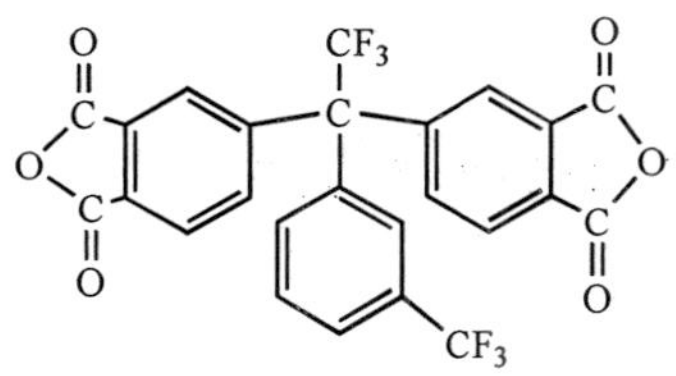

TFDA

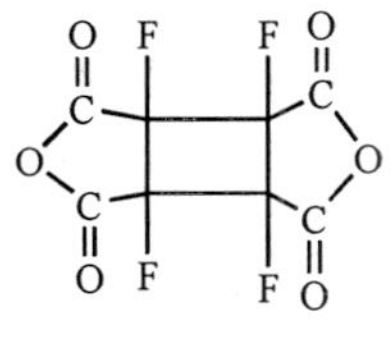

TFCBDA

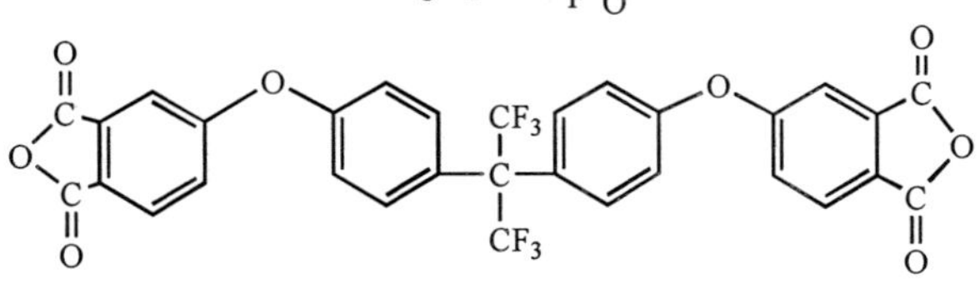

BFDA

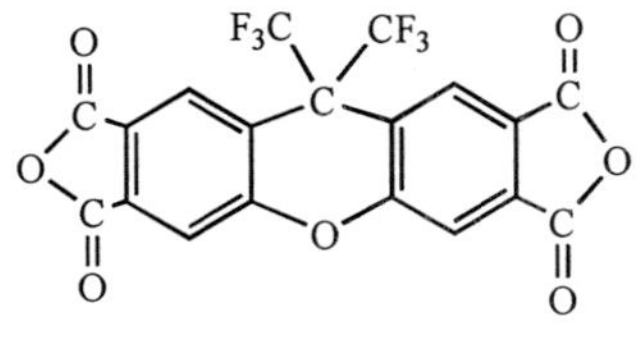

6FXDA

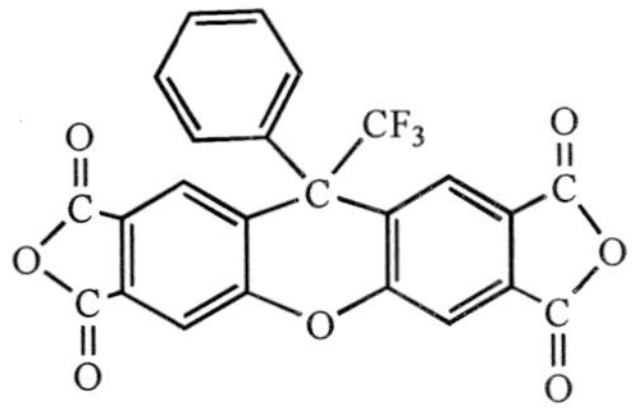

3FXDA

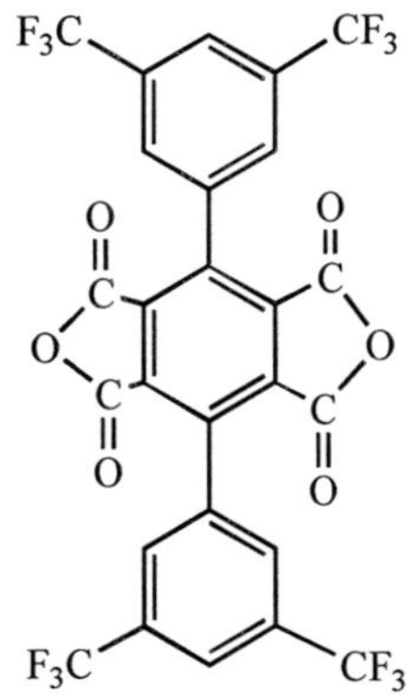

12FPMDA

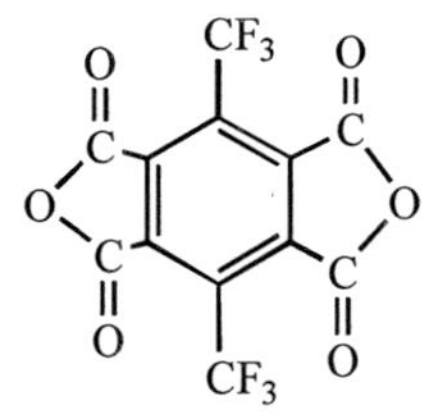

P6FDA

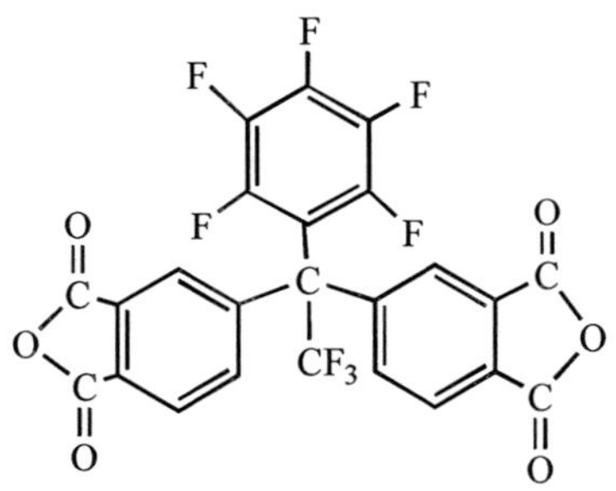

8FDA

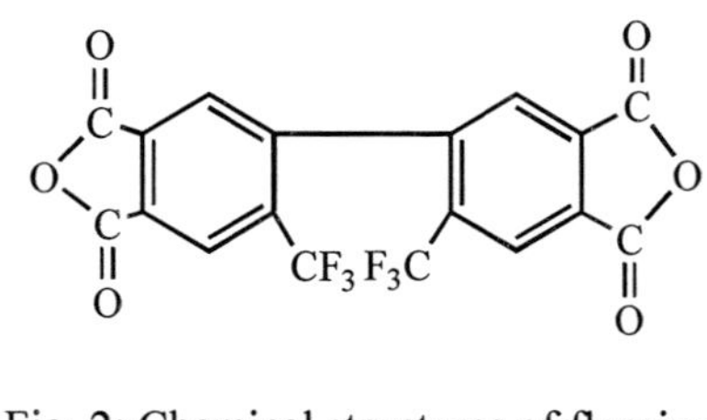

HFBPDA

Fig. 2: Chemical structures of fluorinated dianhydrides

A large number of fluorinated PIs have so far been synthesized using either fluorine containing dianhydrides or diamines and the chemical structures of selected dianhydrides and diamines are given in figures 2 and 3 respectively. Preparation of 2,2-bis(3,4-dicarboxyphenyl) hexafluoropropane dianhydride (6FDA) [13], 4,4'-(2,2,2-trifluoro-1-phenylethylidene)diphthalic anhydride (3FDA) [14], 4,4'-[2,2,2-trifluoro-1-(3-trifluoromethylphenyl) ethylidene]diphthalic anhydride (TFDA) [15], tetrafluorocyclobutane tetracarboxylic dianhydride (TFCBDA) [16], 4,4'-perfluoroisopropylidienediphenoxy dianhydride (BFDA) [17] are reported elsewhere. DuPont [18] has reported the synthesis of PIs based on 9,9-bis(trifluoromethyl)-2,3,6,7-xantheneteracarboxylic dianhydride (6FXDA). 9-phenyl-9-trifluoromethyl-2,3,6,7-xantheneteracarboxylic dianhydride (3FXDA) [19], 3,6-di[3',3'-bis(trifluoromethyl)phenyl]pyromellitic dianhydride (12FPMDA) [20], 1,4-

bis(trifluoromethyl)-2,3,5,6-benzenetetracarboxylic dianhydride (P6FDA) [21], 4,4'-(2,2,2-trifluoro-1-pentafluorophenylethylidene)diphthalic anhydride (8FDA) [22], 2,2'-bis(trifluoromethyl)-4,4',5,5'-biphenyltetracarboxylic dianhydride (HFBPDA) [23] are some of the fluorinated dianhydrides used in the preparation of fluorinated PIs.

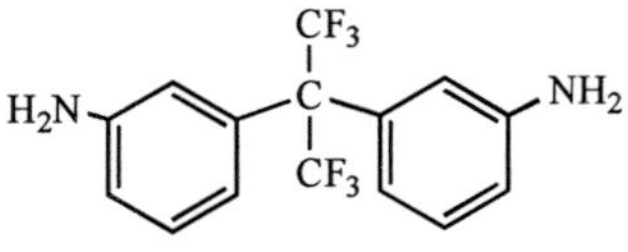

3-6FpDA

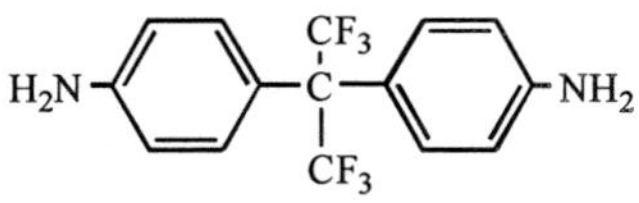

4-6FpDA

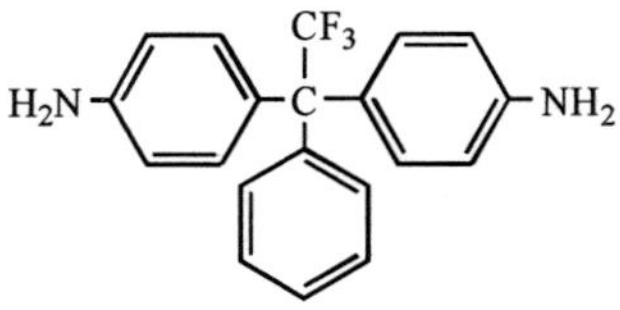

3FDAM

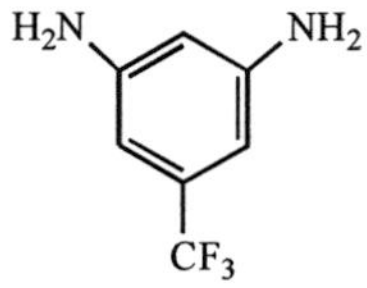

3, 5 - DBTF

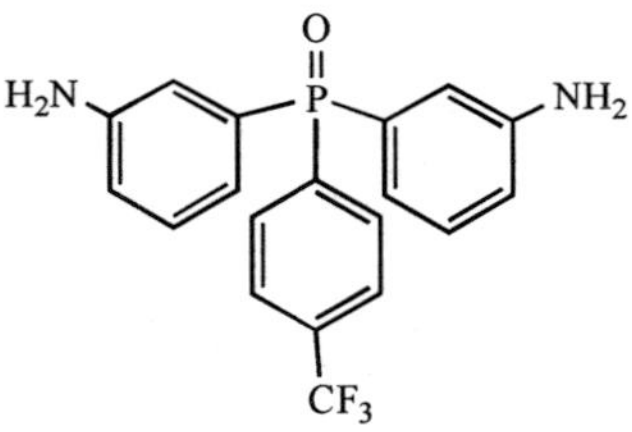

mDA3FPPO

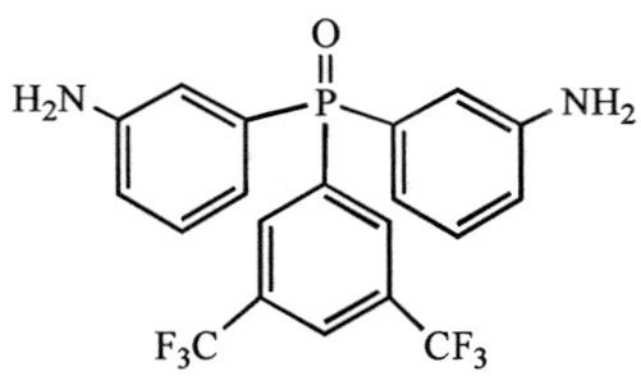

mDA6FPPO

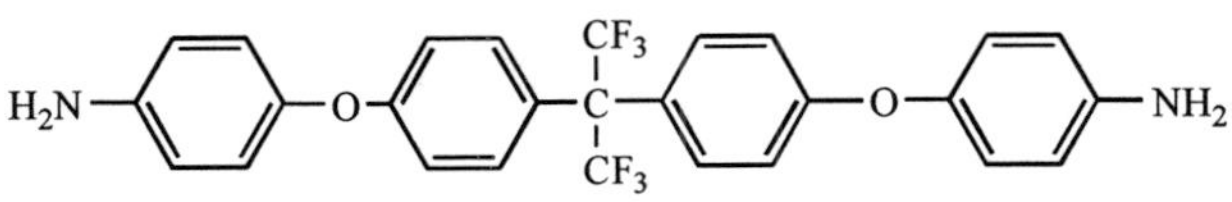

4-BDAF

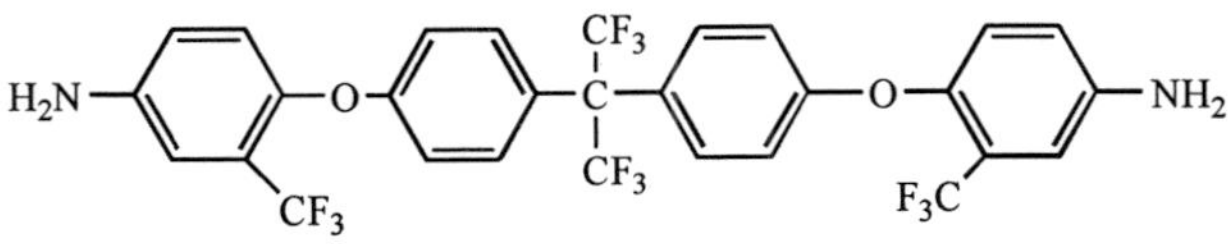

12FBDAF

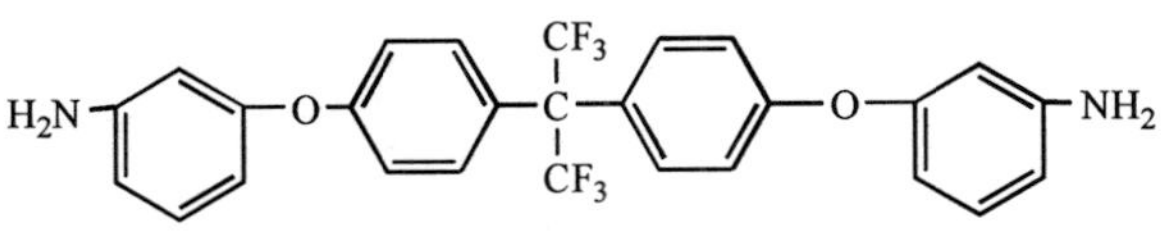

3-BDAF

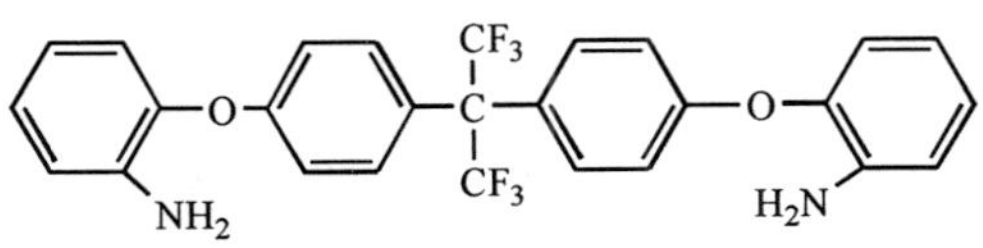

2-BDAF

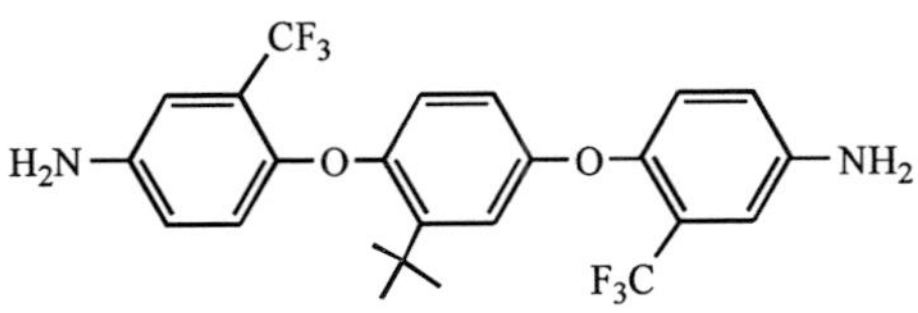

6FBAP-tBB

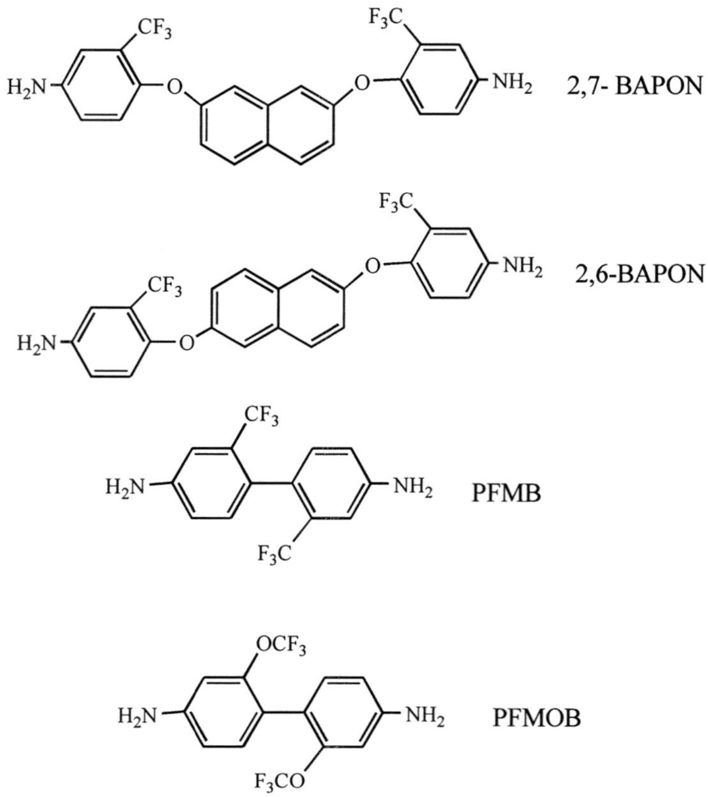


Fig. 3: Chemical structures of fluorinated diamines

6FDA based PIs based on 2,2-bis(3-aminophenyl)-1,1,1,3,3,3-hexafluoropropane (3-6FpDA) and 2,2-bis(4-aminophenyl)-1,1,1,3,3,3-hexafluoropropane (4-6FpDA) [24] were reported in the literature. Preparation of 1,1-bis(4-aminophenyl)-1-phenyl-2,2,2-trifluoroethane (3FDAM) [25] 3,5-diaminobenzotrifluoride (3,5-DBTF) [26], bis-(3-aminophenyl)-4-(trifluoromethyl)phenyl phosphine oxide (mDA3FPPO) [27], bis(3-aminophenyl)-3,5-bistrifluoromethyl)phosphine oxide (mDA6FPPO) [28], 2,2-bis[4-(4-aminophenoxy)phenyl]hexafluoropropane (4-BDAF) [29], 2,2-bis[4-(4-amino-2-trifluoromethylphenoxy)-phenyl]1,1,1,3,3,3-hexafluoropropane (12FBDAF) [30], 2,2-bis[4-(3-aminophenoxy)phenyl]hexafluoropropane (3-BDAF) [31], 2,2-bis[4-(2-aminophenoxy)phenyl]hexafluoropropane (2-BDAF) [32], 2,5-bis(4-amino-2-trifluoromethylphenoxy)-tert-butylbenzene (6FBAP-tBB) [33], 2,7-bis(4-amino-2-trifluoromethylphenoxy)naphthalene (2,7-BAPON) [34], 2,6-bis(4-amino-2-trifluoromethylphenoxy)naphthalene (2,6-BAPON) [35], 2,2'-bis(trifluoromethyl)-4,4'-diaminobiphenyl (PFMB) [36], 2,2'-(trifluoromethoxy)benzidine (PFMOB) [37], were reported in the literature. Review on the entire list of fluorinated PIs is beyond the scope of this chapter and the primary focus is only on 6FDA based fluorinated PIs.

Search on ISI web of Science database indicate that 3,538 PI articles were published during the period starting from 1988 to 2004. Out of which, 344 articles (about 10%) were based on 6FDA based PIs. This indicate the interest, it has created to the researchers all over the World because of the potential it has for various applications and the easy availability as a commercial sample. Various 6FDA manufacturers are listed below:

| Name and address | Phone /Fax Nos./ Email / Web |
|---|---|
| ABCR GmbH KG<br>PF 210135<br>Karlsruhe, D-76151 Germany | Phone: 49-(0)721-95061-0<br>Fax: 49-(0)721-95061-80<br>Email: info@abcr.de |
| Aldrich Chemical Company, Inc.<br>1001 West Saint Paul Avenue, Milwaukee, WI, 53233 USA | Phone: 1 800 558 9160<br>Ph: 414 273 3850<br>Fax: 1 800 962 9591<br>Fax: 414 273 4979<br>Email: aldrich@sial.com<br>Web: www.sigma-aldrich.com |
| Ambinter<br>46 quai Louis Bleriot<br>Paris, F-75016<br>France | Phone: (33-1) 45 24 48 60<br>Fax: (33-1) 45 24 62 41<br>Email: ambinter@compuserve.com<br>Web: www.ambinter.com |
| Apollo Scientific Ltd.<br>Whitefield Rd.<br>Bredbury, Stockport, Cheshire,<br>SK6 2QR UK | Phone: 44(0)870 128 7302<br>Fax: 44(0)870 128 7303<br>Email: sales@apolloscientific.co.uk<br>www.apolloscientific.co.uk |
| Central Glass International Inc.<br>Kow-Hitotsubashi Bldg., 7-1, Kanda-Nishikicho, 3-chrome<br>Chiyoda-ku, Tokyo, 101-0054 Japan | Phone: 81-3-3259-7133<br>Fax: 81-3-3259-7363<br>Email: fcsd@cgco.co.jp<br>Web: www.cgco.co.jp |
| Chriskev Company, Inc.<br>5109 W. 111th Terrace<br>Leawood, KS, 66211-1742 USA | Phone: (913) 491-4911<br>Fax: (913) 491-9451 |
| Clariant Division LSE<br>Stroofstrasse 27<br>Frankfurt-am-Main, 65933<br>Germany | Phone: 49 (69) 3800-2109<br>Fax: 49 (69) 3800-2203<br>Email: dieter.regnat@clariant.com<br>Web: www.lse.clariant.com |
| Colour-Chem Limited, Mumbai-Agra Rd,<br>Balkum, Thane, 400 608 India | Web: www.colour-chem.com |
| Fluka Chemical Corp.<br>1001 West St. Paul Avenue, Milwaukee, WI, 53233 USA | Phone: 1 800 558 9160<br>Phone: 414 273 3850<br>Fax: 1 800 962 9591<br>Fax: 414 273 4979<br>Email: Fluka-RdH@sial.com<br>Web: www.sigma-aldrich.com |
| Fluorochem Ltd.<br>Wesley Street, Old Glossop, Derbyshire, SK13 7RY UK | Phone: (01457) 868921<br>Fax: (01457) 869360<br>Fax: (01457) 860927<br>Email: enquiries@fluorochem.co.uk<br>Web: www.fluorochem.net |
| Indofine Chemical Company, Inc.<br>121 Stryker Lane, Bldg 30, Suite 1<br>Hillsborough, NJ, 08844 USA | Phone: (908) 359-6778<br>Phone: (888) INDOFINE (463-6346)<br>Fax: (908) 359-1179<br>Email: chemical@indofinechemical.com<br>Web: www.indofinechemical.com |
| Interchim<br>213 Avenue Kennedy, BP 1140, Montlucon,<br>Cedex, 03103 France | Phone: (33) (0) 4 70 03 88 55<br>Fax: (33) (0) 4 70 03 82 60<br>Web: www.interchim.com |
| LaboTest<br>Falkenberger Str. 4<br>Niederschona, 09600 | Phone: 49 35209 21501<br>Fax: 49 35209 21502<br>Email: sales@labotest.com |

| Name and address | Phone /Fax Nos./ Email / Web |
|---|---|
| Germany | Web: www.labotest.com |
| Lancaster Synthesis Ltd.<br>Newgate, White Lund<br>Morecambe,<br>Lancashire, LA3 3BN UK | Phone: 0800 262336<br>Fax: 0800 616440<br>Email: Lancaster_UKSales@clariant.com<br>Web: www.lancastersynthesis.com |
| Matrix Scientific<br>P O Box 25067<br>Columbia, SC,<br>29224-5067 USA | Phone: (803) 788-9494<br>Fax: (803) 788-9419<br>Email: sales@matrixscientific.com<br>Web: www.matrixscientific.com |
| Oakwood Products, Inc<br>(Fluorochem USA)<br>1741 Old Dunbar Rd.<br>West Columbia, SC, 29172 USA | Phone: 803-739-8800<br>Fax: 803-739-6957<br>Email: sales@oakwoodchemical.com<br>Web: www.oakwoodchemical.com |
| Ryan Scientific, Inc.<br>P O Box 845<br>Isle of Palms, SC, 29451 USA | Phone: 1-843-884-4911<br>Fax: 1-843-884-5568<br>Web: www.ryansci.com |
| SynQuest Laboratories, Inc.<br>P O Box 309<br>Alachua, FL,<br>32616-0309 USA | Phone: 1-877-4-FLUORO<br>Phone: 1-386-462-0788<br>Fax: 1-386-462-7097<br>Email: info@synquestlabs.com<br>Web: www.synquestlabs.com |
| Tokyo Kasei Kogyo Co., Ltd.<br>4-10-1, Nihonbashi-Honcho Chuo-ku<br>Tokyo, 103-0023, Japan | Phone: 81-3-5651-5172<br>Fax: 81-3-5640-8022<br>Email: international@tokyokasei.co.jp<br>Web: www.tokyokasei.co.jp |
| TimTec, Inc.<br>100 Interchange Blvd.<br>Newark, DE, 19711 USA | Phone: (302) 292-8500<br>Fax: (302) 292-8520<br>Email: info@timtec.net<br>Web: www.timtec.net |
| Wako Chemicals USA, Inc.<br>1600 Bellwood Road<br>Richmond, VA, 23237<br>USA | Phone: 800-992-WAKO<br>Phone: 804-271-7677<br>Fax: 804-271-7791<br>Email: Information@wakousa.com<br>Web: www.wakousa.com |

Database: CHEMCATS

The various synonyms of 6FDA are

(i) 2,2-bis(3,4-anhydrodicarboxyphenyl) hexafluoropropane
(ii) 2,2-bis(3,4-dicarboxyphenyl)-1,1,1,3,3,3-hexafluoropropane dianhydride
(iii) 4,4'-(hexafluoroisopropylidene)bis-phthalic anhydride
(iv) 4,4'-(hexafluoroisopropylidene)diphthalic anhydride
(v) 4,4'-[2,2,2-trifluoro-1-(trifluoromethyl)ethylidene]diphthalic anhydride
(vi) 5,5'-[2,2,2-trifluoro-1-(trifluoromethyl)ethylidene]bis-1,3-isobenzofuranedione
(vii) 2,2,2-trifluoro-1-(trifluoromethyl)ethylidene diphthalic anhydride.

Some of the commercially available 6FDA based fluorinated PIs are given in figure 4. DuPont's Avimid N is a copolyimide, derived from 6FDA and 95 mol % of 1,4-phenylene diamine and 5 mol percent of 1,3-phenylene diamine [38]. It is used as a composite matrix resin and retains mechanical properties after 100 h aging in air at 371 °C. The limitation of this material is that it requires high pressure (>1000 psi) for processing into a composite. It has favorable dielectric properties making it well suited for missile and turbine engine applications.

Another series, vinyl capped Avimid N oligomer called VCAP-II was prepared by capping anhydride terminated Avimid N oligomer with p-amino styrene. Because of the presence of end double bond, it could be thermally crosslinked to produce a thermoset polymer having a $T_g$ of 388 °C [39].

Both AF-700A and AF-700B are the modified version of Avimid N having one nadic end-capped group at one end another end is having amine and anhydride group respectively. [40]. In another modification of Avimid N structure, two nadic end groups were introduced and designated as PMR-II series and has improved processability. Theoretical prepolymer molecular weight of PMR-II-30 and PMR-II-50 are 3000 and 5000 respectively.

In 1988, Hoechst Celanese commercialized Sixef-44$^{TM}$ and was prepared by reacting with 4,4'-6F diamine (4-6FpDA). It was found to be soluble in THF, chloroform and acetone and had $T_g$ at 320 °C. It exhibited a very low dielectric constant of 2.58 [41].

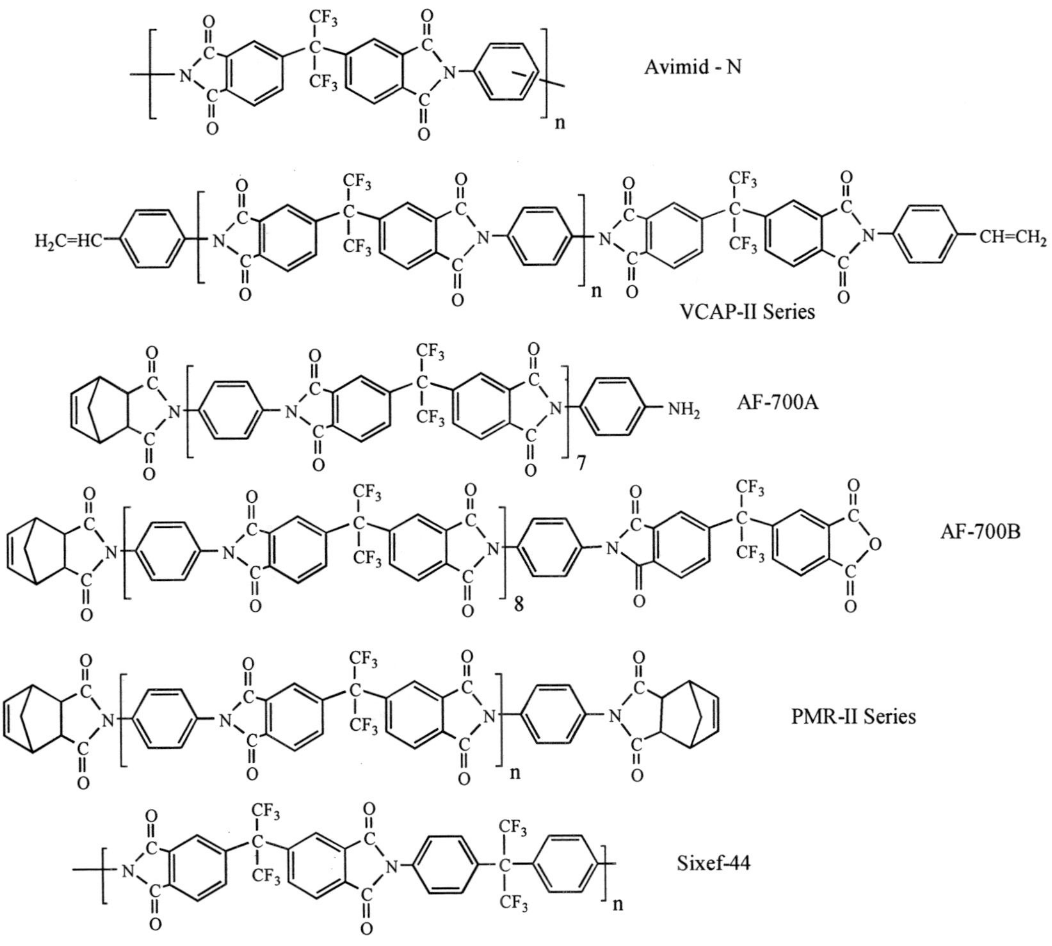


Fig. 4: Commercial polymers

# Synthesis

## Monomer

Reaction scheme for the synthesis of 6FDA from o-xylene is given in figure 5. In the first step, hexafluoroacetone (HFA) was condensed with o-xylene in the presence of hydrofluoric acid (HF) at 100-200 °C to give 4,4'-(hexafluoroisopropylidene)-bis(o-xylene). In the second step, 4,4'-(hexafluoroisopropylidene)-bis(o-xylene) was oxidized using potassium permanganate to obtain4,4'-(hexafluoroisopropylidene) bis(phthalic acid). In the final step, bis(phthalic acid) was refluxed in xylene using Dean-Stark separator to remove water and when the collection of water stoped, the xylene was evaporated to get crude 6FDA [13,14]. This was further purified by recrystallization from acetic anhydride followed by sublimation. The reported melting point of 6FDA was in the range of 250-4 °C.

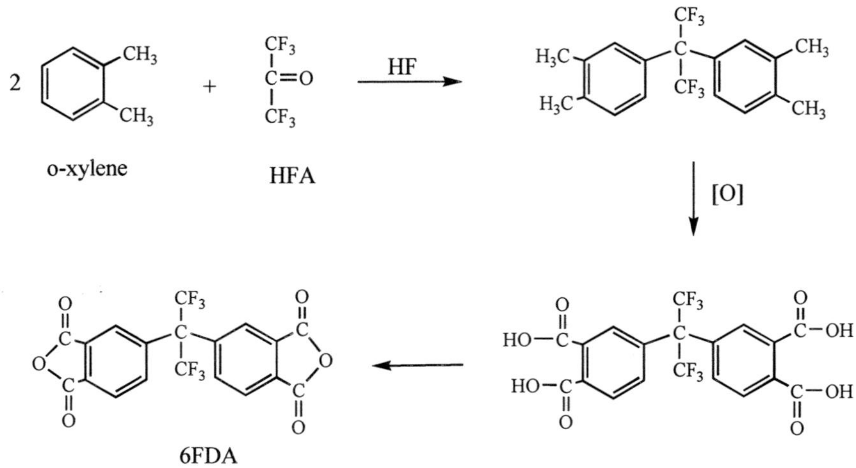


Fig. 5: Synthesis of 6FDA

Molecular model of 6FDA is given in figure 6. Based on the molecular Orbital Package (MOPAC) method [42], the calculated dihedral angle of the bonds in the 1,1'-positions of 6FDA and 3FDA are 47.9° and 84.7° respectively, which indicates that the degree of out-of-plane twisting is lower for 6FDA than it is for 3FDA. The solvent accessible surface area, molecular surface and solvent excluded volume of 6FDA are 540.54 $A^{o2}$, 294.07 $A^{o2}$ and 271.7 $A^{o3}$ respectively [43]. Molar dielectric polarization based on Lorentz and Lorentz theory ($P_{LL}$) and molar dielectric polarization based on Vogel theory ($P_V$) of 6FDA are 121.1 and 752.4 respectively [41].

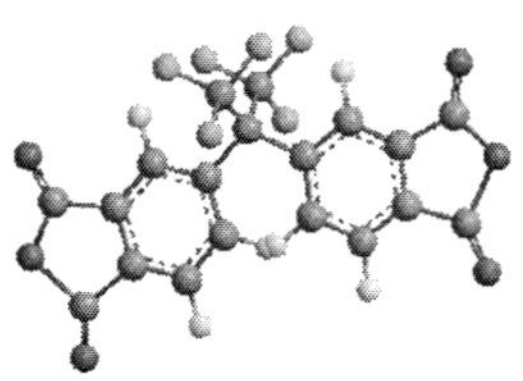

Fig. 6: Molecular model of 6FDA

## Polymer

In 1960s, Rogers et al first synthesized 6FDA based fluorinated PI from 4,4'-6F (4-6FpDA) [44]. Thereafter, a large number of diamines were used to synthesize 6FDA based fluorinated PIs and the chemical structure of some of them are given in figure.7. 2,2-bis(4-aminophenyl)norbornane (BANB) [45], 2,2-bis(4-aminophenyl)adamantane (BAAD) [45], 5,7-diamino-1,1,4,6-tetramethylindan (DAI) [46], 5(6)-amino-1-(4-amino-1-(4-aminophenyl)1,3,3-trimethylindane (DAPI) [47], trans-1,4-diaminocyclohexane (DACH) [48], bis(p-aminocyclohexyl)methane (PACM) [48], bis(3-methyl-4-aminocyclohexyl)methane (PMACM) [48], 4,4'-diaminodiphenylacetylene (p-intA) [49], bis(3-aminophenyl)acetylene (m-intA) [50], 3,7- diaminophenothiazinium chloride (thionine) [51], 1,7-bis(4-aminophenoxy)naphthalene (1,7-BAPON) [52], 1,1-bis(4-aminophenyl)cyclohexane (BACH) [53], 2,2'-dichloro-4,4'-diaminobiphenyl (DCB) [36], 4-phenyl-2,6-bis(4-aminophenyl)pyridine (PAPPY) [54], 4-cumyl-2,6-bis(4-aminophenyl)pyridine (CAPPY) [54], 4-anisyl-2,6-bis(4-aminophenyl)pyridine (AAPPY) [54], 4-naphthyl-2,6-bis(4-aminophenyl)pyridine (NAPPY) [54], 4-tolyl-2,6-bis(4-aminophenyl)pyridine (TAPPY) [54], 3,6-diaminocarbazole (CDA) [55,56], 3,6-diamino-N-phenylcarbazole (PCDA) [55], N-ethyl-3,6-diaminocarbazole (ECDA) [56] and N-(4'-aminophenyl)-3-aminocarbazole (APCDA) [55] are used to prepare 6FDA based PIs.

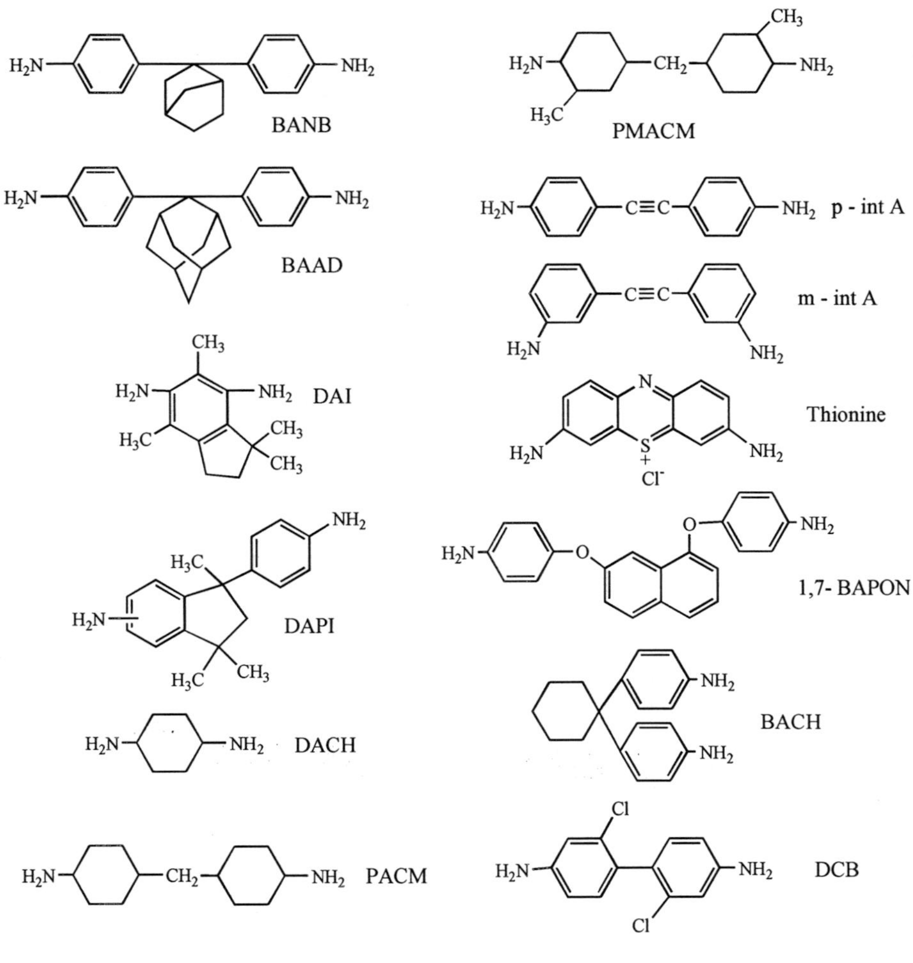

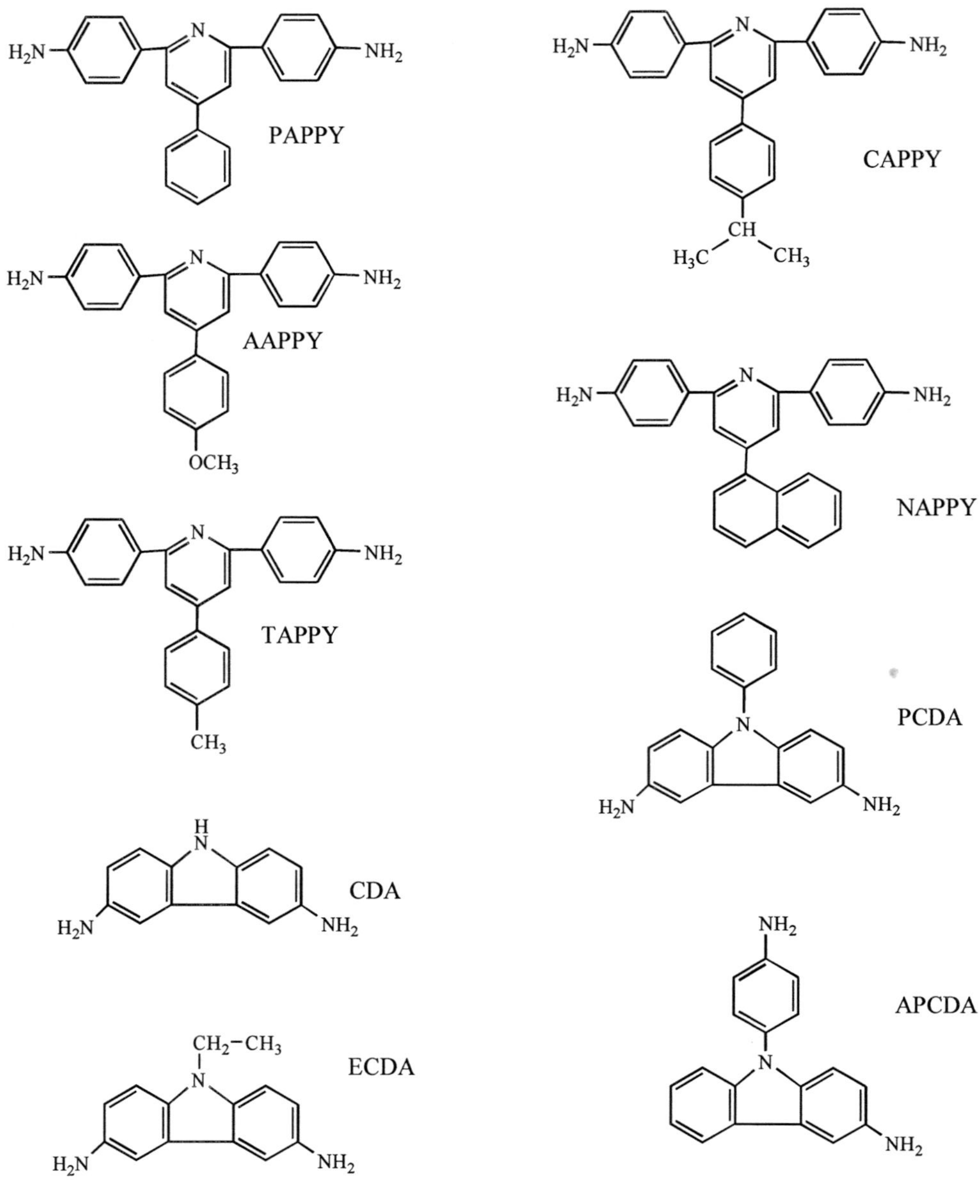


Fig. 7: Chemical structures of diamines

PIs can be synthesized by one or two- step methods. Most of these PIs were synthesized using a traditional two-step synthetic method namely a polyaddition and polycondensation step (figure 8). During the first step, solid 6FDA was added in installments to an equimolar amount of diamine pre-dissolved in an aprotic solvent such as N-methyl-2-pyrrolidone (NMP). Other aprotic solvents that are normally used for PI synthesis are N,N-dimethyl acetamide (DMAc), N,N-dimethyl formamide (DMF), γ-butyrolactone (BLO) and dimethyl sulfoxide (DMSO). The reaction was carried out under nitrogen at RT to obtain a soluble PAA. The precursor PAA is initially formed via nucleophilic attack by the diamine at a carbonyl carbon of the dianhydride. The rate of this reaction is largely depended on the electron-donating potential of the diamine and on the electron affinity of dianhydride. In general, the high electron negativity of fluorine reduces the basicity of diamines whereas it increases the electron affinity of dianhydrides. In other words, the incorporation of fluorine atoms in the dianhydride increases the rate of the reaction whereas the fluorination of diamines suppresses the reaction. 6FDA is highly reactive than fluorinated diamines. PAAs were known to be unstable upon storage. Kreuz [57] had studied the hydrolysis of PAA derived from 6FDA.

Imidization of PAA could be carried out by either chemical or thermal means. For chemical imidization, acetic anhydride / propionic anhydride was used as a dehydrating agent and triethyl amine/3-picoline/pyridine/2-ethyl-4-methyl imidazole was used as a catalyst. The solution was stirred for about 16 h and then precipitated in methanol. Solid PI was dried at 150 °C for overnight in air circulating oven. Alternatively, solution imidization of PAA could be performed at 165 °C for 24 h using an 8:2 mixture of co-solvent NMP and an azeotrope solvent o- dichlorobenzene. Other solvents such as phenol, o-chlorophenol, 2,4-dichlorophenol, m-methoxy phenol, m-cresol [58] or nitrobenzene can also be used for high temperature imidization. Proton NMR was used to determine number average molecular weight ($M_n$) by assessing the concentration of end groups i.e., t-butyl phthalimide in PI [59]. Proper chromatographic conditions for the determination of weight average molecular weight ($M_w$) and molecular weight distribution (MWD) by size exclusion chromatography (SEC) has been reported in the literature and the proposed procedure was verified for 6FDA based PI endcapped with m-ethynylaniline (MEA) [60,61]. The effect of rhodamine B on the completeness of the imidization process of 6FDA and DAB (diaminobenzophenone) [62] and its scintillation characteristics [63] have been reported.

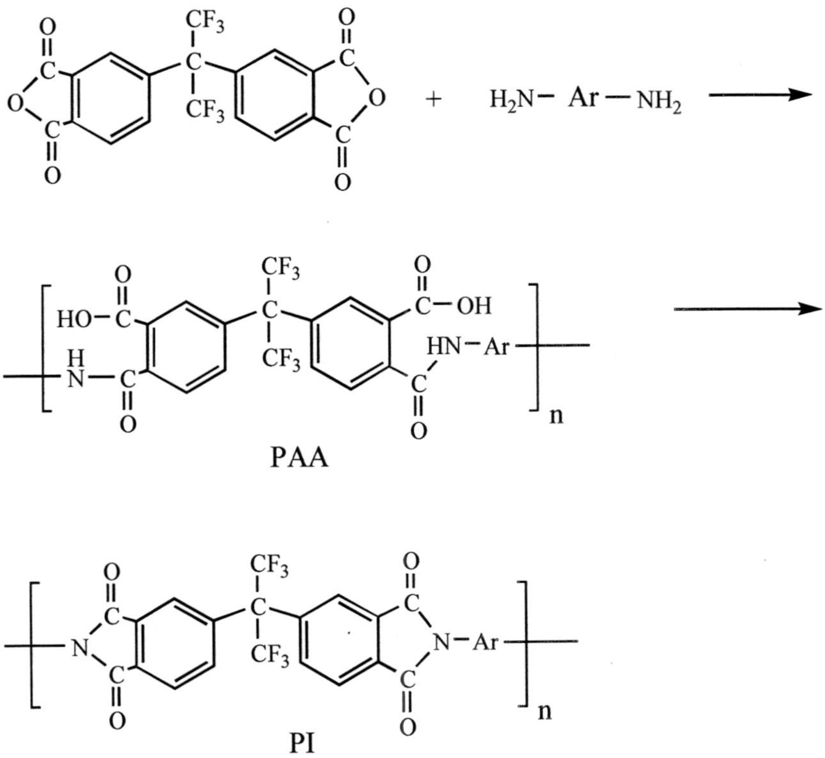


Fig. 8: Synthesis of 6FDA PIs

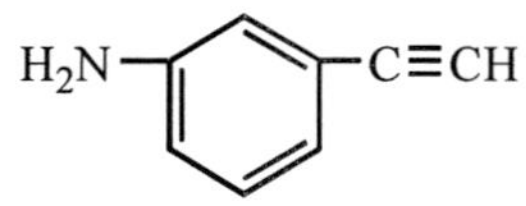


MEA

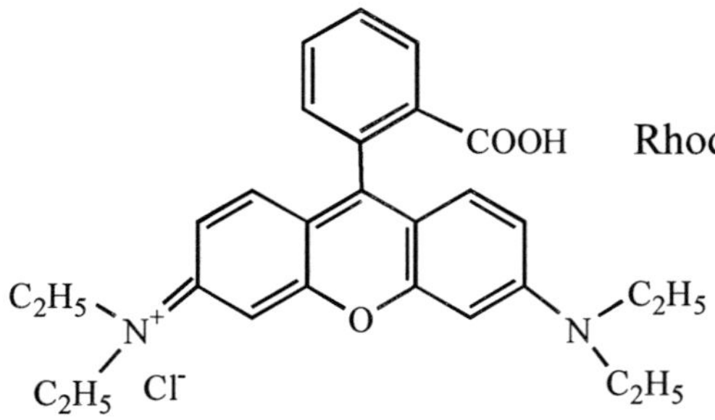


Rhodamine B

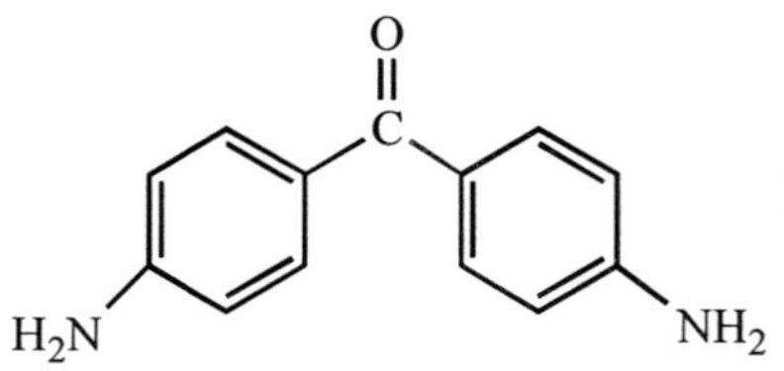


DAB

Fig. 9: Chemical structures

For making films, PAA solution is casted on a glass plate using Doctor's plate and is subjected to a programmed heating rate. A typical curing cycle used is one hour each at 55, 80, 110, 150, 200, 250 and 300°C [64]. In this approach, solvent removal occurs concurrently with imide formation. PI fibers can be produced by extrusion into various aqueous organic media. It has been reported that PI fibers obtained from PAA fibers after thermal imidization had round or oval cross sections, whereas fibers obtained from PI resin exhibited dog bone, C-shaped or ovall cross section [65].

## Copolyimides

Copolyimides are usually prepared using either a diamine and two different dianhydrides or a dianhydride and two different diamines. Most researchers have used two different diamines and a dianhydride for the preparation of copolyimides. 6FDA based copolyimdes containing various mol percents of 1,4-phenylene diamne (pPDA), [66], or 1,3-phenylene diamine (mPDA), [67], or 2,6-diamino toluene (DAT) [67] and 2,3,5,6-tetramethyl-1,4-phenylene diamine (durene diamine, DDA) were synthesized and studied for thermal properties. Series of 6FDA based copolyimides were synthesized using mPDA, DAT and 2,4,6-trimethyl-1,3-phenylene diamine (TMemPDA) to study the effect of methyl group substitution in the diamine and copolymer composition on their properties [68]. 6FDA based copolyimides using a combination of DDA, 4-6FpDA and 3,5-diamino benzoic acid (DABA) as monomers were studied for the membrane-based separation of benzene/cyclohexane mixtures [69]. Another copolyimde derived from 4-6FpDA, DABA and 4',4"(5")-diamino-dibenzo-15-crown-5 (15-crown-5) were studied for propylene/propane separation [70].

Photoluminescence and electroluminescence properties of copolyimide derived from 6FDA, 3,6-diamino acridine (proflavine, ACR) and 4,4'-(9H-fluoren-9-ylidene)bisphenylamine (FBPA) was studied and single layer device fabricated out of this copolyimide emitted green light [71].

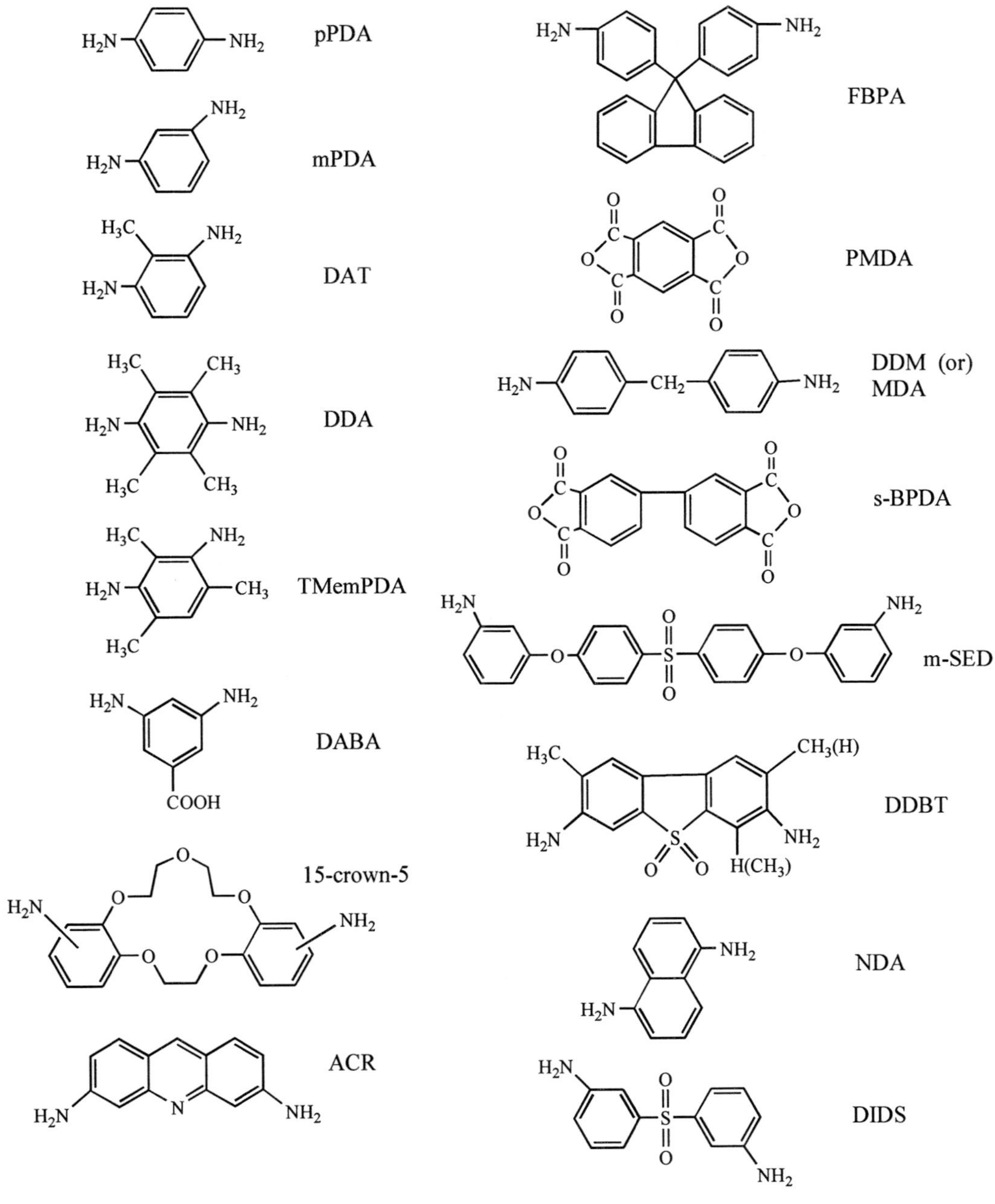


Fig. 10: Structure of various comonomers

Effect of dianhydride addition order on the properties of copolyimides derived from 6FDA, pyromellitic dianhydride (PMDA) and 4,4'-diaminodiphenyl methane (DDM / MDA) was reported [72]. Soluble copolyimides of both random and block type were prepared by reacting 6FDA and 3,3',4,4'-biphenyltetracarboxylic dianhydride (s-BPDA) as dianhydride components and 4,4-bis(3-aminophenoxy)diphenyl sulfone (m-SED) as a diamine component. Random type was prepared by reacting both the dianhydrides simultaneously with m-SED whereas to prepare block type 6FDA was reacted with m-SED initially for few hours followed by s-BPDA. Random type was found to be more effective in improving the brittleness of cyanate ester resin than block type [73].

Similar copolyimide was prepared using 3,7-diamino-2,8(6)-dimethyldibenzothiophene5,5-dioxide (DDBT) instead of m-SED and its performance for olefin/paraffin separation was investigated [74]. Block copolymers of liquid crystalline polyamide and amorphous PI were prepared from a two-pot polycondensation reactions [75]. Hydrocarbon (C-2 and C-3) separations in copolyimide dense membranes derived from 6FDA, DDA and 1,5-naphthalene diamine (NDA) was studied [76]. Gas transport properties of 6FDA, DDA and 3, 3'-diaminodiphenyl sulfone (DIDS) was reported [77].

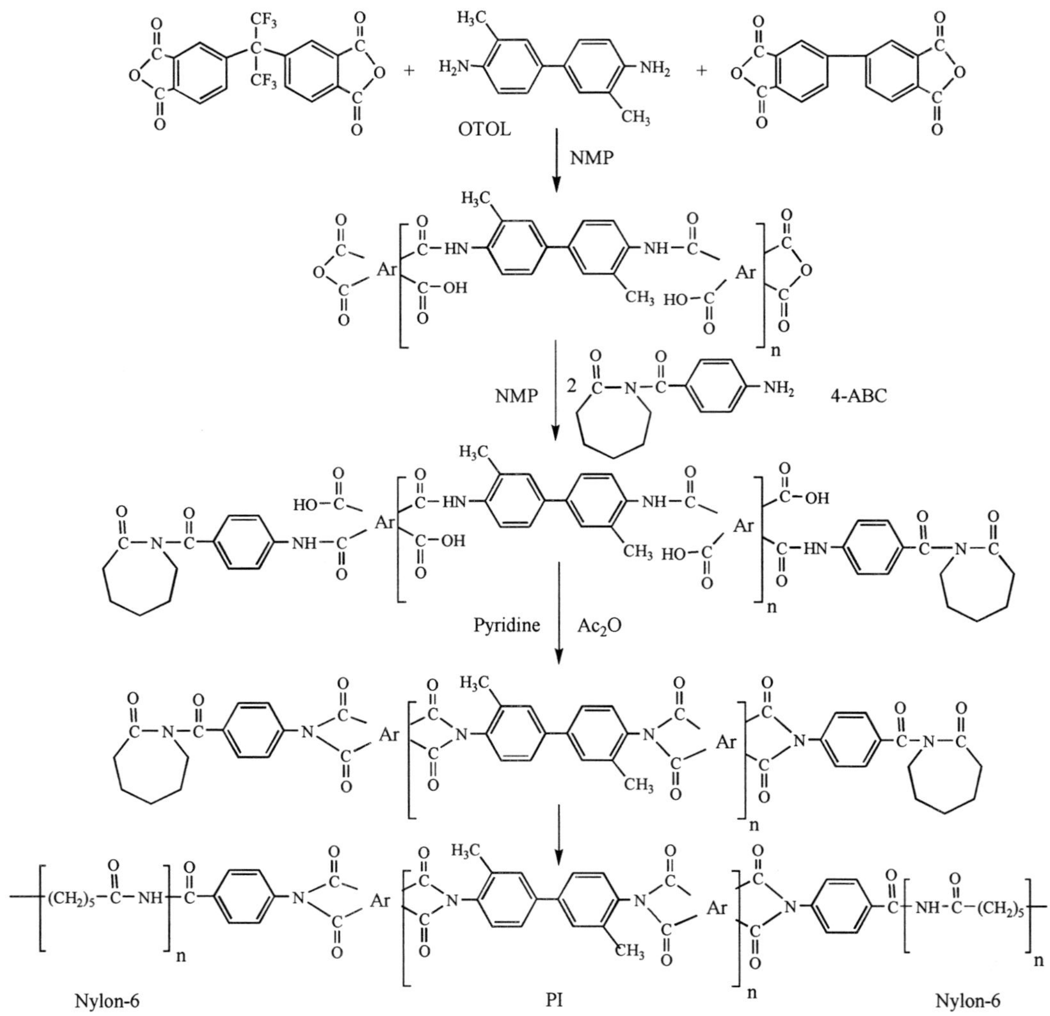


Fig. 11 Synthesis of PI nylon-6 block copolymers

PI-Nylon-6 block copolymers were synthesized (Figure 11) using 6FDA, s-BPDA, 3,3'-dimethyl-4,4'-diaminobiphenyl (OTOL) and 4-aminobenzoyl caprolactam (4-ABC) by polycondensation and subsequent anionic ring-opening polymerization methods [78].

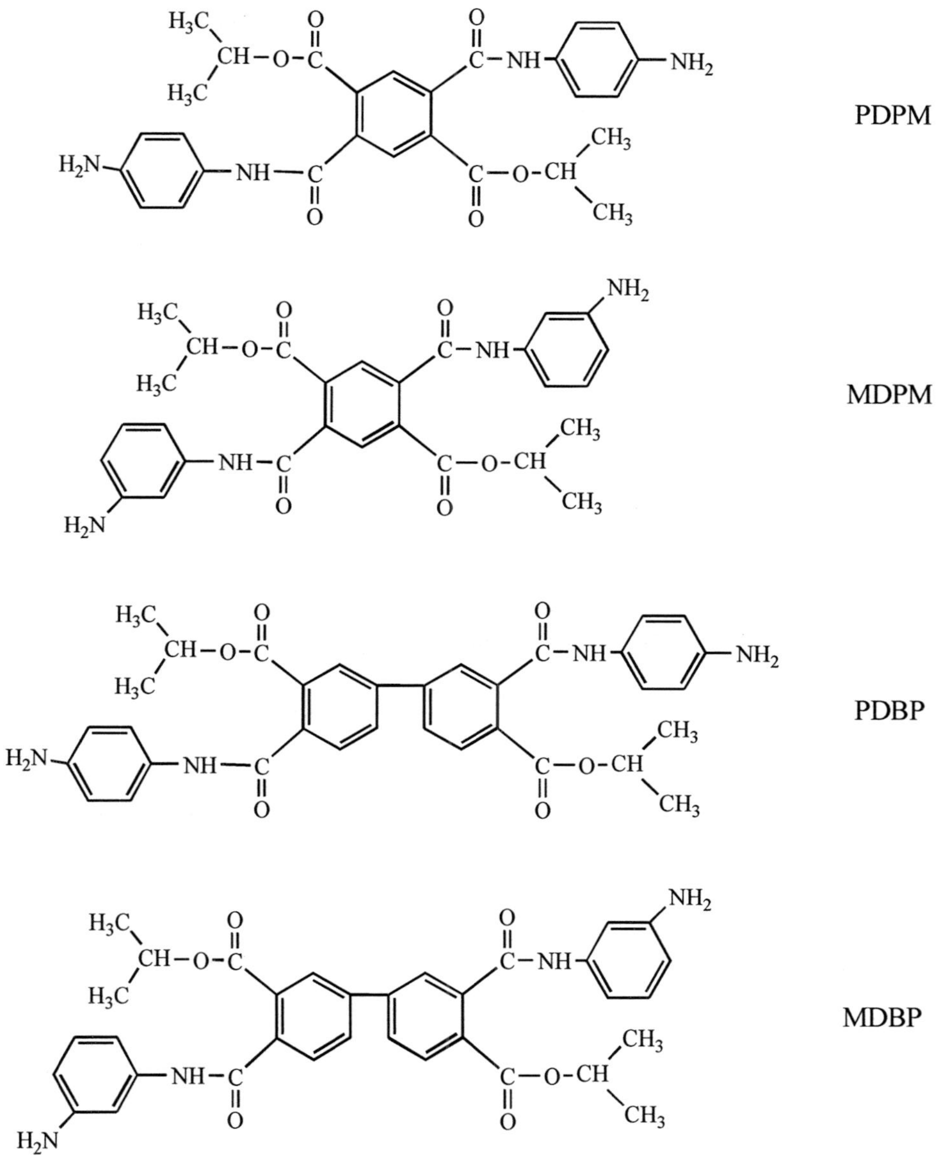


Fig. 12 Chemical structures of amido-amines

Various precursors for amido diamine [79] (figure 12) such as N,N'-bis(4-aminophenyl)-2,5-bis[(isopropyloxy)carbonyl]benzene-1,4-dicarboxamide (PDPM), N,N'-bis(3-aminophenyl)-2,5-bis[(isopropyloxy)carbonyl]benzene-1,4-dicarboxamide (MDPM), N,N'-bis(4-aminophenyl)-4,4'(4,3' or 3,3')-bis[(isopropyloxy) carbonyl]-biphenyl-3,3'(3,4' or 4,4')-dicarboxamide (PDBP), N,N'-bis(3-aminophenyl)-4,4'(4,3' or 3,3')-bis[(isopropyloxy)carbonyl]-biphenyl-3,3'(3,4' or 4,4')-dicarboxamide (MDBP) was reacted with 6FDA to prepare alternating copolyimide via alternating copoly(amic acid ester) intermediate [80]. Alternating copolyimide was found to have enhanced solubility when compared with the corresponding random isomers.

## Other Polymerization Process

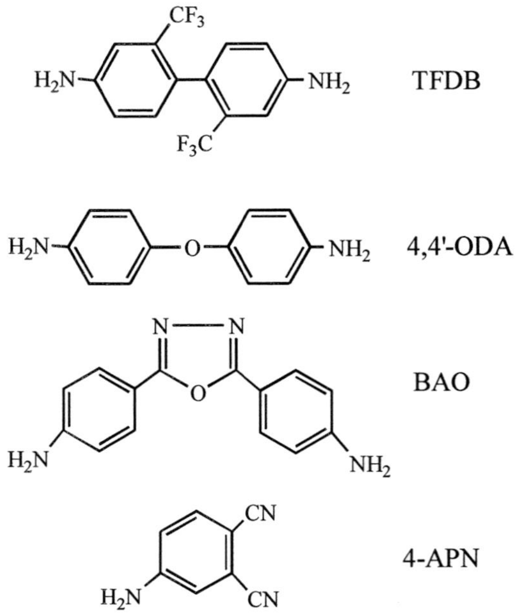


Fig. 13 Chemical structures

In vapor deposition polymerization (VDP), coevaporation of dianhydride and diamine was utilized to prepare fluorinated PI films. 6FDA and 2,2'-bis(trifluoromethyl)-4,4'-diaminobiphenyl (TFDB) monomers were used to prepare films by this method. It has been reported that 6FDA/TFDB pressure ratio of 1:10 was best for obtaining stoichiometric 6FDA/TFDB PI [81]. Using 6FDA and 4,4'-oxydianiline (4,4'-ODA) as monomers, this polymerization method was used to prepare composite membrane and its separation performance for water-ethanol systems in pervaporation mode and $CO_2$-$N_2$ system in the gas separation mode was studied [82]. Similarly, by this method, 2,5-bis(4-aminophenyl)-1,3,4-oxadiazole (BAO) and FBPA were reacted with 6FDA to prepare light emitting PI thin films [83].

Melt polymerizable bisimido-bisphthalonitrile polymer precursors were synthesized by the reaction of 4-aminophthalonitrile (4-APN) with 6FDA. The synthesized monomer showed a melting point at about 270 °C, which upon melt polymerization gave thermally – stable tough polymers. [84]. 6FDA based PIs can be prepared via one-step synthesis using gamma radiation from a cobalt-60 source [85].

Melt processable trifluorovinylether-terminated imide oligomers were prepared (Figure 14) by reacting 6FDA with 4-(trifluorovinyloxy)aniline (TFVA) which underwent thermal cyclopolymerization to afford PIs containing perfluorocyclobutane rings [86]. These polymers possess a unique combination of properties of well suited for optical applications such as high temperature stability, low moisture absorption, excellent melt and solution processability, high thermo-optic coefficient and low absorption at 1.3 and 1.55 μm.

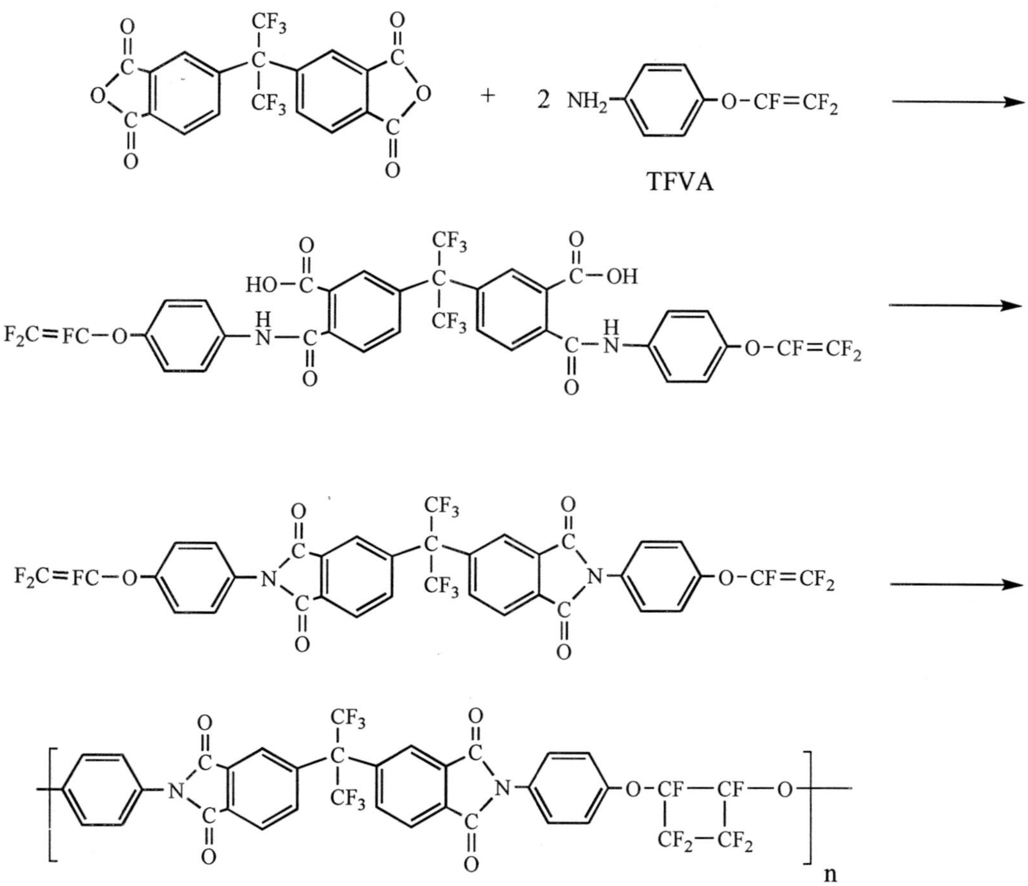


Fig. 14 Reaction scheme of PI containing perfluorocyclobutane

## Graft Copolymerization

Grafting is an effective approach for incorporating specific properties into a material, while retaining desirable properties of the parent polymer. Compared to parent polymer, graft copolymers often exhibit improvements such as enhanced compatibility with other polymers, adhesion to metallic and inorganic substrates and dye retention. Graft copolymerization can be induced thermally on ozone pretreated polymers. Ozone pretreatment produces peroxide and hydroperoxide species onto polymer chains and surfaces. Under thermal induction the peroxide functional groups on the main chains undergo decomposition. The resulting reactive sites serve as initiation sites for the free-radical polymerization of comonomers namely vinyl monomers. Thermal graft copolymerization of acrylic acid (AA) or 4-vinyl pyridine (4-VP) or N-isopropylacrylamide (NIPA) with the ozone pre-activated 6FDA PI has been reported [87, 88]. These graft copolymers containing poly(acrylic acid) (PacA)/poly(vinyl pyridine) (PVPy) and poly(N-isopropylacrylamide) (PNIPA) side chains are shown to be promising materials for fabricating microfiltration membranes with pH and temperature sensitive permeability to aqueous solutions respectively.

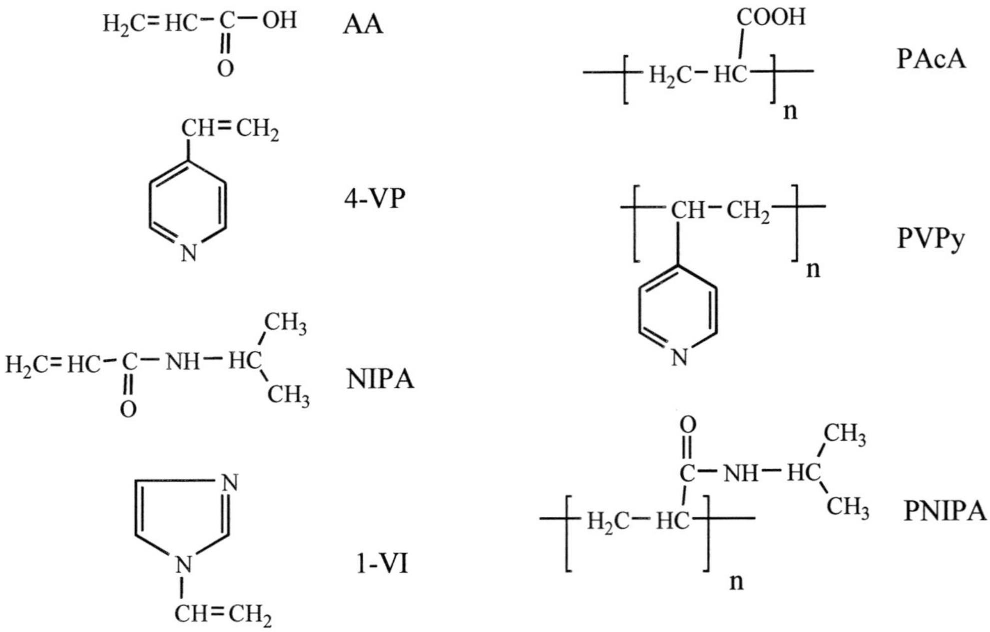


Fig. 15 Chemical structures

Ultraviolet induced graft copolymerization of 1-vinyl imidazole (1-VI) and 4-VP was reported on the argon plasma pretreated 6FDA based PI [89, 90]. These graft copolymerized PI surfaces were found to be more susceptible to the electroless deposition of copper via tin-free process than the pristine PI and argon plasma pretreated ungrafted PI surfaces and were having high adhesion strength values than those of the electrolessly deposited copper with the pristine and argon plasma pretreated ungrafted PI surfaces.

## Poly(ether-imide) (PEI)

Poly(ether-imide)s (PEI)s are unique polymers, which exhibit superior physical and chemical properties, including high heat resistance, exceptional strength and excellent processability. These polymers can be used as wire coatings and are particularly suited for injection molding applications. PEIs are prepared from diamines containing ether linkages. Two colorless PEIs developed by NASA Langley Research Center are LaRC™ CP1 and LaRC™ CP2 and are produced under license to SRS Technologies. These materials may be used to make transparent, thin polymer films for building large space reflector/collector, inflatable anteannas, solar arrays and radiometers. LaRC™ CP1 is derived from 3-6FpDA and 4-BDAF and LaRC™ CP2 from 3-6FpDA and 1,3-bis(3-aminophenoxy)benzene [91]. LaRC™ CP1 has a $T_g$ of 263 °C whereas LaRC™ CP2 has 209 °C [92]. Dielectric constant of LaRC™ CP1 and LaRC™ CP2 at 10 GHz is 2.4 and 2.7 respectively. It is not commercially available now.

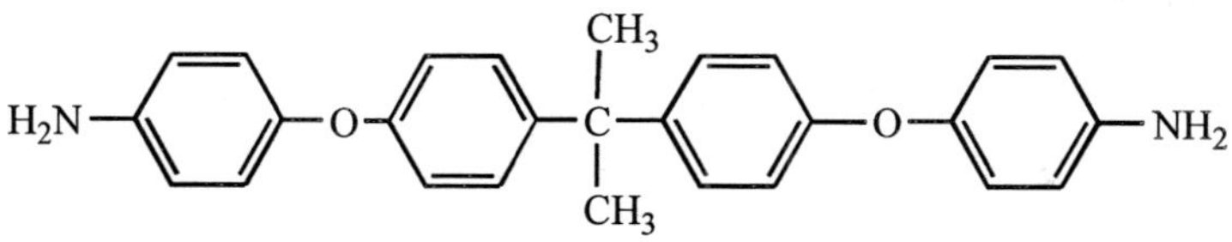
BPADE

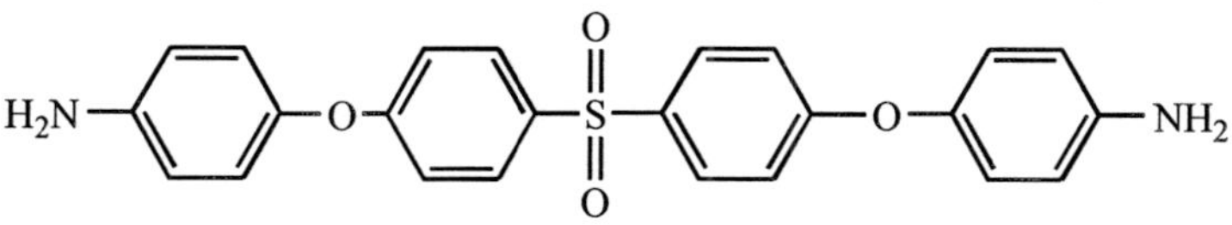
p-SED

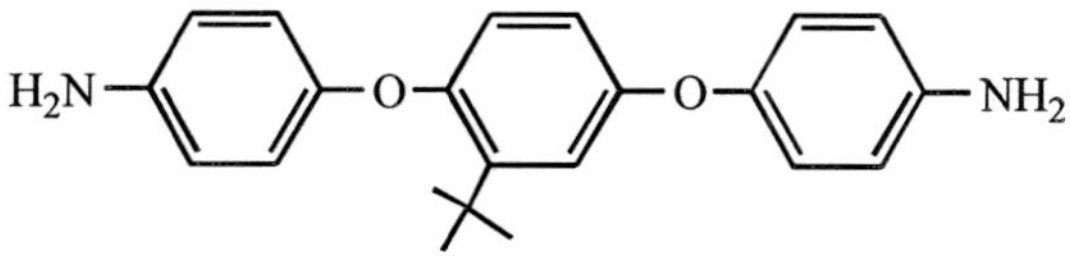
BAP-tBB

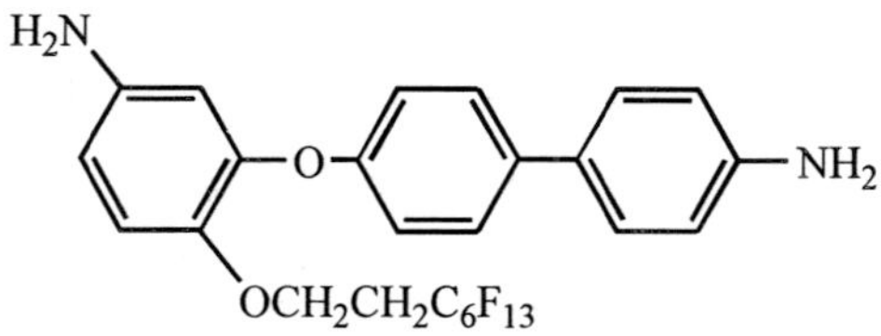
13FAPAB

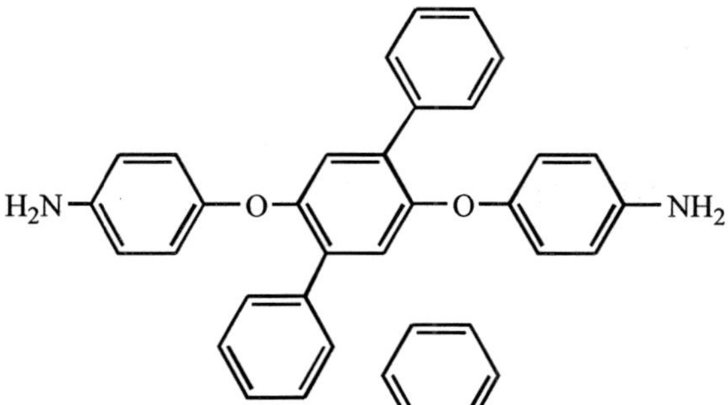
APOTP

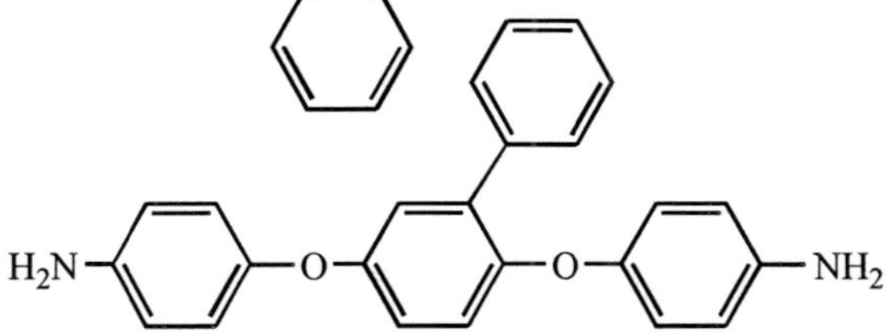
APOBP

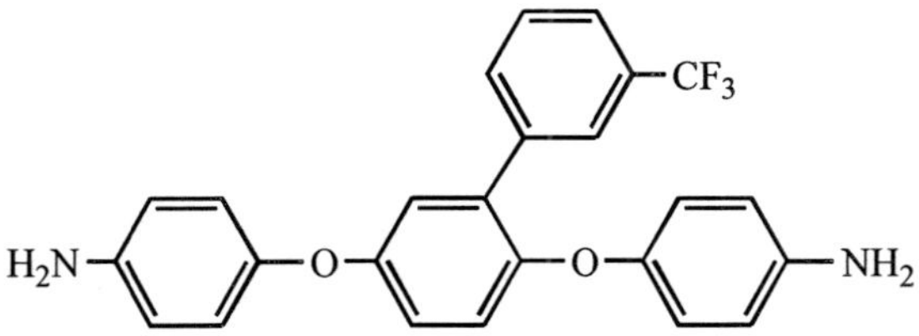
TFPPA

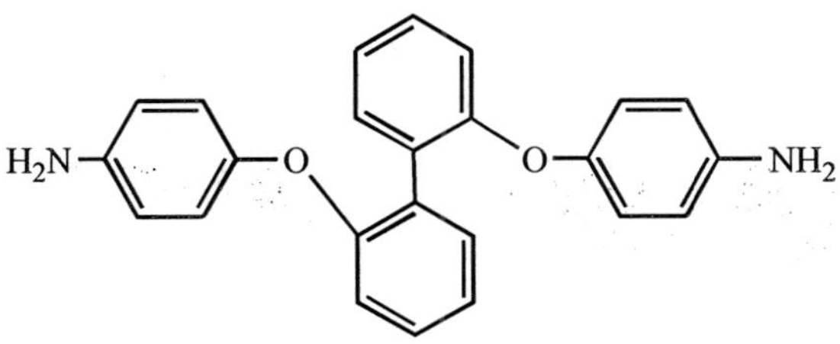
1-APOBP

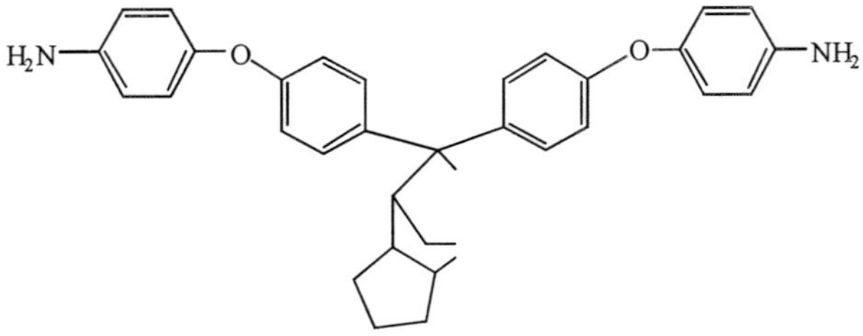


BAPPMI

n

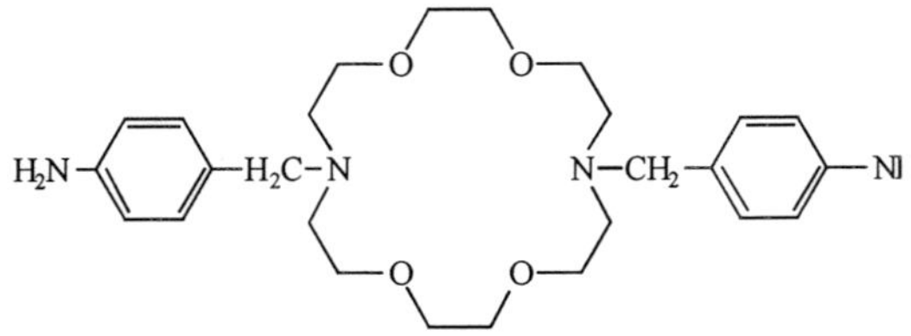


AB18C6

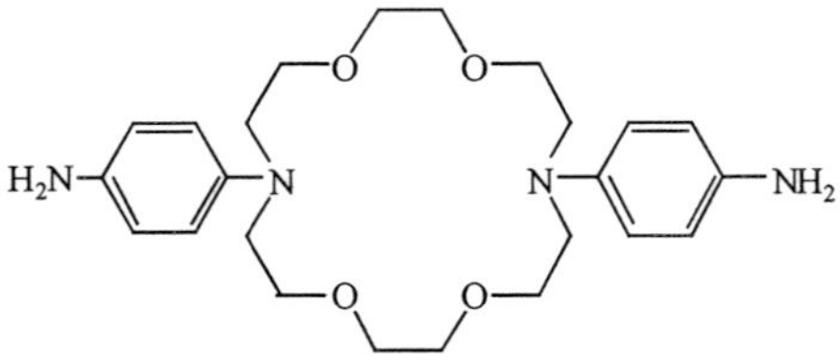


AP18C6

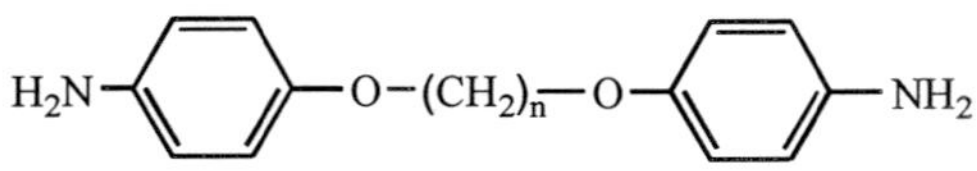


DA-n
n = 4,6,10

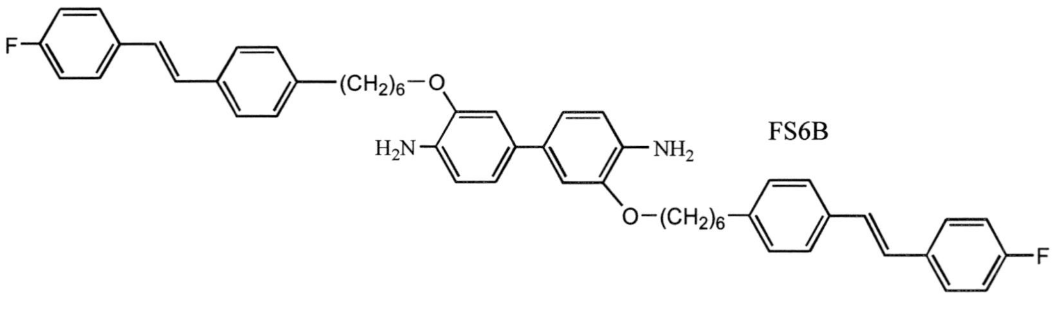


FS6B

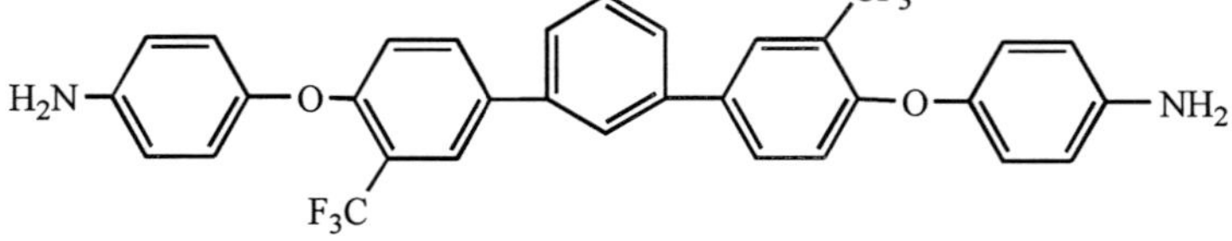


TABB

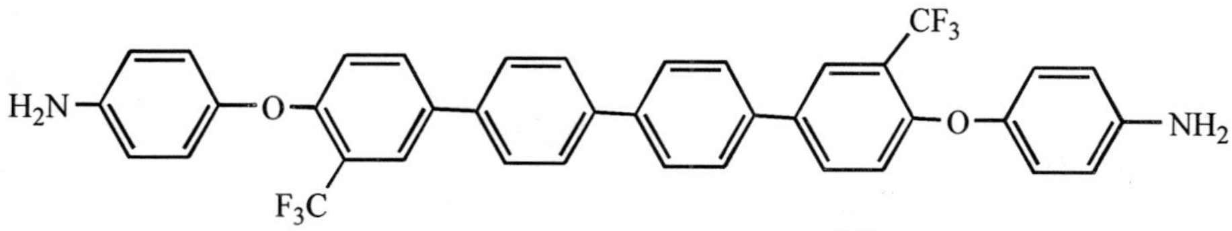

TABBP

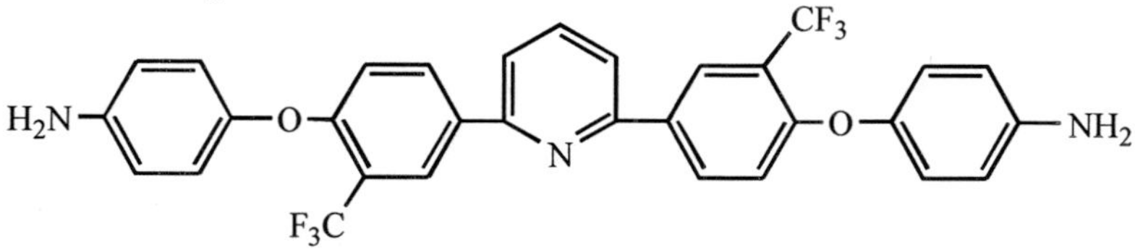

TABP

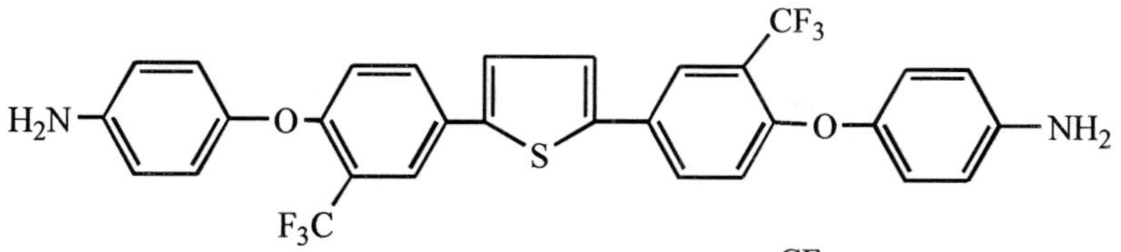

TABT

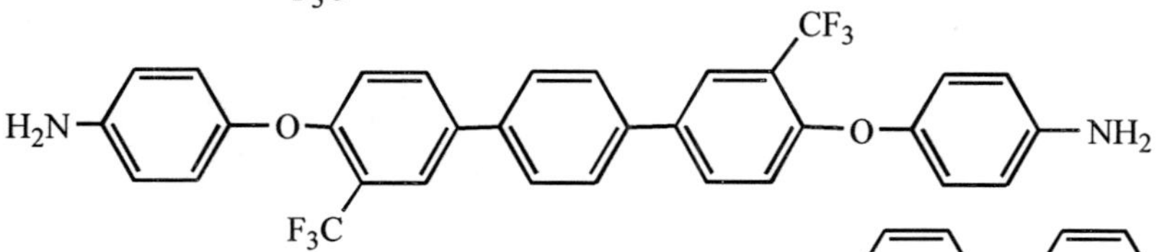

ATFT

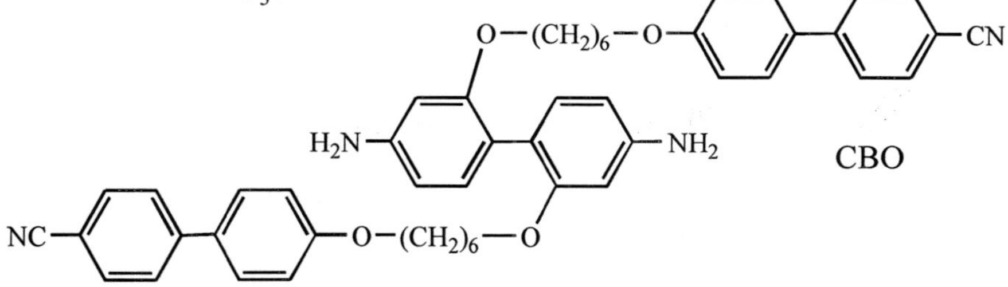

CBO

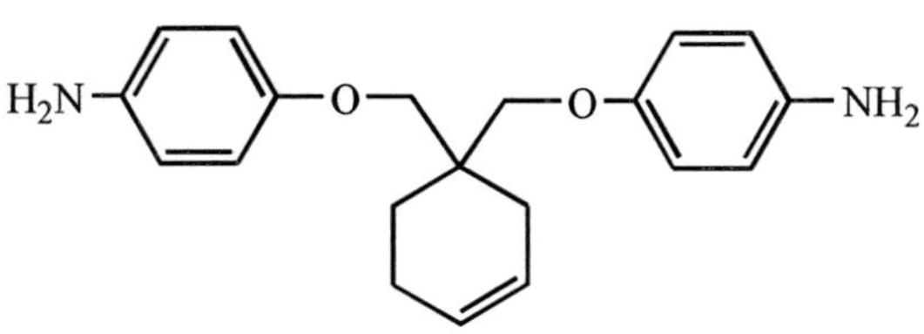

CHEDA

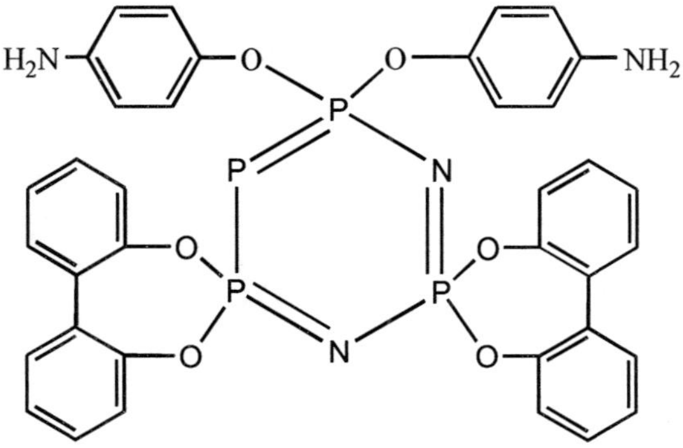

GDA

Fig. 16: Chemical structure of Aromatic diamines containing ether linkages

Some of the ether linked diamines used for making PEIs are 2,2'-bis[4-(4-aminophenoxy)phenyl] propane (BPADE) [93, 94], 2,2'-bis[4-(4-aminophenoxy)diphenyl] hexafluoropropane (4-BDAF) [93, 94], 4,4-bis(4-aminophenoxy)diphenyl sulfone (p-SED) [94] and m-SED [94], 1,4-bis(4-aminophenoxy)-2-tert-butyl benzene (BAP-tBB) [95], 1-(3,3,4,4,5,5,6,6,7,7,8,8,8-tridecafluorooctan-1-oxy)-2-(4-aminophenoxy)-4-aminobenzene (13FAPAB) [96], 2',5'-bis(4-aminophenoxy)-1[1,1',4',1"]terphenyl (APOTP) [97], 2,5-bis(4-aminophenoxy)biphenyl (APOBP) [97], 1,4-(2'-trifluoromethyl-4'-aminophenoxy)-2-(3'-trifluoromethylphenyl)benzene (TFPPA) [98], 1,1'-bis(p-aminophenoxy)-2,2'-biphenyl (1-APOBP) [99], 5,5'-bis[4-(4-aminophenoxy)phenyl]-4,7-methanohexahydroindan (BAPPMI) [100], N,N'-bis(4-aminobenzyl)-4,13-diaza-18-crown-6 (AB18C6) [101], N,N'-bis(4-aminophenyl)-4,13-diaza-18-crown-6 (AP18C6) [102], 4,4'-(alkylenediyldioxy)dianilines (DA-n) [102], 3,3'-bis[(4'-fluoro-4-stilbenyl)oxyhexyloxy]-4,4'-biphenyldiamine (FS6B) [103].

Novel diamine monomers such as 1,3-bis[3'-trifluoromethyl-4'(4"-aminobenzoxy)benzyl]benzene (TABB) [104], 4,4-bis[3'-trifluoromethyl-4'(4-amino benzoxy)benzyl]biphenyl (TABBP) [104], 2,6-bis(3'-trifluoromethyl-p-aminobiphenyl ether)pyridine (TABP) [105], 2,5-bis(3'-trifluoromethyl-p-aminobiphenyl ether)thiophene (TABT) [105], 4,4"-bis(aminophenoxy)-3,3"-trifluoromethyl terphenyl (ATFT) [106], 2,2'-bis(4'-cyanobiphenyl-4-yloxy)-4,4'-diaminobiphenyl (CBO) [107], PFMB [108], 4,4-bis(p-aminophenoxymethyl)-1-cyclohexene (CHEDA) [109], 2,2-bis(4'-aminophenoxy)-4,4,6,6-bis[spiro(2',2"-dioxy-1',1"-biphenylyl)] cyclotriphosphazene (geminal diamine, GDA) [110] were synthesized to prepare PEIs.

## Photosensitive Polyimide

Photosensitive polyimide (PSPI)s are widely used in semiconductor manufacturing because the number of processing steps is reduced by avoiding the use of classical photoresist in the microlithography. They are used as protection and insulation layers of VLSI, multi-chip modules for computers, telecommunication, photosensors and thermal heads. They can easily give fine-patterned films with excellent characteristics of PIs by photolithographic procedure. They are of two types. (i) positive and (ii) negative working. Positive working PI is the one, which is soluble in the development solution after irradiation whereas negative working is insoluble after irradiation. PI prepared from 2,2-bis(3-amino-4-hydroxyphenyl)hexafluoropropane (AHHFP) [111] was reported to be positive working. 3,3'-diamino chalcone (3DAC) and 4,4'-diaminochalcone (4DAC) were used to prepare negative type PIs [112]. Benzhydrol PI (derived from 60 mol percent of benzhydroltetracarboxylic dianhydride (BHTDA) and 40 mol percent of 6FDA with and DDM / MDA was reacted with methacryloyl isocyanate (MAI) to obtain another negative type PSPI [113]. Another PSPI with a cinnamoyl pendant group was prepared by reacting cinnamic acid (CA) with chloromethyl groups in chloromethylated PI [114].

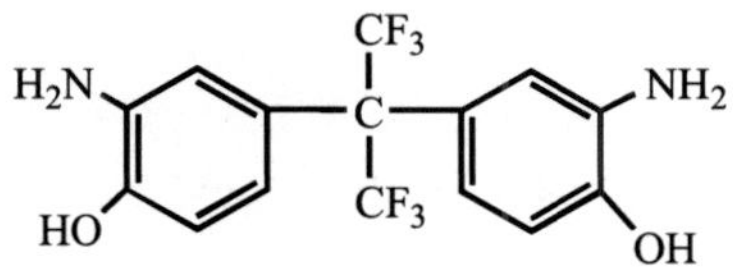

AHHFP

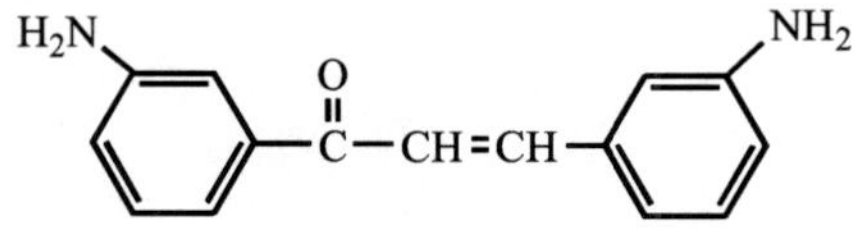

3DAC

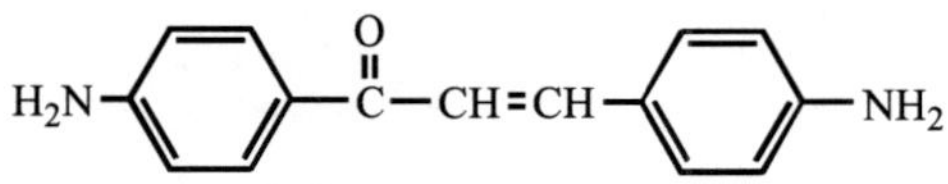

4DAC

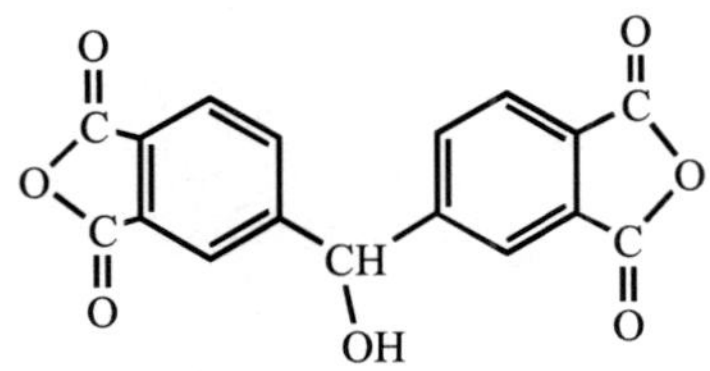

BHTDA

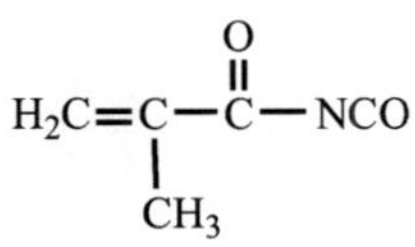

MAI

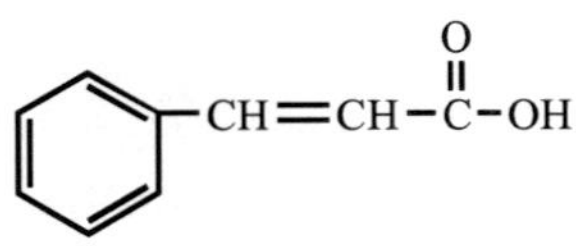

CA

Fig. 17: Chemical structures

## Hyberbranched Polyimide

Hyberbranched polymers are dendritic polymers. However, hyperbanched polymers do not have well-defined architectures as dendrimers and they are generally comprised of three parts: linear (L), dendritic (D) and terminal (T) units. Since they can be easily prepared, they are more of significance than dendrimers from the view point of industrial applications. The presence of large number of reactive groups at the ends (T units) distinguishes them from linear polymers. In contrast with conventional linear polymers, hyperbranched polymers possess good solubility in organic solvents, decreased viscosity and a low level of interchain entanglement. They are simply prepared by one-step polymerization of multifunctional monomers and the monomer concentration was kept low to prevent the formation of gel. Generally, they are synthesized by self-polymerization of an $AB_2$-type monomer [115,116] or by copolymerization of an $A_2$-type monomer with $B_3$-type monomer [117] and have poor film-forming ability due to the lack of chain entanglement [117,118].

Tris(4-aminophenyl)amine (TAPA) [119], 2,4,6-triaminopyrimidine (TAP) [120], tris[4-(4-aminophenoxy)phenyl) ethane (TAPE) [121] and 1,3,5–tris(4-aminophenoxy)benzene

(TAPOB) [122] were used to prepare 6FDA based hyberbranched PIs. Depending upon the molar ratio or addition manner of monomers, two types of hyberbranched PIs (amine terminated and anhydride terminated) were obtained (figure 18). The addition of a dianhydride to triamine with the monomer molar ratio of 1:1 yielded amine terminated polymers, while reverse monomer addition order with the molar ratio of dianhydride over triamine of 2:1 gave anhydride terminated polymers. SEC measurement revealed that both the amine terminated and anhydride terminated 6FDA-TAPA PIs had moderate number-averaged molecular weights, but the latter had very broad MWD [123]. Physical and gas transport properties of 6FDA-TAPOB was compared with its linear type analogues namely 6FDA-1,4-bis(4-aminophenoxy) benzene (TPEQ) and 6FDA-1,3-bis(4-aminophenoxy) benzene (TPER) [122].

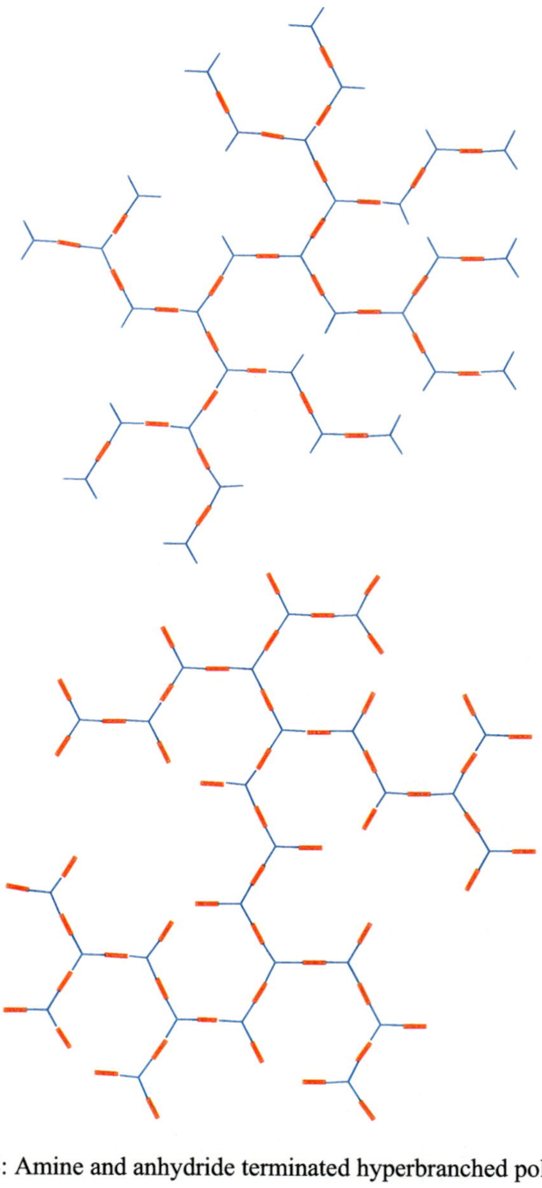

Fig. 18: Amine and anhydride terminated hyperbranched polymers

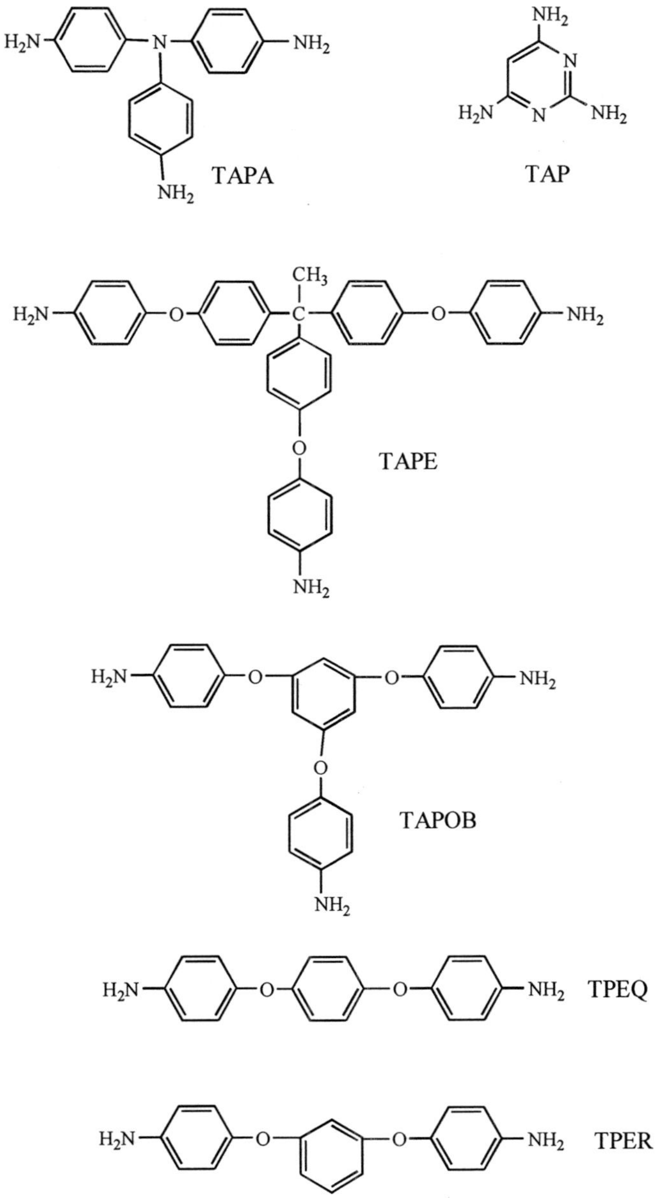


Fig. 19: Chemical structures

## Addition Polyimides

Addition (Thermosetting) PIs are used as matrix resins for structural composites in aircrafts and thermal insulation materials [124]. They are synthesized by reacting a dianhydride and the diamine in the presence of monofunctional endcapper such as maleic anhydride, nadic anhydride, methyl nadic anhydride, 3-ethynyl aniline. They are classified by the chemical nature of their reactive end groups. The addition PIs containing various endgroups are listed in figure 20. The reactive endgroups can undergo homo- and /or copolymerisations by thermal or catalytical means. Thermosetting PIs are easier to process than their thermoplastic counterparts because they use low molecular weight, low viscosity monomers and/or prepolymers as starting materials. Furthermore, they have excellent shelf life and there are no volatiles generated during cure, if the thermosetting PIs are preimidized. Crosslinked PIs possess desirable properties such as improved solvent resistance, good stress crack behavior and high modulus.

AF-700A, AF-700B and PMR-II resins discussed earlier belong to nadicimide resins as they have nadimide group at one or both the ends. Thermid FA-700, an ethynyl endcapped resin, was marketed by National Starch and Chemical Corporations, but is no longer available. It is based on 6FDA and 1,3-bis(3-aminophenoxy) benzene and is endcapped with MEA. Its chemical structure is given in figure 21. This resin is very attractive because it has low $T_g$ when uncured and high $T_g$ for cured resin. The resin is soluble in a variety of common solvents and can be crosslinked by heating at 250-275 °C without the evolution of volatile products [125]. Another resin VCAP-II discussed earlier also belong to addition PIs and has vinyl as end group. N-CYCAP oligomer (amiNe substituted CYClophane Addition Polyimide) has aromatic endgroups and is prepared from using endcapper aminosubstituted 2,2-paracyclophane [126].

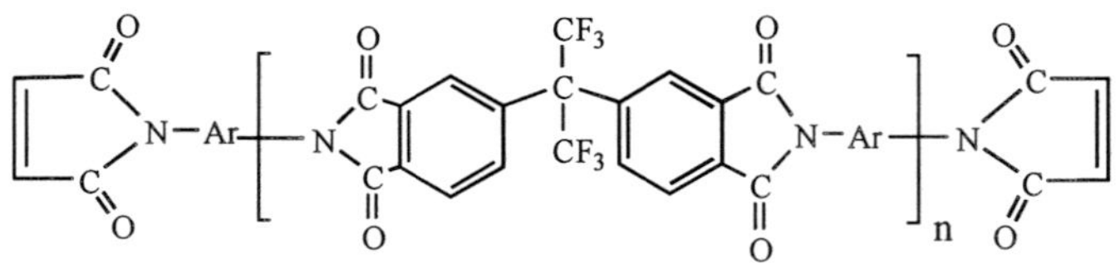

Bismaleimide resin

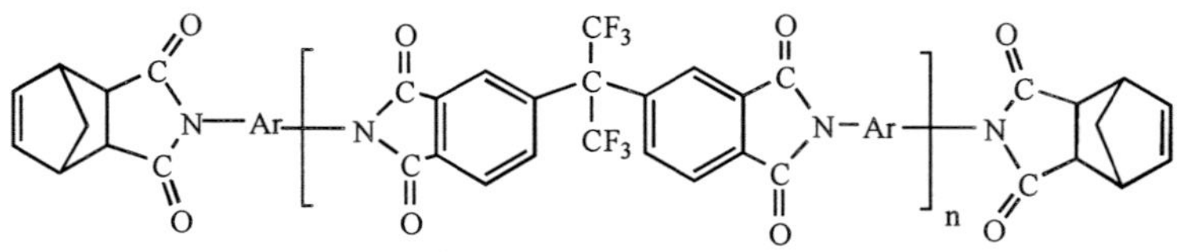

Bisnadicimide resin

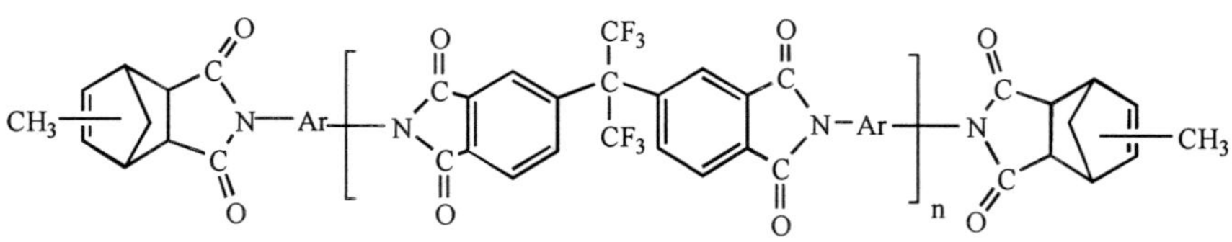

Bis(methyl)nadicimide resin

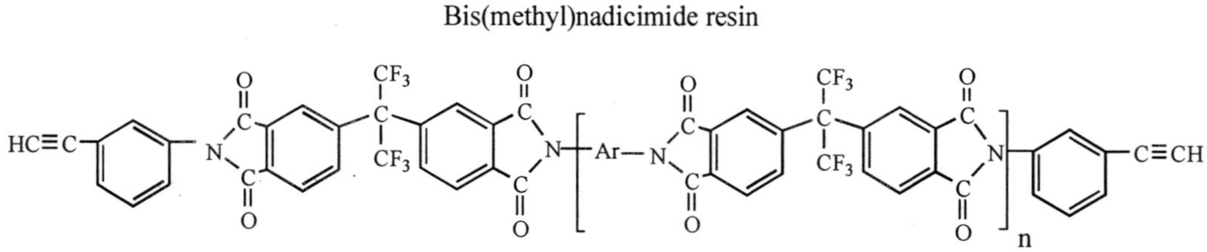

Ethynyl end-capped resin

Fig. 20: Chemical structure of addition polyimides

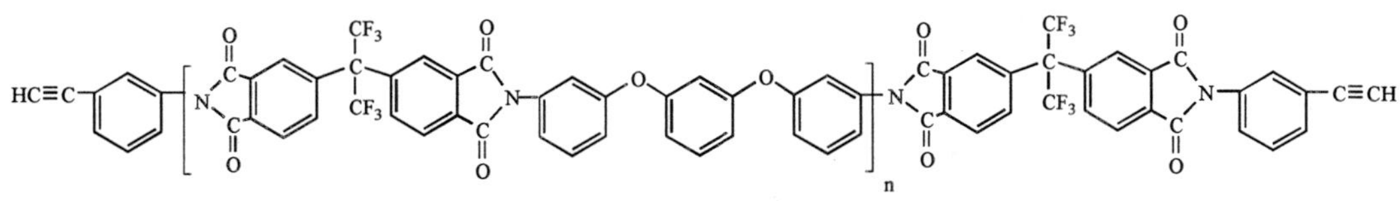

Thermid FA 700

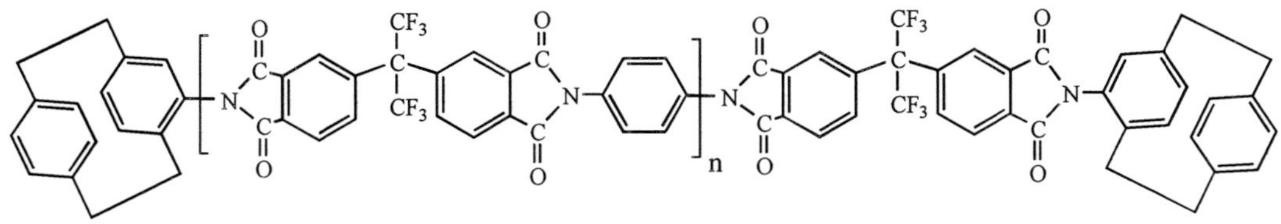

N-CYCAP oligomer

Fig. 21: Chemical structure of Thermid FA-700 and N-CYCAP oligomer

A great variety of structurally distinct thermosetting PIs have been synthesized and characterized. These PI oligomers have been derived from different aromatic diamines, dianhydrides, as well as varying reactive endgroups. A typical synthetic scheme for maleimide terminated imide oligomer is given in figure 22.

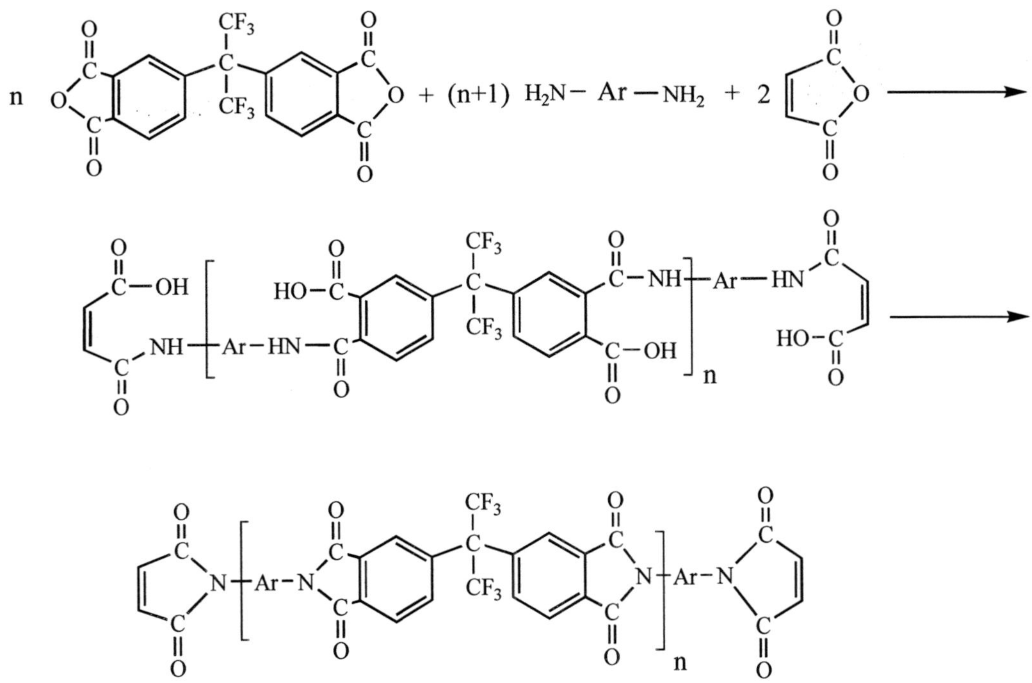

Fig. 22: Preparation of maleimde-terminated imide oligomer

Aromatic diamine, 6FDA and maleic anhydride were reacted in N, N-dimethyl acetamide/xylene at 50 °C to form amic acid oligomer, which was subsequently cyclodehydrated by refluxing in the presence of pyridine as a catalyst. Water is removed azeotropically and the oligomer is isolated by precipitation in water or a non-solvent [127]. These oligomers have been used for molding, adhesive and composite applications. Similarly mixed end-capped such as ethenyl and ethynyl terminated imide resins can be prepared [128, 129].

Nadicimide end-capped oligomers were usually prepared in two steps namely preparation of amine terminated nadicimide and chain extension with a dianhydride. Appropriate quantities of nadic or methyl nadic anhydride and diamine in glacial acetic acid were refluxed for several hours and the amine terminated nadicimide was recovered by precipitation. Then chain extension with 6FDA was carried out in acetone at 60 °C. Chemical imidization of the amic acid to imide was carried out using sodium acetate and acetic anhydride [130,131]. Instead of diamines, triamines can also be used [132–134]. Chemical structure of nadicimide end-capped resins prepared from di and triamine (Figure 23) is given below.

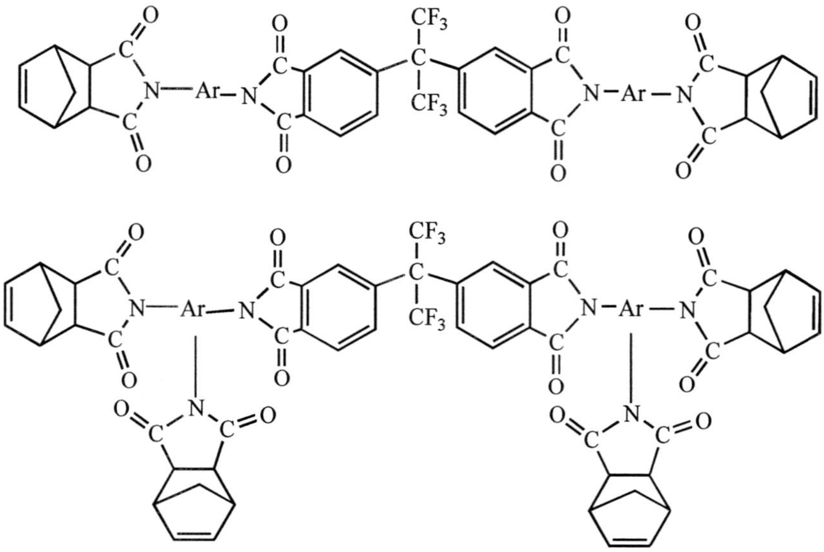


Fig. 23: Chemical structures of nadicimide resins

## Poly(amide-imide)

High softening/melting temperatures limit the applications of PIs. To overcome these drawbacks, PIs structures are often modified as poly(amide-imide)(PAI). PAIs have the advantages of both PA and PIs and possess thermal stability balanced with processability. These polymers can be synthesized from various aromatic monomers containing anhydrides, carboxylic acids and aromatic diamines by condensation. PAIs are used in wire coatings, laminates, molded products, films and fibers. PAIs were mainly prepared under anhydrous and isothermal conditions (<35 °C) by reacting trimellitic anhydride acid chloride, 6FDA and aromatic diamine in N, N-dimethyl acetamide solvent and in the presence of triethylamine as the catalyst. The intermediate was then cyclized by chemical imidization method. Finally, the product was filtered to remove the by-product chloride salt and the filtrate was precipitated in water followed by washing in methanol and drying at 110 °C in an oven (Figure 24). [135]. PAIs are relatively hydrophilic but the presence of fluorine atoms in 6FDA based PAI reduces the apparent moisture uptake when compared to the non fluorinated commercially available PAIs namely Torlon.

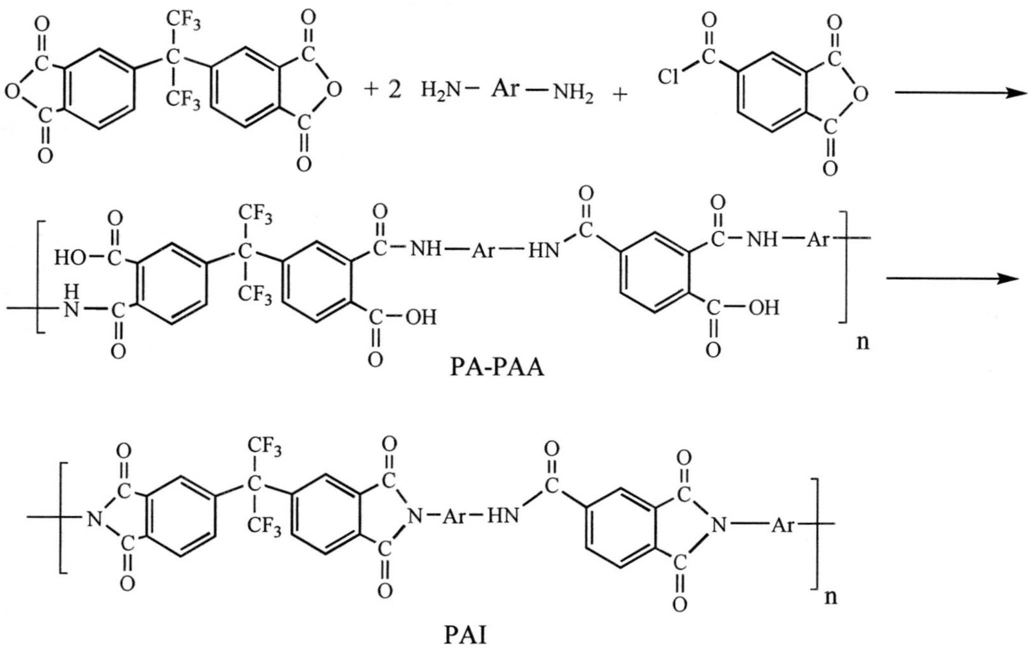


Fig. 24: Synthesis of PAI

Aromatic PIs can be synthesized through the polycondensation of imide containing monomers and aromatic diamines/dicarboxylic acids. A new type of PAI (Figure 25) with good processability was prepared by reacting 6FDA with various aromatic diamines in 1: 2 molar ratio to prepare imide ring preformed diamines which was then directly polycondensed with equimolar amount of terephthalic acid [136].

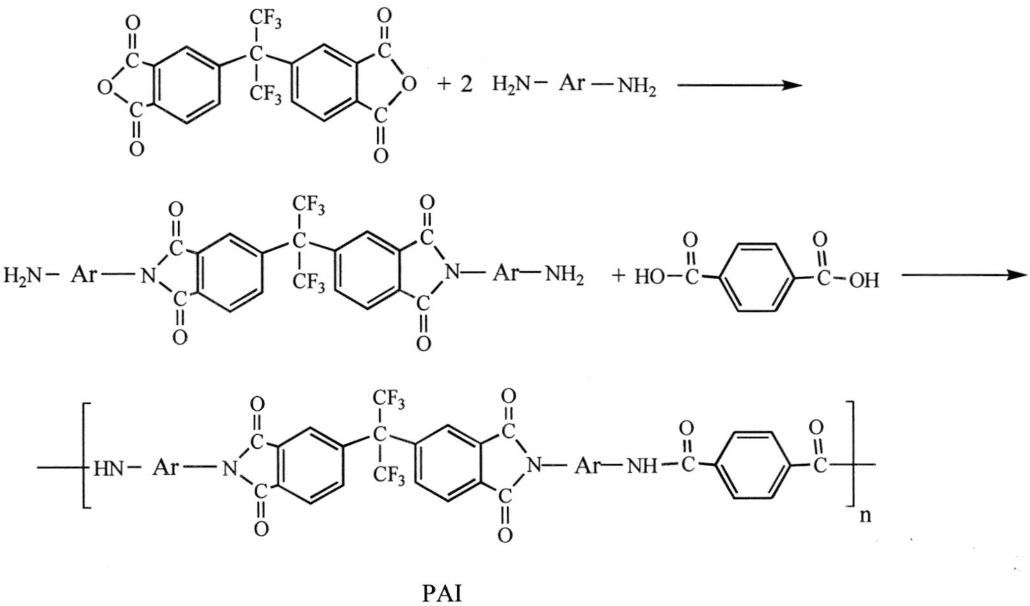


Fig. 25: Synthesis of PAI using terephthalic acid

## Poly(urethane-imide)

Polyurethane (PU) is known to degrade above 200 °C and this limits its applications. This necessitated the introduction of PI or oligoimide unit into PU and the resulting poly(urethane-imide) (PUI) has improved thermal stability when compared to PU. The most convenient method for the preparation of PUI is by the reaction of isocyanate terminated PU prepolymer with 6FDA (Figure 26) [137,138]. The reaction between isocyanate and anhydride yields imide and carbon dioxide. The use of isocyanate instead of conventionally used amine is advantageous in that imidization is achieved in a single step at lower temperature. Because of the higher reactivity of isocyanate, it is necessary to protect the isocyanate against moisture and self-addition reactions. These problems can be overcome by blocking isocyanate with aromatic alcohols through the formation of carbamate having a labile bond which can dissociate at a higher temperature to regenerate isocyanate functionality.

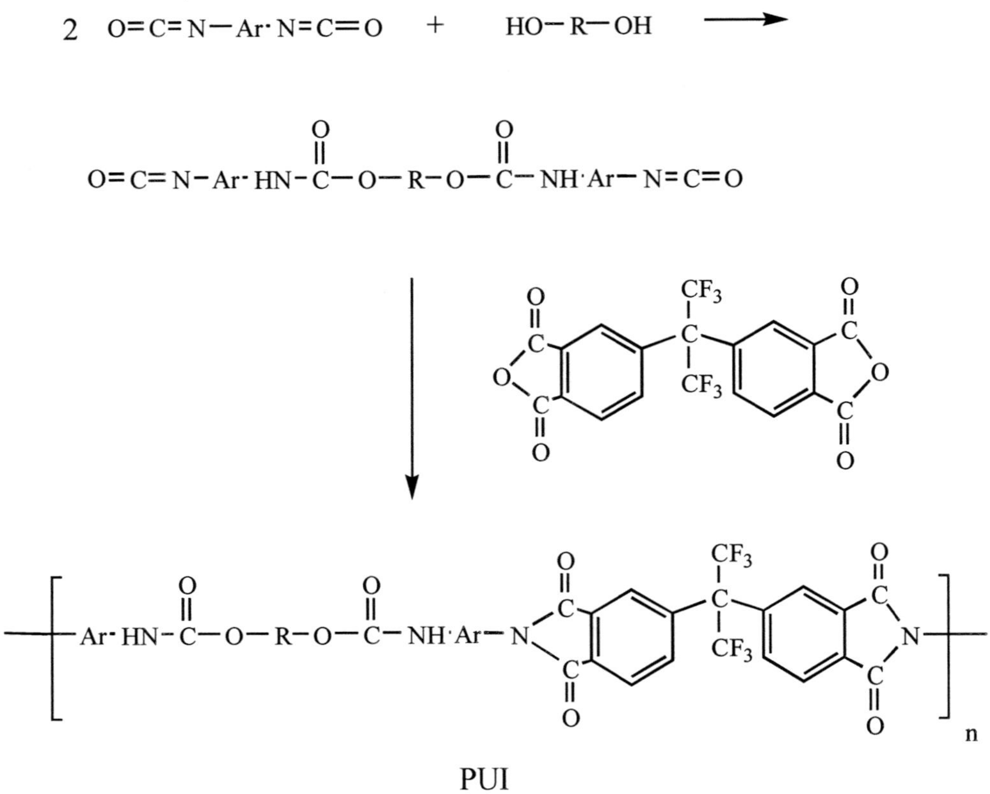


Fig. 26: Synthesis of PUI

In another method, phenol terminated PU prepolymer (PTPP) was heat treated with PI containing hydroxyl groups (derived from 6FDA and 3,3'-diamino-4,4'-dihydroxybiphenyl (AHBP)) to obtain PUI [139]. During heat treatment, end-capped phenol was released and isocyanate reacts with the hydroxyl groups of PI giving PUI. Alternatively, PTPP was thermally treated with PAA derived from 6FDA and pPDA [140] or ODA [141] to prepare PUI. Here, thermal treatment releases isocyanate, which reacts with carboxylic acid groups of PAA giving PUI [142].

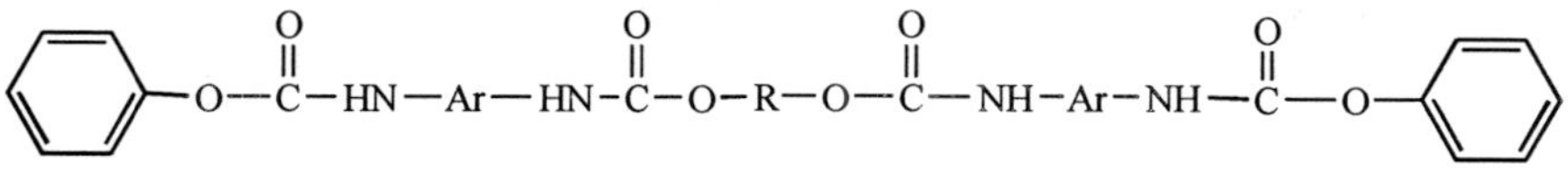

PTPP

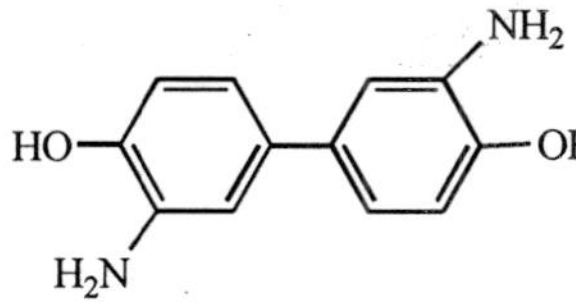

AHBP

Fig. 27: Chemical structures

## Poly(ester-imide)

Poly(ester-imide) was prepared in two steps. In the first step, 6FDA was reacted with 5-amino-1-naphthol to obtain diimide-dinaphthols. This underwent high temperature polycndensation reactions with aromatic and aliphatic diacid chlorides to produce poly(ester-imide) [143] (Figure 28).

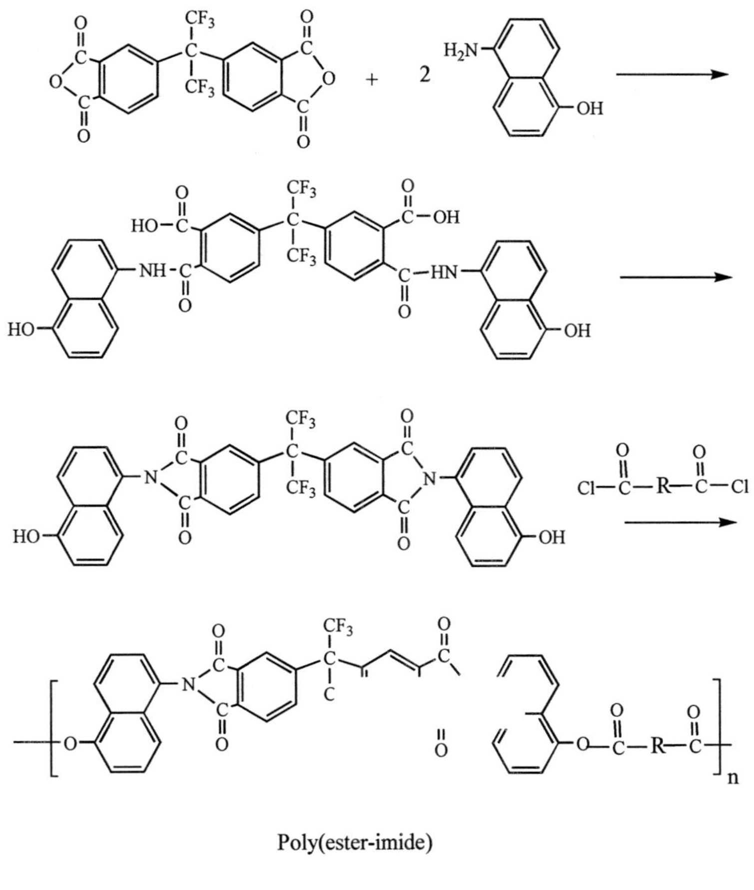

Poly(ester-imide)

Where R = Aromatic (or) aliphatic

Fig. 28: Reaction scheme of poly(ester-imide)

## Poly(siloxane-imide)

PI exhibit poor adhesion to glass, silica and silicon which limits its use. Poly(siloxane-imide) overcomes this limitation. Apart from superior adhesion, it also has low dielectric constant. Hence, it is recommended for microelectronic applications such as protective overcoating and used as dielectric layers. Poly(siloxane-imide)s were synthesized by reacting 6FDA with amine terminated poly(dimethyl siloxanes) (PDMS) having different molecular weights [144] (Figure 29). 6FDA was reacted with AHHFP and 1,3-bis(3-aminopropyl)tetramethyldisiloxane [145] to prepare hydroxyl containing poly(siloxane-imide). PI-PDMS graft copolymer was synthesized by polycondensation of 3-(3,5-diaminobenzyloxy)propyl-terminated polydimethylsiloxane with ODA and 6FDA followed by chemical imidisation and its application as a separation membrane was investigated. [146].

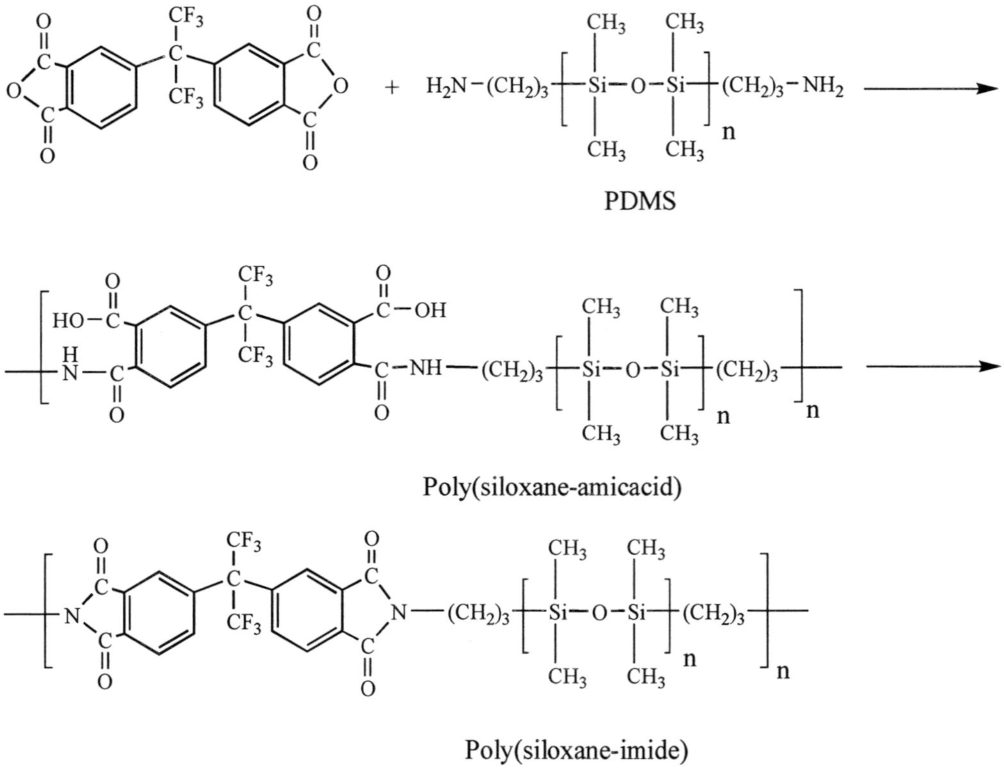


Fig. 29 Reaction scheme of poly(siloxane-imide) from PDMS

## Poly(epoxy-imide)

PI containing epoxy as side chain group was prepared by reacting epichlorohydrin with poly(hydroxy imide) in the presence of benzyl(trimethyl)ammonium chloride at 110–120 °C. Earlier, poly(hydroxy imide) was derived from 6FDA and 2,2-bis(3-amino-4-hydroxyphenyl)hexafluoropropane (AHHFP). Reaction scheme is given in fig. 30 [147]. Its photochemical reactivity was studied in the presence of diphenyl-iodonium-9,10-dimethylanthracene-2-sulfonate as photo-acid generator [148].

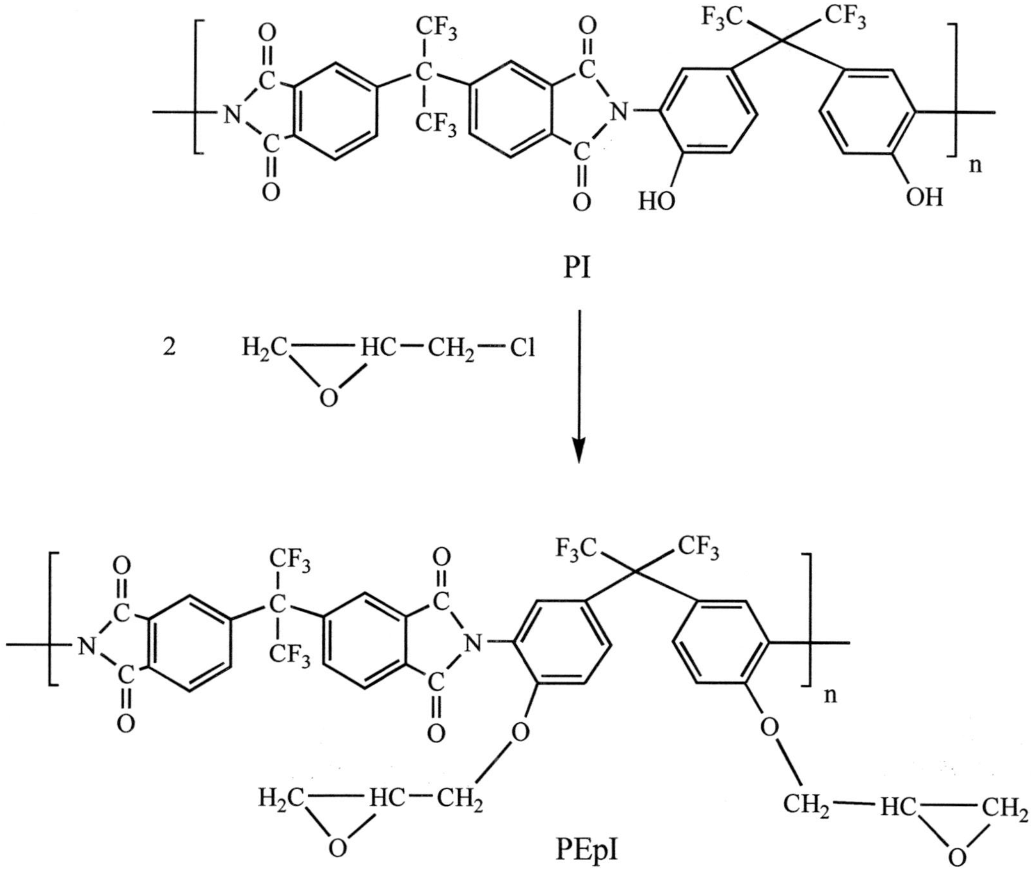


Fig. 30: Reaction scheme of poly(epoxy-imide)

## Nanocomposites

Nanocomposites are a new class of particle-filled composites in which at least one dimension of the dispersed particles is within 100 nm. Because of the dispersion of nanosize clay particles, polymer-clay nanocomposites exhibit improved moduli and strength, decreased thermal expansion coefficient, decreased gas permeability, increased swelling resistance, better thermal stability and enhanced ionic conductivity when compared to the pristine polymers or microscale composites [149–151]. They find increased applications in various fields such as automobile, packaging, electronic, coating and aerospace industries [152,153].

The preparation of a series of 6FDA based PI-clay nanocomposites and their film properties were reported in ref. [154]. Studies on the effect of transformer oil on these films [155] and their theoretically calculated glass transition temperature based on Monte-Carlo simulation [156] were documented. These nanocomposites has the potential to prevent the degradation of ultra-large-scale integration (ULSI) and giga-scale integration (GSI) devices due to the ability of silicate layer in these nanocomposites to retard the diffusion of copper at the interface [157].

Oxo-lanthanide (III) nanocomposites derived from 6FDA based PI were found to have low CTE and better solvent sensitivity than the those of the parent polymer and it is one of the many approaches to overcome the limitation of fluorinated PIs [158, 159]. 6FDA based PI-silica hybrid films containing silica particles in nanoscale were prepared successfully by the sol-gel method and these materials have excellent thermal stability and enhanced miscibility between silica and polymer. [160–162]. These hybrid materials have shown potential for microelectronic and optoelectronic applications [163].

## Non-linear Optical Polyimides

Nonlinear optical (NLO) materials are expected to play a major role in the fields of optical information processing, optical sensing and telecommunications [164–166]. Recent developments on NLO materials portend exciting new possibilities for low cost integrated devices for the telecommunication and data communication industries. NLO materials include inorganic and organic crystals, polymers, organic-inorganic hybrids and Langmuir-Blodgett compounds. Of these, polymers are considered the most promising materials because of their ease of processability, excellent thermal stability, environmental resistance, good mechanical strength and flexibility in molecular designs. Some of the issues that limit the use of polymers are instability of the dipole orientation, inadequate optical nonlinearity and high optical loss limit [167]. Numerous approaches have been developed to tackle the above issues and some of them are reviewed in this section with respect to 6FDA based PI. The use of 6FDA, results in highly soluble NLO PIs, as a result, the polymer solutions could be spin coated on the ITO glass or other substrates to form the optical quality thin films. Other advantages of using 6FDA are the dielectric constant and optical losses of the resulting PI are reduced significantly.

NLO moieties are introduced either by doping NLO chromophores into amorphous polymer matrices (guest-host systems) [168] or covalently attaching to polymer backbone as side chain or in main chain [169]. NLO chromophores are one- or two-dimensional. One-dimensional NLO chromophores are designated as push-pull compounds and are represented as D-π-A where the electron donor (D) and electron-acceptor (A) groups are liked by a π–conjugated bridge [170]. These chromophores exhibit large first–order hypersusceptibility (β). But they provide less phase-matching behavior owing to their small off-diagonal component [171] whereas two-dimensional chromophores provide phase-matching behavior better than one-dimensional chromophores because of their larger off-diagonal components [172].

6FDA based PI containing NLO chromophores as side chain could be prepared either through the introduction of chromophores to poly(hydroxy imide)s [173] or by reacting 6FDA with NLO containing diamine. Diaminophenol derivatives are needed for the synthesis of poly(hydroxy imide)s. Most of the diaminophenol derivatives are not sufficiently stable as monomers due to the oxidation of free diamine to the nitroso or nitro group except few hydroxy diamines. For hydroxy diamines which are not stable to oxidation, their dihydrochloride salts could be used for the synthesis of poly(hydroxy imide)s [174]. The introduction of NLO chromophores to poly(hydroxy imide)s at the last stage of synthesis, as shown in figure 31, prevents the exposure of harsh chemical conditions of the imidization process to the chromophores since a very mild Mitsunobu condensation conditions (using diethyl azodicarboxylate and triphenyl phosphine at RT) are used here. Further, the loading level of chromophore can be easily controlled.

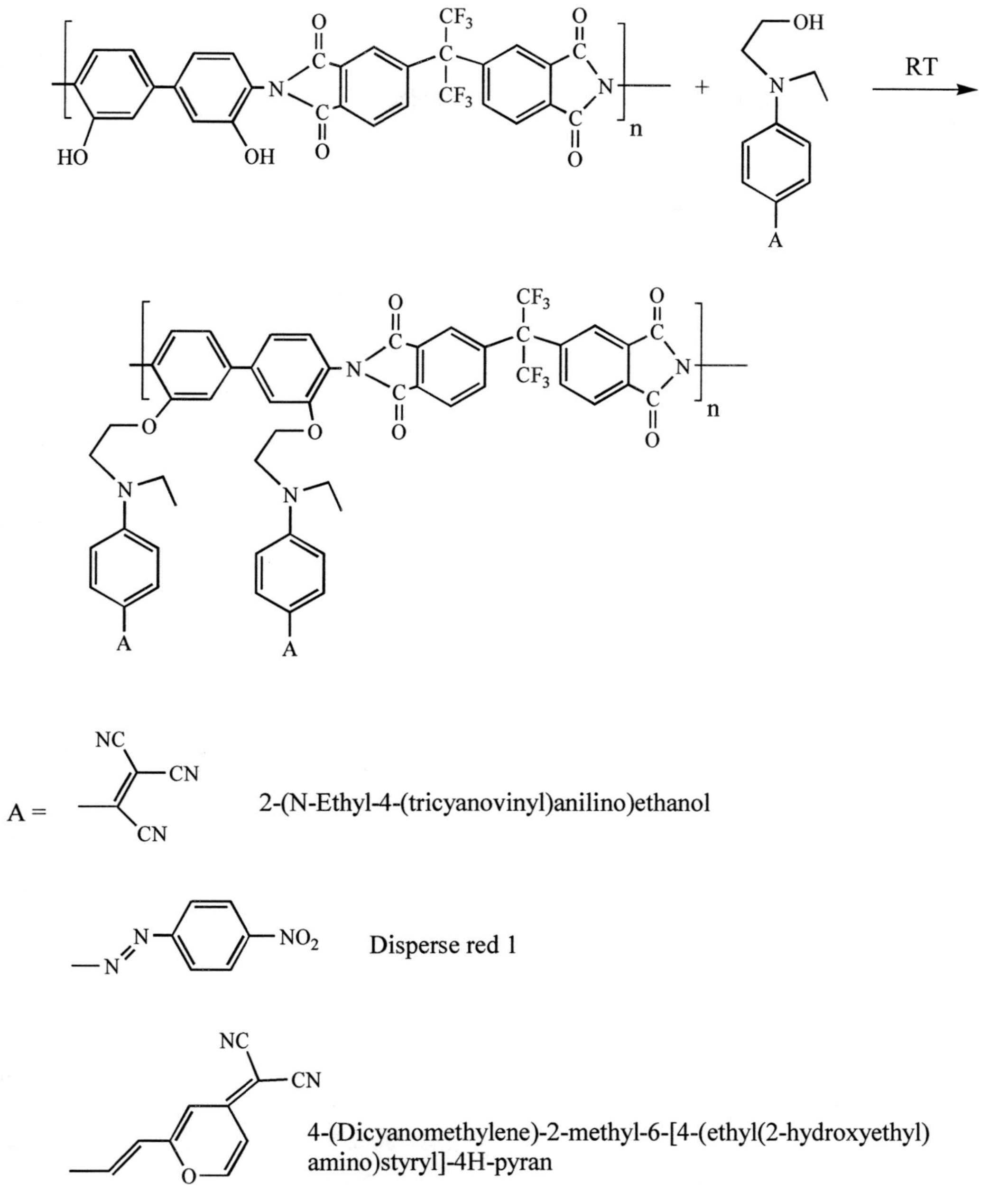


Fig. 31: Synthesis of NLO containing PIs [173]

Some of the hydroxydiamine monomers used in the synthesis of poly(hydroxy imide)s are listed in figure 32. The listed monomers are 2,2-bis(3-amino-4-hydroxyphenyl)hexafluoropropane [175, 176], 4-(4-amino-2-hydroxy)phenoxyaniline [176, 177] and 2,4-diamino-phenol dihydrochloride [174] containing hydroxyl groups [178]

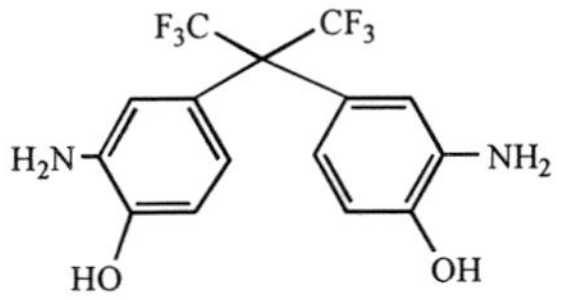


2,2-bis(3-amino-4-hydroxyphenyl)hexafluoropropane

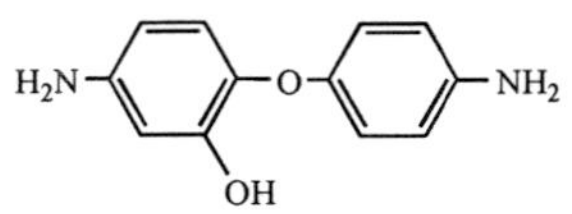


4-(4-amino,2-hydroxy)phenoxyaniline

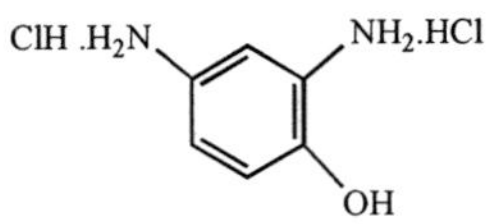


2,4-diamino-phenol dihydrochloride

Fig. 32: Chemical structure of hydroxyl diamine monomers

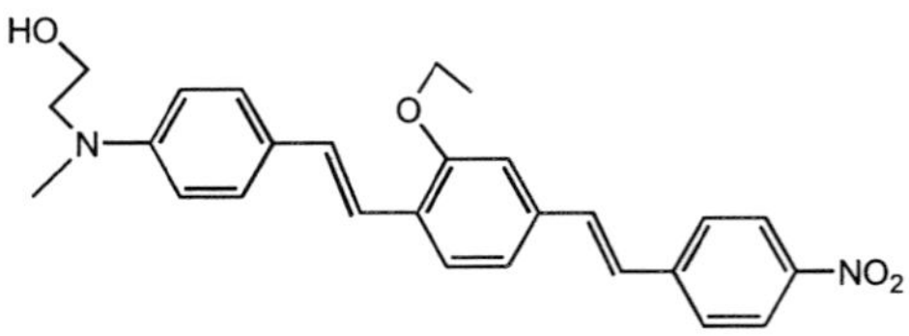


2-{[4-(2-{2-Ethoxy-4-[2-(4-nitro-phenyl)-vinyl]
-phenyl}-vinyl)-phenyl]-methyl-amino}-ethanol

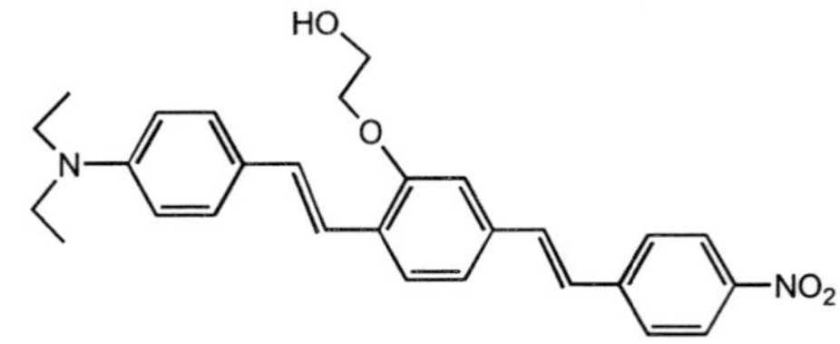


2-{2-[2-(4-Diethylamino-phenyl)-vinyl]
-5-[2-(4-nitro-phenyl)-vinyl]-phenoxy}-ethanol

Fig. 33: Chemical structure of NLO chromophores

Other NLO chromophores, which can be introduced to poly(hydroxy imide) through Mitsunobu reaction [178] are listed in figure 33 and are 2-{[4-(2-{2-Ethoxy-4-[2-(4-nitro-phenyl)-vinyl]-phenyl}-vinyl)-phenyl]-methyl-amino}-ethanol [178] and 2-{2-[2-(4-Diethylamino-phenyl)-vinyl]-5-[2-(4-nitro-phenyl)-vinyl]-phenoxy}-ethanol [178].

Another alternative and convenient method for the synthesis of NLO-PIs is through a one-step reaction between di(hydroxyalkyl) chromophores and diimide under Mitsunobu conditions. Earlier dimide was prepared by heating a mixture of

6FDA and urea for few hours [179, 180]. Some of the NLO chromophores [179, 180] used in this method are 2-[3-(2-{4-[bis-(2-hydroxy-ethyl)-amino]-phenyl}-vinyl)-5,5-dimethyl-cyclohex-2-enylidene]-malononitrile [179], 4-[N,N-bis(hydroxyethyl)amino]-4'-nitrostilbene [179], 4-[N,N-bis(hydroxyethyl)amino]-4'-nitroazobenzene [179], 4-[N,N-bis(hydroxyethyl)amino]-2'-methyl-4'-nitroazobenzene [179], 4-[N,N-bis (hydroxyethyl) amino]-4'-cyanoazobenzene [179],4-[N,N-bis(hydroxyethyl)amino]-4'-formylazobenzene [180] and 2-[[4-(4-[1,3]dioxolan-2-yl-phenylazo)-phenyl]-(2-hydroxy-ethyl)-amino]-ethanol [180] and their chemical structures are given in figure 34.

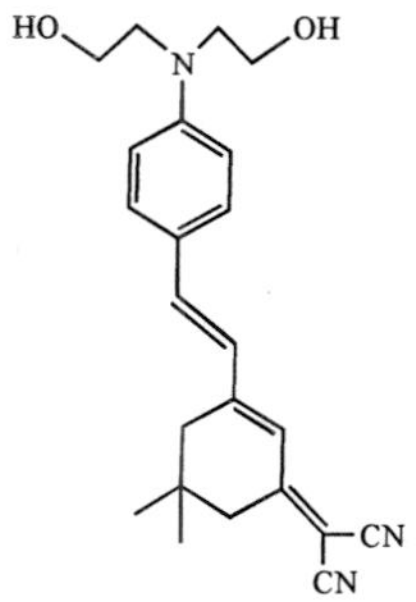

2-[3-(2-{4-[bis-(2-hydroxy-ethyl)-amino]-phenyl}
-vinyl)-5,5-dimethyl-cyclohex-2-enylidene]-malononitrile

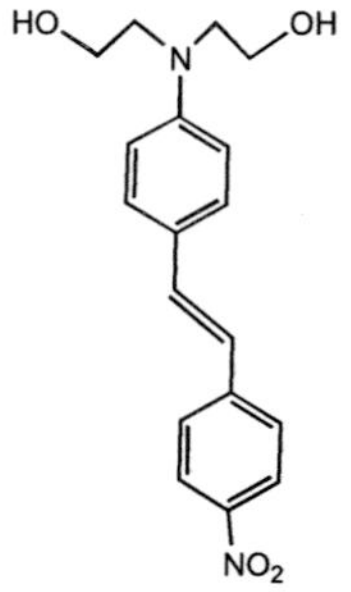

4-[N,N-bis(hydroxyethyl)amino]-4'-nitrostilbene

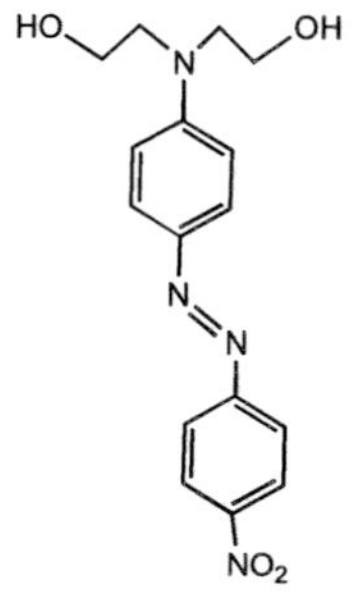

4-[N,N-bis(hydroxyethyl)amino]-4'-nitroazobenzene

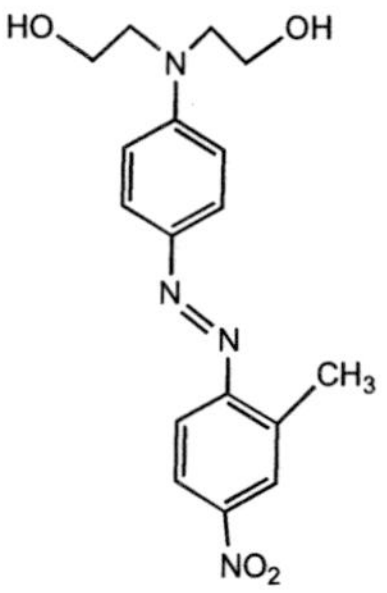

4-[N,N-bis(hydroxyethyl)amino]
-2'-methyl-4'-nitroazobenzene

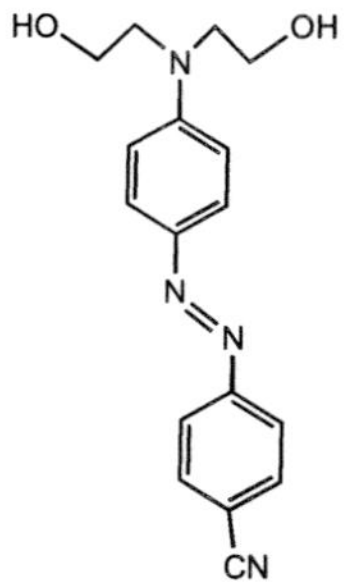

4-[N,N-bis(hydroxyethyl)amino]
-4'-cyanoazobenzene

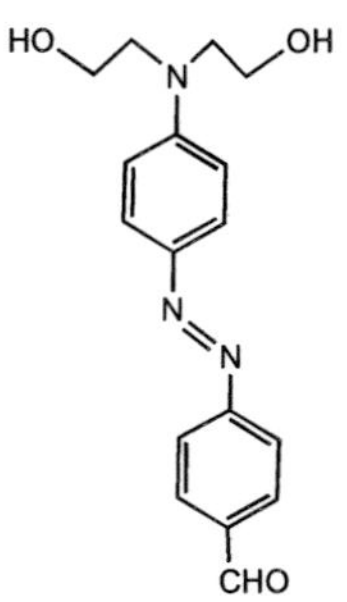

4-[N,N-bis(hydroxyethyl)amino]
-4'-formylazobenzene

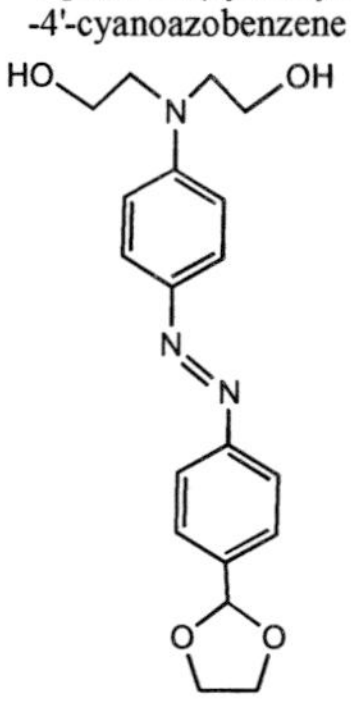

2-[[4-(4-[1,3]dioxolan-2-yl-phenylazo)-
phenyl]-(2-hydroxy-ethyl)-amino]-ethanol

Fig. 34: Chemical structure of dihydroxy alkyl NLO chromophores

In another approach, NLO chromophore diamines [181-187] are used for the preparation of 6FDA based PIs. Some of them are 2-(2-{ethyl-[4-(4-nitro-phenylazo)-phenyl]-amino}-ethoxy)-benzene-1,4-diamine [181,182], 2-[2-(ethyl-{4-[2-(4-nitro-phenyl)-vinyl]-phenyl}-

amino)-ethoxy]-benzene-1,4-diamine [181,182], 2-[2-(ethyl-{4-[5-(4-nitro-phenyl)-thiophen-2-yl]-phenyl}-amino)-ethoxy]-benzene-1,4-diamine [181,182], 2-(2-{ethyl-[4-(4-methanesulfonyl-phenylazo)-phenyl]-amino}-ethoxy)-benzene-1,4-diamine [181,182], 4-(4'-nitrophenyl-diazenyl)phenyl-1,3-diamine [183], 2,4-diamino-4'-(4-nitophenyl-diazenyl) azobenzene [183], 2,4-diamino-4'-[(6-nitrobenzothiazol-2-yl)diazenyl]azobenzene [184], [(6-nitrobenzothiazol-2-yl)diazenyl]phenyl-1,3-diamine [184], bis(4-aminophenyl)[4-(2-(6-nitrobenzothiazol-2-yl)vinyl)phenyl]amine [185], 2-[4-((4-(bis(4-aminophenyl)amino) phenyl)diazenyl)phenyl]-2-phenyl-1,1-dicyanoethylene[185], bis(4-aminophenyl)[4-(2-(4-nitrophenyl)vinyl)phenyl]amine [186], 4-nitro-4'-[N-(4,6-di-4-aminophenylamino)-1,3,5-triazin-2-yl]aminoazobenzene [187], 2,4-di-p-aminophenylamino-6-p-nitrophenylamino-1,3,5-triazine [187], 4-nitro-4'-[N-(4,6-di-β-aminoethylamino)-1,3,5-triazin-2- yl] aminoazobenzene [187], 2,4-di-β-aminoethylamino-6-p-nitrophenylamino-1,3,5-triazine [187] and their chemical structures are given in figure 35. Here, a traditional two-step synthetic step is followed that included a polycondensation step to form polyamic acid followed by a chemical imidization step to form PI.

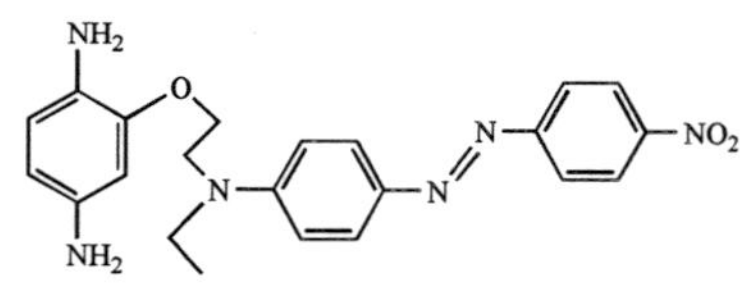

2-(2-{ethyl-[4-(4-nitro-phenylazo)-phenyl]-amino}-ethoxy)-benzene-1,4-diamine

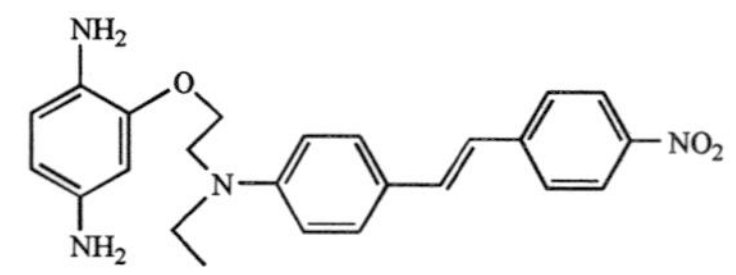

2-[2-(ethyl-{4-[2-(4-nitro-phenyl)-vinyl]-phenyl}-amino)-ethoxy]-benzene-1,4-diamine

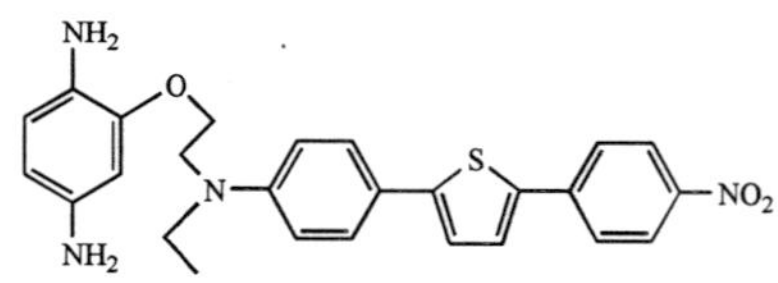

2-[2-(ethyl-{4-[5-(4-nitro-phenyl)-thiophen-2-yl]-phenyl}-amino)-ethoxy]-benzene-1,4-diamine

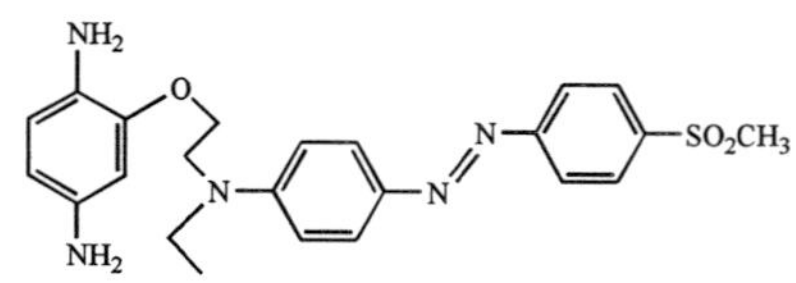

2-(2-{ethyl-[4-(4-methanesulfonyl-phenylazo)-phenyl]-amino}-ethoxy)-benzene-1,4-diamine

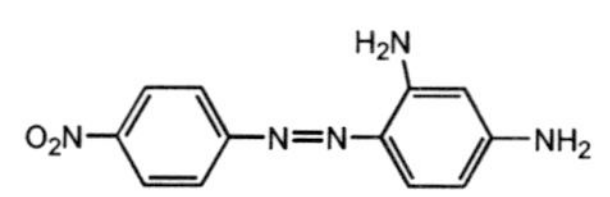

4-(4'-nitrophenyl-diazenyl) phenyl-1,3-diamine

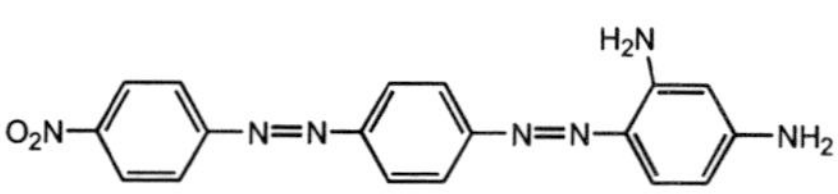

2,4-diamino-4'-(4-nitrophenyl-diazenyl)azobenzene

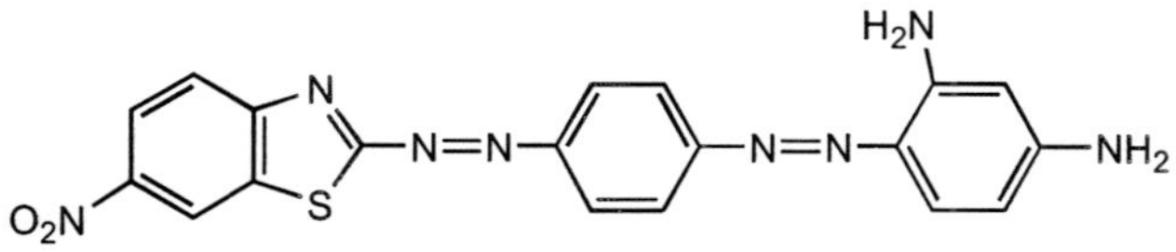

2,4-diamino-4'-[(6-nitrobenzothiazol -2-yl)diazenyl]azobenzene

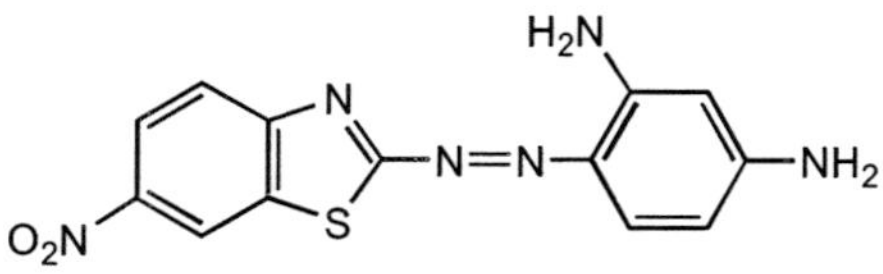

[(6-nitrobenzothiazol-2-yl)diazenyl] phenyl-1,3-diamine

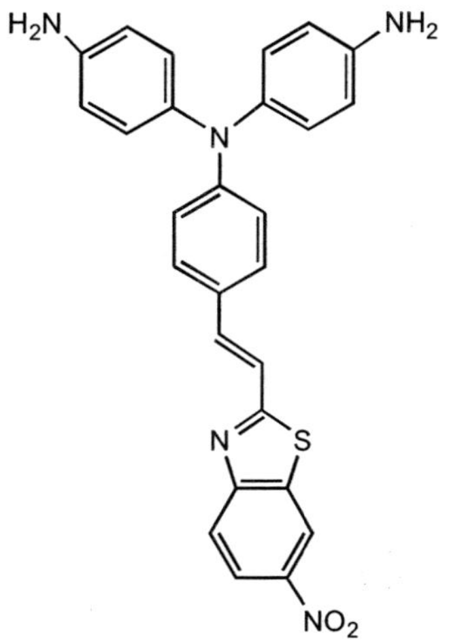

Bis(4-aminophenyl)[4-(2-(6-nitro benzothiazol-2-yl)vinyl)phenyl]amine

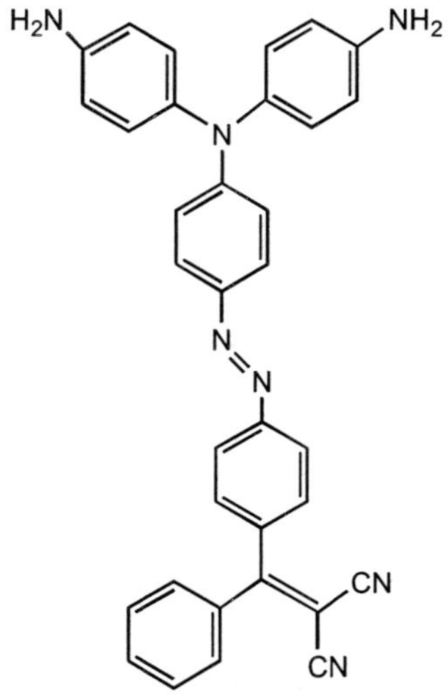

2-[4-((4-(bis(4-aminophenyl)amino)phenyl) diazenyl)phenyl]-2-phenyl-1,1-dicyanoethylene

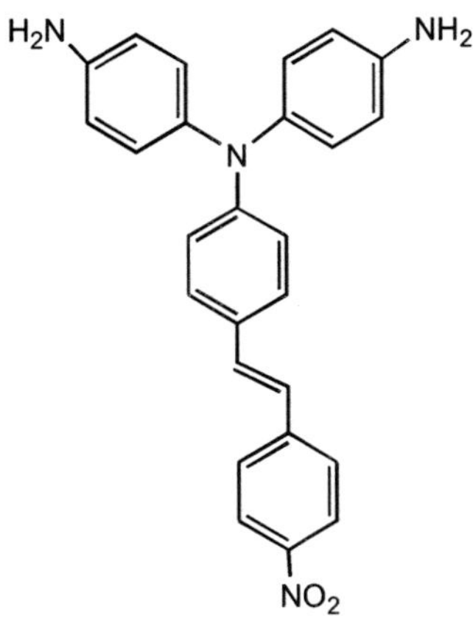

Bis(4-aminophenyl)[4-(2-(4-nitro phenyl)vinyl)phenyl]amine

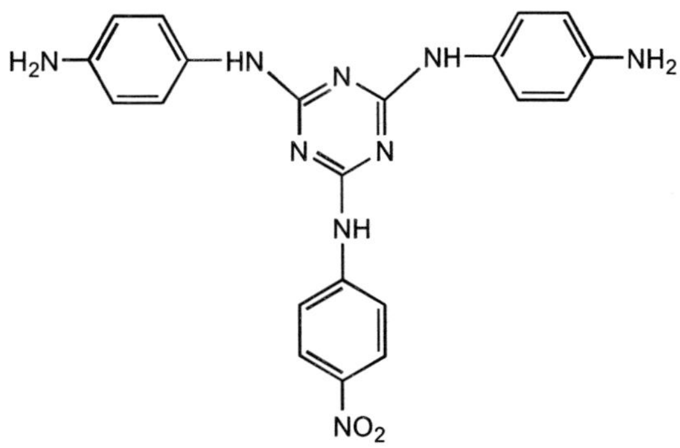

2,4-Di-p-aminophenylamino-6-p-nitrophenylamino-1,3,5-triazine

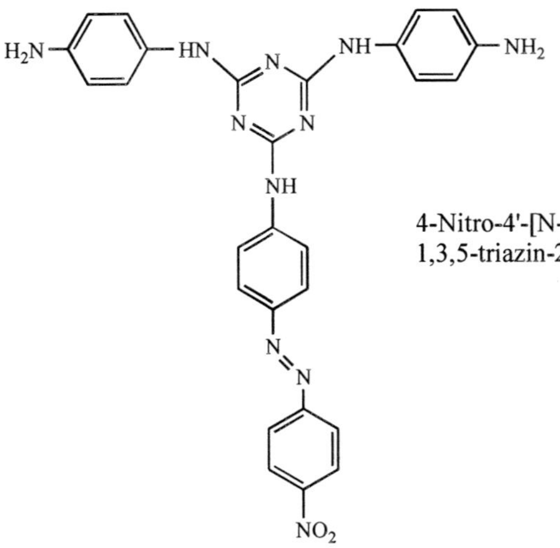


4-Nitro-4'-[N-(4,6-di-4-aminophenylamino)-1,3,5-triazin-2-yl]aminoazobenzene

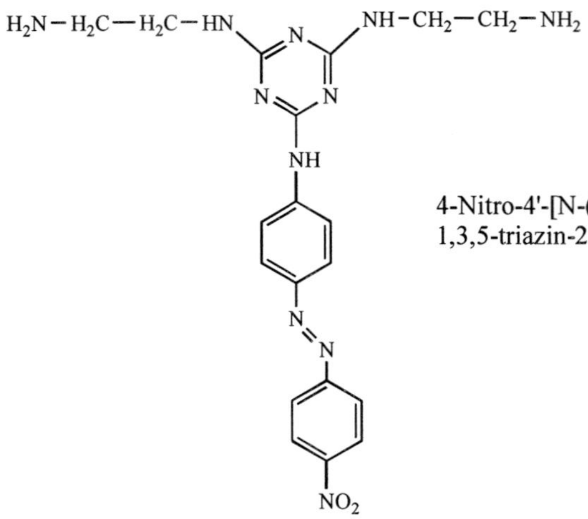


4-Nitro-4'-[N-(4,6-di-b-aminoethylamino)-1,3,5-triazin-2-yl]aminoazobenzene

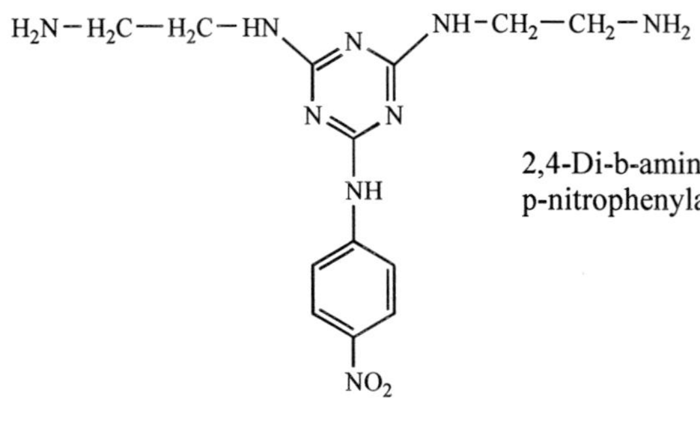


2,4-Di-b-aminoethylamino-6-p-nitrophenylamino-1,3,5-triazine

Fig. 35: Chemical structures of NLO chromophore diamines

A series of guest-host NLO polymers were prepared by doping a two-dimensional chromophore, 3,6-di(2'-(4"-ethylsulfonylphenyl)-1'-ethenyl)-9-hexyl 9H-carbazole (Cz2PhSO$_2$) to a PEI derived from 6FDA and 1-APOBP. The second harmonic coefficients (d33) for the guest-host system range from 5 to 22 pm/V, depending upon the doping level and the guest-host system possess excellent temporal stability. There was no aggregation of NLO chromophores, till doping level reaches 38 weight percent [188].

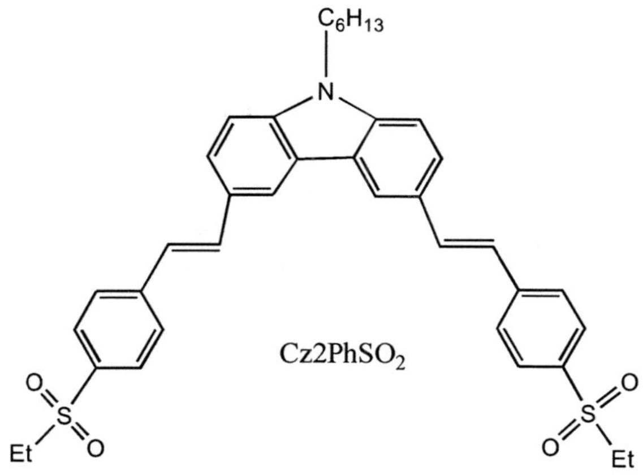


Fig. 36: Chemical structure of NLO chromophore

Third order NLO materials have potential applications such as optical switches, modulations and other nonlinear optical devices [189]. Polydiacetylenes [190], polyacetylenes [191], poly(phenylene-vinylene)s [192] and polythiophenes [193] have large third-order nonlinear susceptibility $\chi^{(3)}$ ranging from $10^{-12}$ to $10^{-9}$ esu. Metallophthalocyanines are well known organometallic chromophore for third-order NLO properties [194]. Metallophthalocyanines are poorly soluble in organic solvents and easily crystallize in matrices. These problems were overcome by attaching metallophthalocyanines to polymer chains. This approach has an advantage that a high concentration of nonlinear chromophores can be incorporated into polymer systems without crystallization, phase separation or the formation of concentration gradients. The chemical structure of a third-order NLO PI with aluminum phthalocyanine unit as a side chain is given in figure 37 and it was prepared by reacting a poly(hydroxyimide) with chloroaluminum phthalocyanine in the presence of silver triflate [195]. The third-order nonlinear susceptibility $\chi^{(3)}$ of PI film with approx. 60 mol percent of aluminum phthalocyanine at 1064 nm wavelength was $5.3 \times 10^{-9}$ esu.

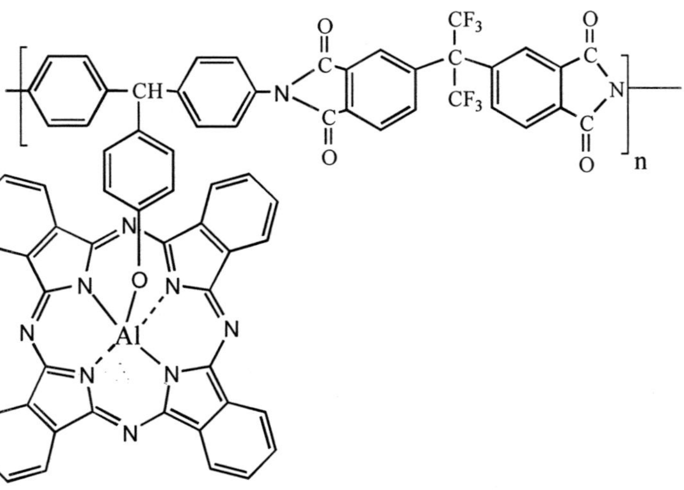


Fig. 37: Third order NLO polymer

## Electronic Applications

PIs are used in microelectronics industry as interdielectric layers, passivation layers and α–particle barriers. The electrical performance of PIs in these applications is dictated by its dielectric constant and can be further improved by reducing the dielectric constant. The propagation velocity of signal in microelectronic devices is inversely proportional to the square of the dielectric constant of the propagating medium. Therefore, signal propagation in microelectronics devices is faster, when the dielectric constant is low. Further, lower dielectric constant materials reduce crosstalk between adjacent circuit lines and transmission delay time.

The dielectric constant can be further reduced by the introduction of fluorine atoms into PIs. The lowering of dielectric constant by fluorine substitution is due to the combination of number of factors. As in the case of 6FDA based PIs, the incorporation of bulky $CF_3$ groups prevented the close packing of polymer chains and reduced interchain transfer of the highly polar dianhydride groups [196]. Further, the large fluorine atoms increased the free volume fraction in the polymer, which in turn reduced the number of polarizable groups in unit volume. Due to the large electronegativity of the C-F bond, fluorine substitution lowered the electronic polarization in the polymer. One should be cautious during fluorine substitution because indiscriminate substitution gives undesired effect. Hougham et al [197] have shown that nonsymmetric substitution of fluorine for hydrogen, in fact increased the average magnitude of the dielectric constant by 0.05 per substituted ring. Another approach to reduce dielectric constant is based on reducing polar imide concentration (weight fraction of imide linkage by molecular weight ($M_w$) in repeating unit), which contribute to make dielectric constant higher [198].

Dielectric constant (κ) of 6FDA PIs are listed below:

| Structure | κ | Freq. | Ref. |
|---|---|---|---|
| 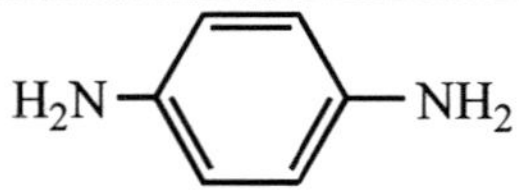 | 3.04 | 1kHz | 41 |
| 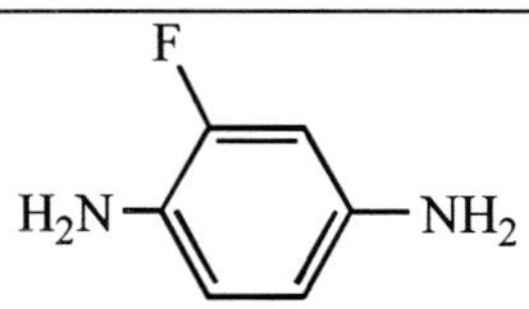 | 2.85†<br>3.19‡ | 1kHz<br>1kHz | 199<br>199 |
| 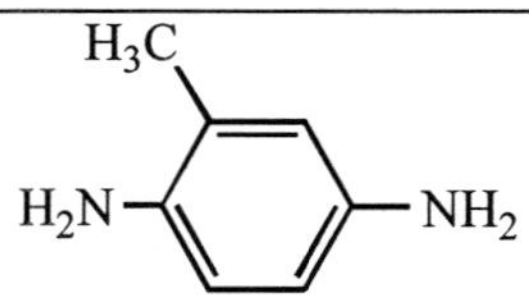 | 2.75†<br>3.16‡ | 1kHz<br>1kHz | 199<br>199 |
| 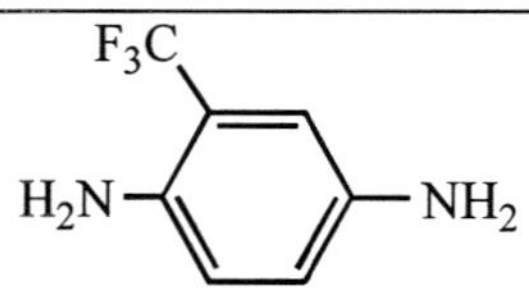 | 2.72†<br>3.05‡ | 1kHz<br>1kHz | 199<br>199 |

| Structure | κ | Freq. | Ref. |
|---|---|---|---|
| 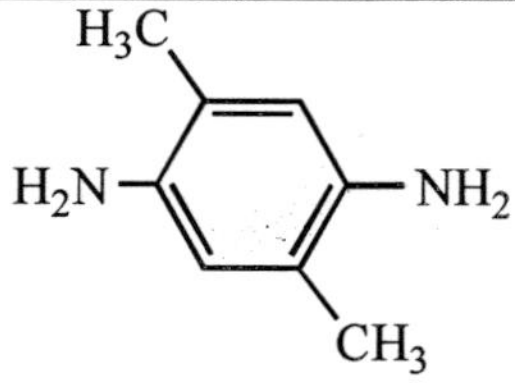 | 2.74†<br>3.21‡ | 1kHz<br>1kHz | 199<br>199 |
| 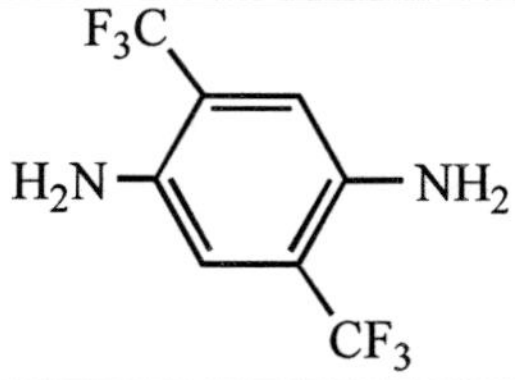 | 2.59†<br>2.87‡ | 1kHz<br>1kHz | 199<br>199 |
| 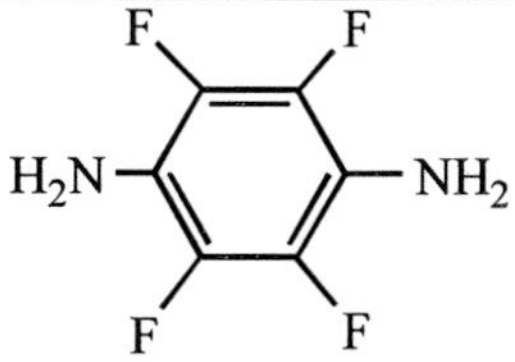 | 2.68†<br>2.91‡<br>2.75 | 1kHz<br>1kHz<br>100kHz | 199<br>199<br>200 |
| 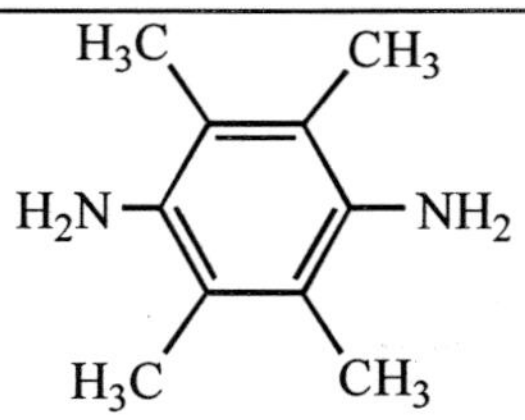 | 2.87 | 1kHz | 41 |
| 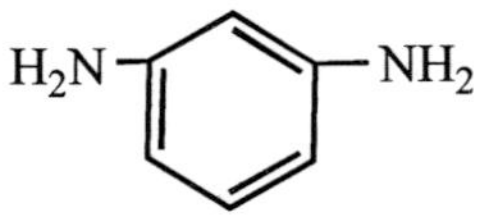 | 3.05 | 1kHz | 41 |
| 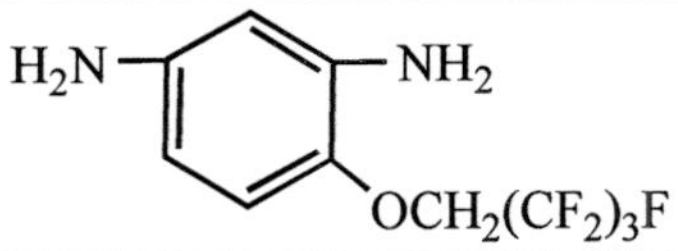 | 2.9 | 1kHz | 201 |
| 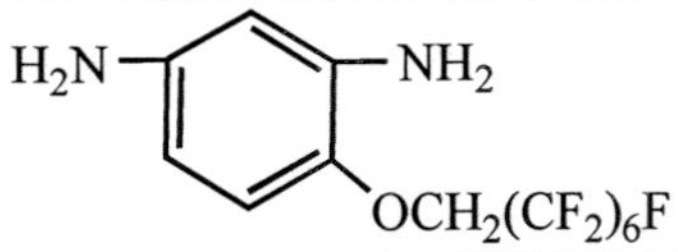 | 2.7 | 1kHz | 201 |
| 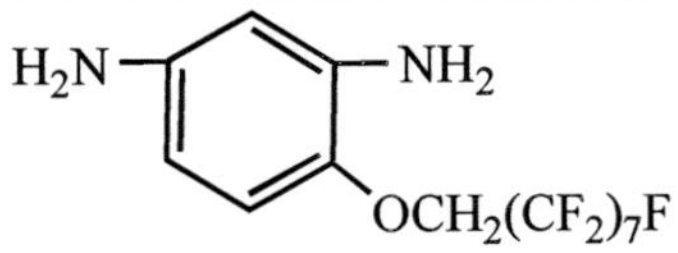 | 2.6 | 1kHz | 201 |
| 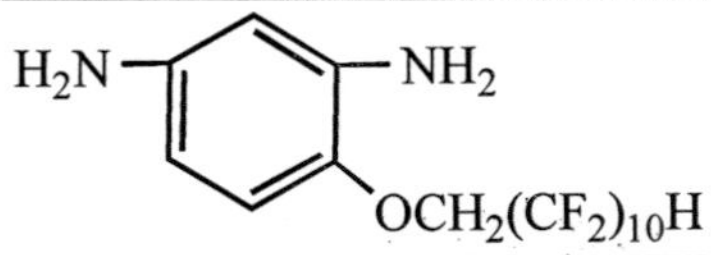 | 2.6 | 1kHz | 201 |

| Structure | κ | Freq. | Ref. |
|---|---|---|---|
| 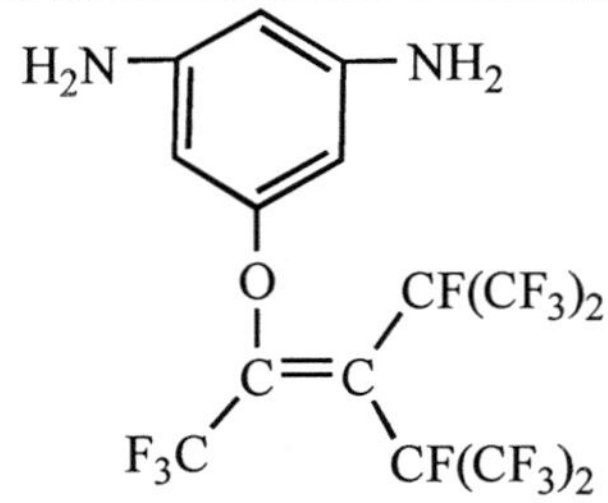 | 2.50 | 10kHz | 202 |
| 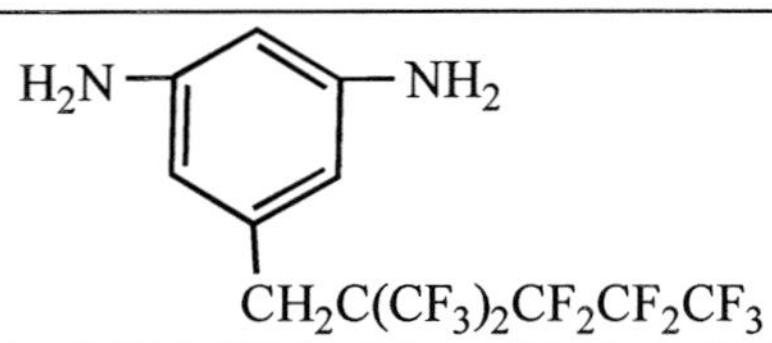 | 2.70 | 1MHz | 203 |
| 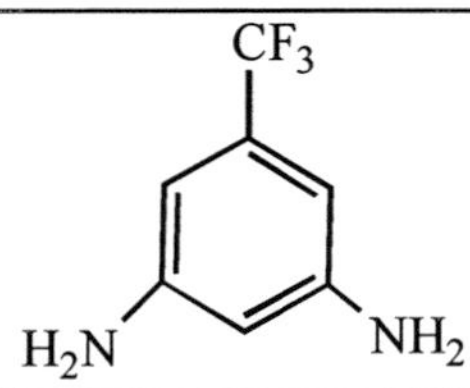 | 2.58 | 10GHz | 26 |
| 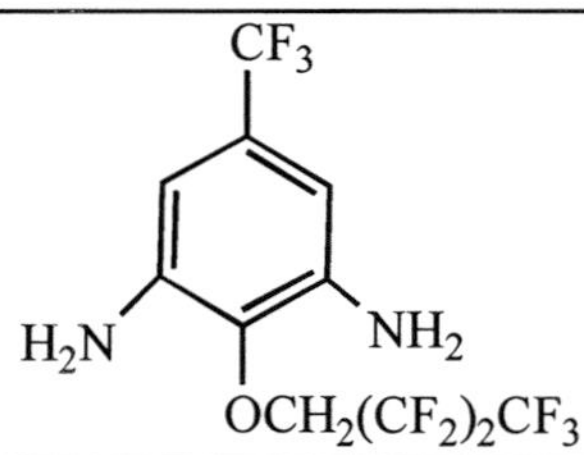 | 2.93 | 100kHz | 204 |
| 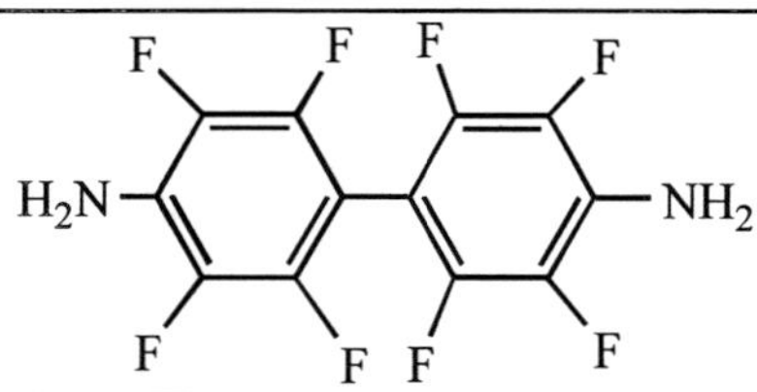 | 2.55†<br>2.73‡ | 1kHz<br>1kHz | 199<br>199 |
| 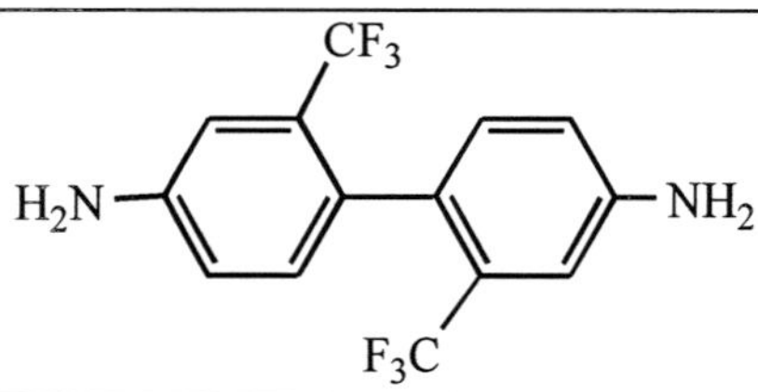 | 2.80 | 1MHz | 205 |
| 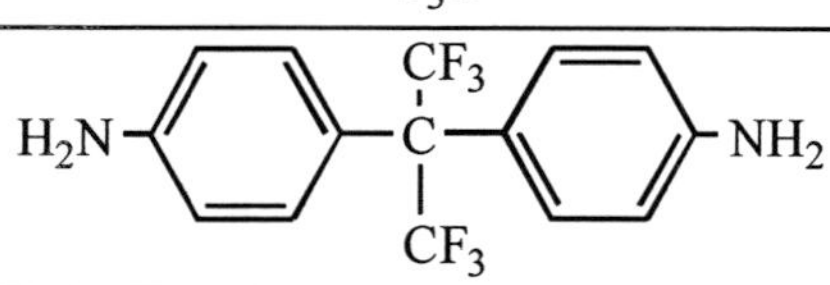 | 2.78<br>2.39 | 1kHz<br>10 GHz | 41<br>206 |
| 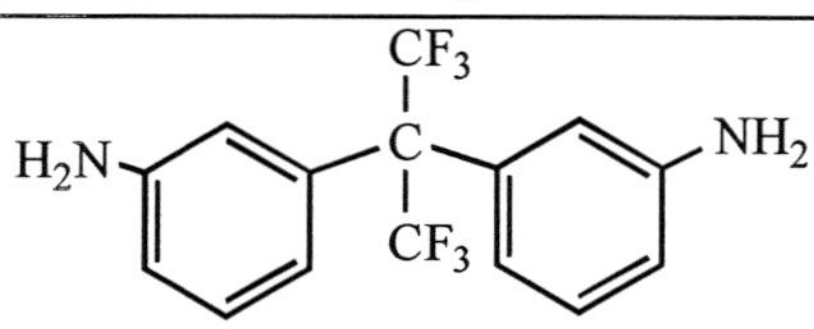 | 2.58 | 1kHz | 207 |

| Structure | κ | Freq. | Ref. |
|---|---|---|---|
| 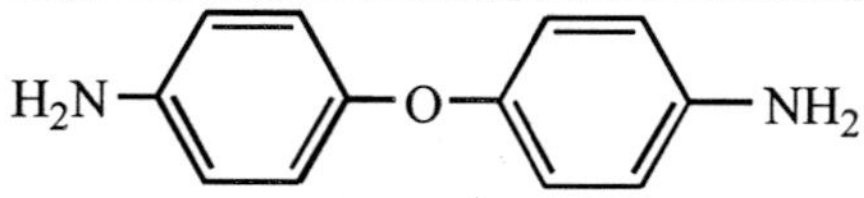 | 2.90<br>2.79 | 1kHz<br>10 GHz | 41<br>206 |
| 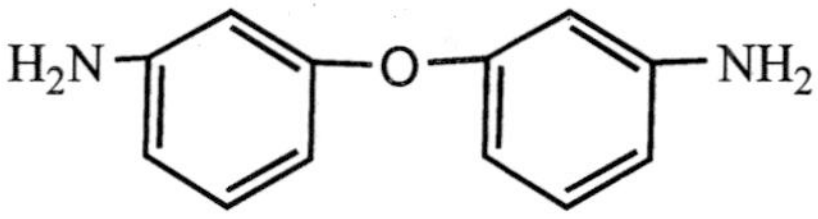 | 2.73 | 10GHz | 206 |
| 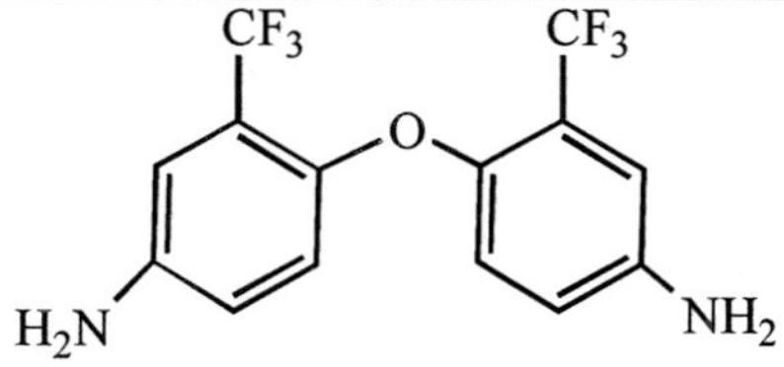 | 2.76 | 1MHz | 208 |
| 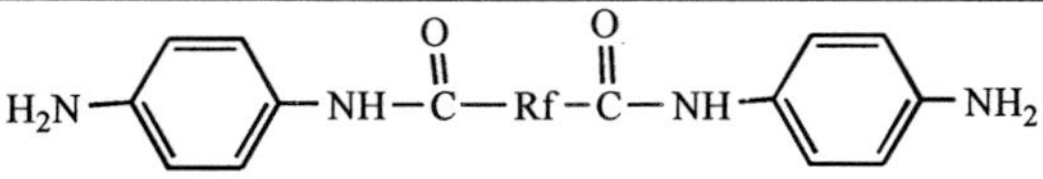<br>Where Rf = $CFCF_3(OCF_2CFCF_3)_mO(CF_2)_5O(CFCF_3CF_2)_nCFCF_3$<br>$m + n = 3$ | 3.00 | 100kHz | 204 |
| 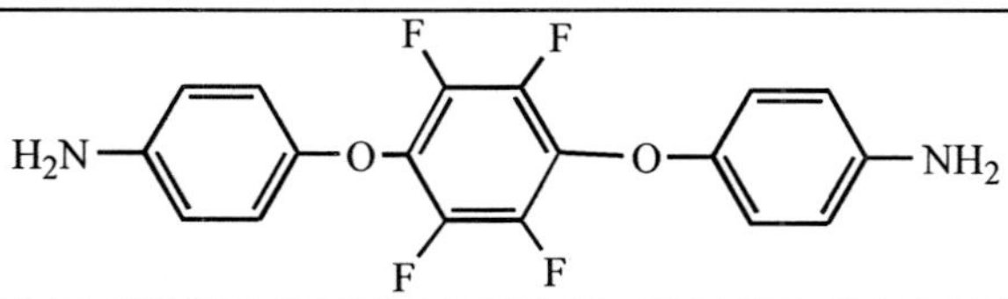 | 2.95 | 100kHz | 204 |
| 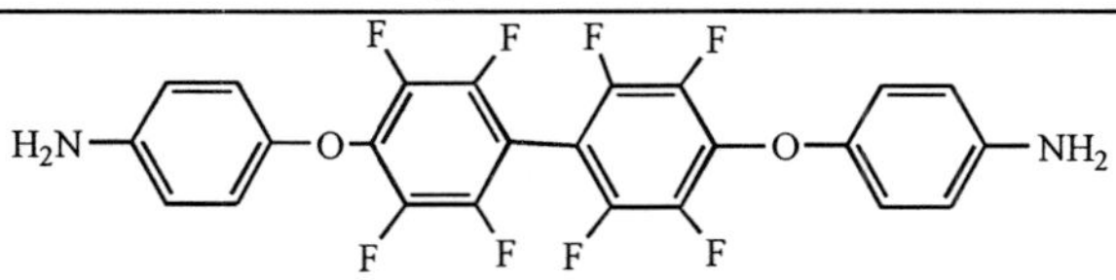 | 2.95 | 100kHz | 204 |
| 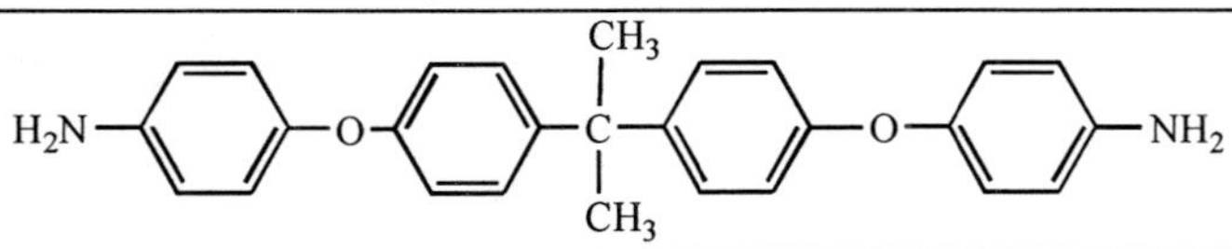 | 2.65 | 10MHz | 207 |
| 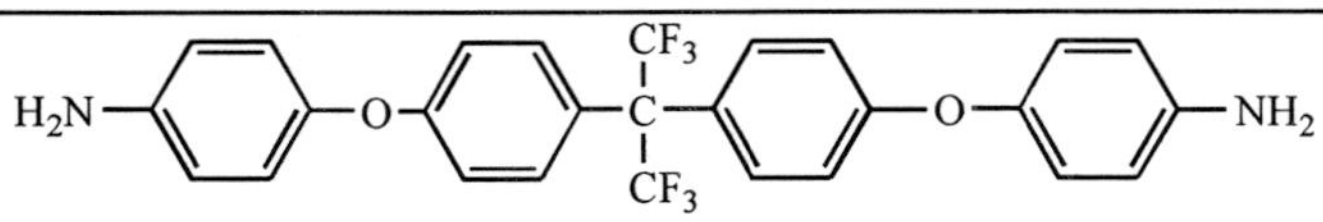 | 2.99<br>2.50 | 1kHz<br>10kHz | 41<br>206 |
| 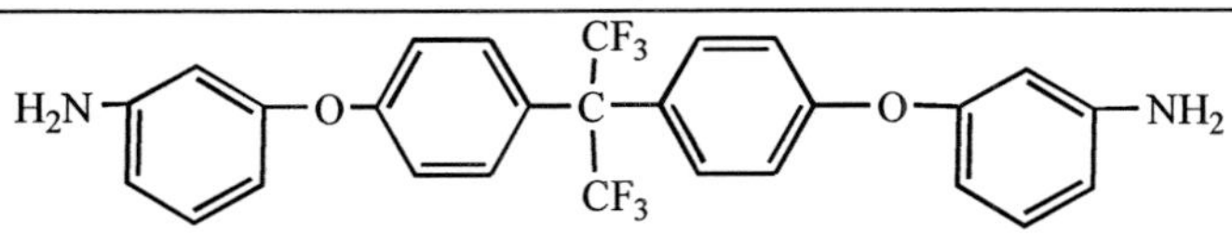 | 2.40 | 10GHz | 206 |
| 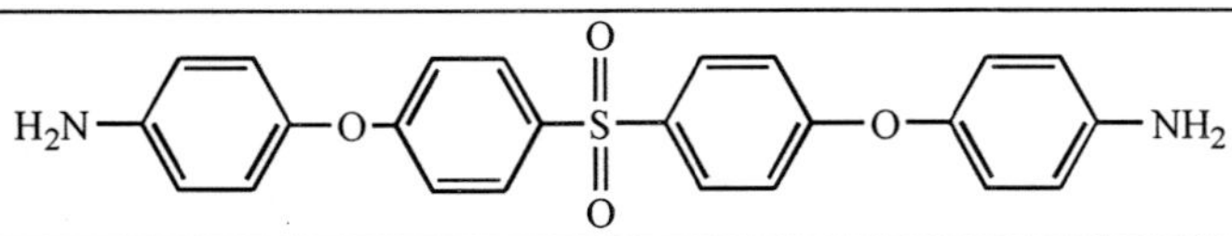 | 2.74 | 10MHz | 207 |

| Structure | κ | Freq. | Ref. |
|---|---|---|---|
| 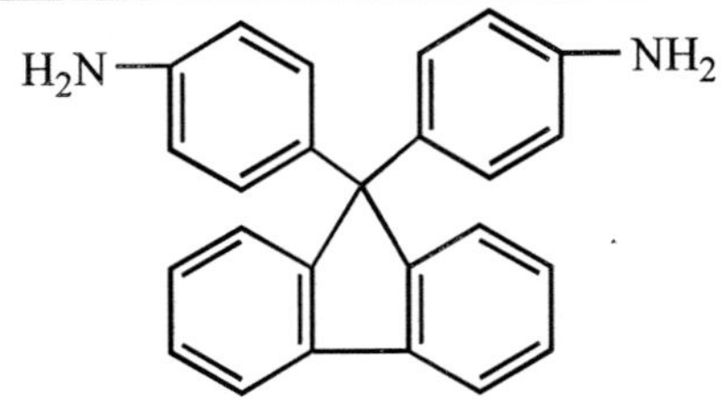 | 2.81 | 1MHz | 198 |
| 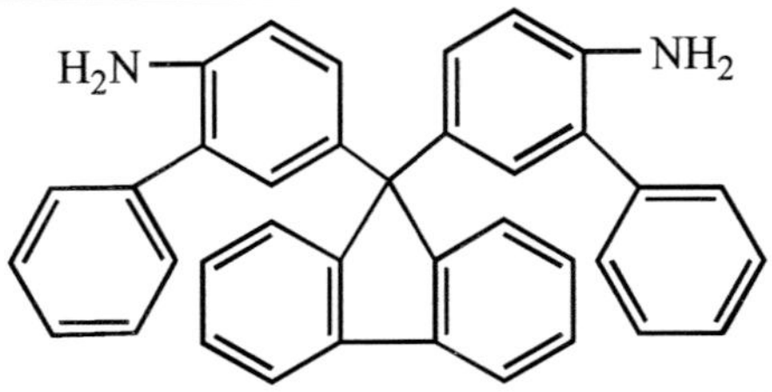 | 2.63 | 1MHz | 198 |
| 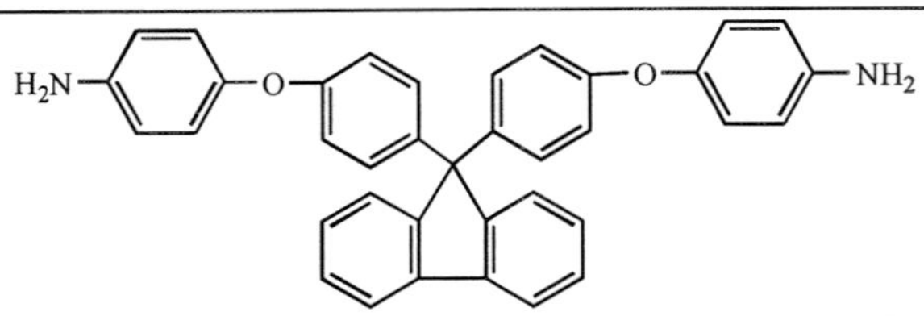 | 2.65 | 1MHz | 198 |
| 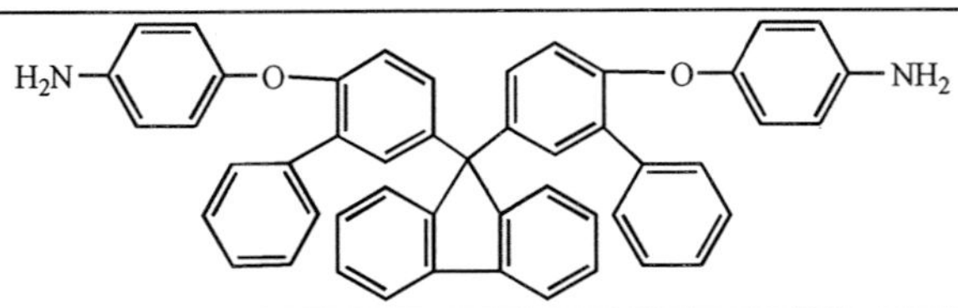 | 2.65 | 1MHz | 198 |
| 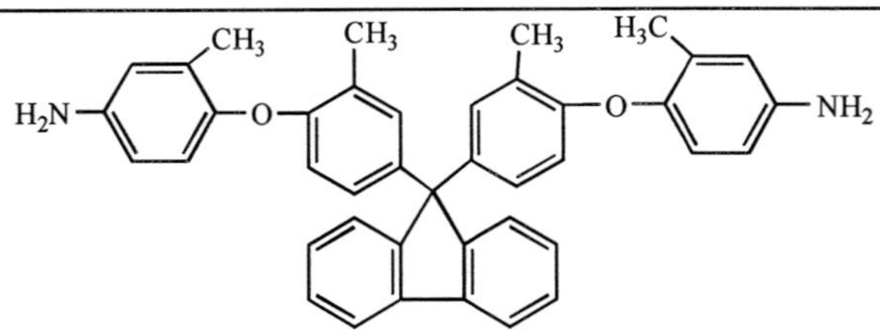 | 2.62 | 1MHz | 198 |
| 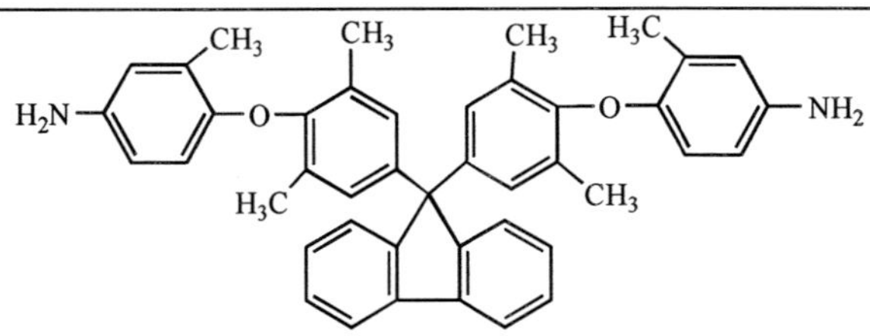 | 2.70 | 1MHz | 198 |
| 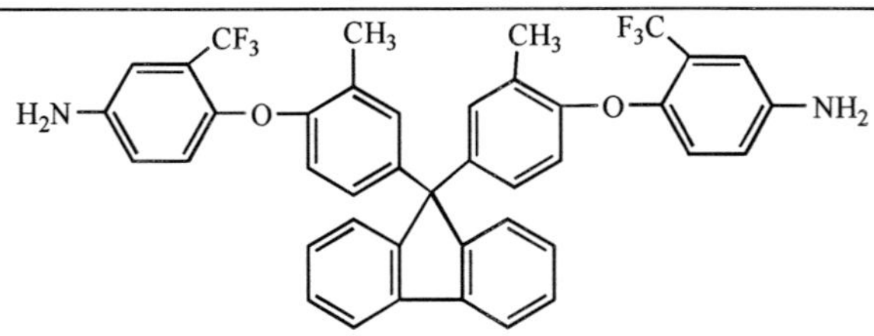 | 2.51 | 1MHz | 198 |

| Structure | κ | Freq. | Ref. |
|---|---|---|---|
| 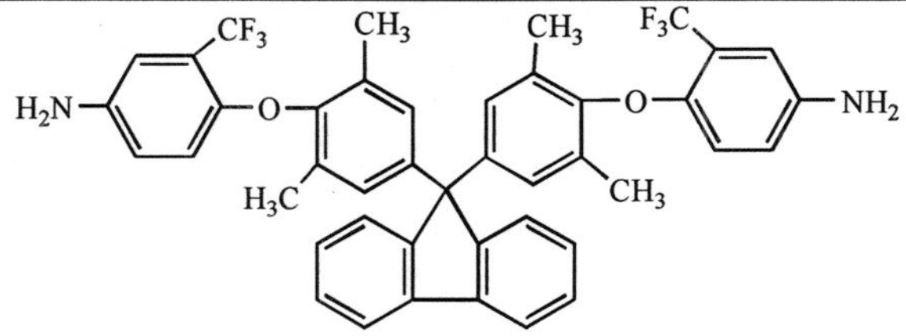 | 2.53 | 1MHz | 198 |
| 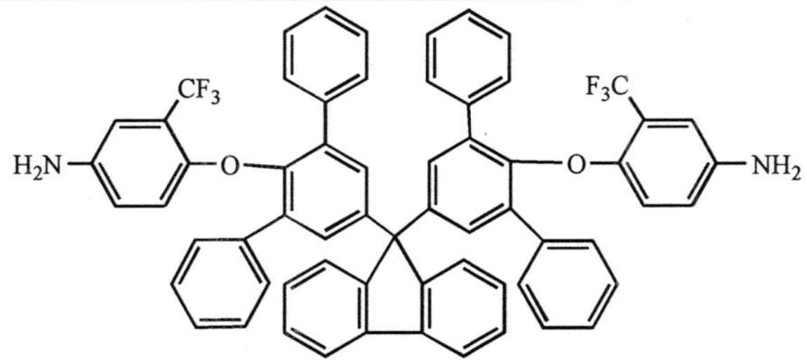 | 2.46 | 1MHz | 198 |
| 0.5 mol % 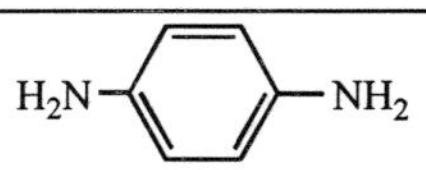 + 0.5 mol % 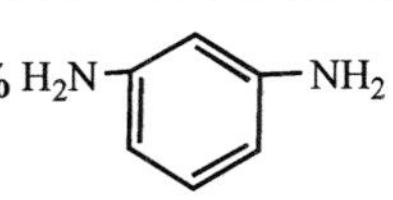 | 3.06 | 1kHz | 41 |
| 0.5 mol % 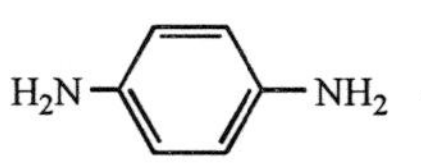 + 0.5 mol % 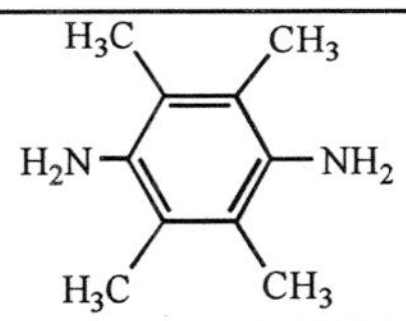 | 2.98 | 1kHz | 41 |
| 0.75 mol % 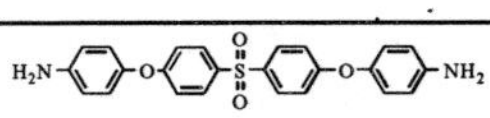 + 0.25 mol % 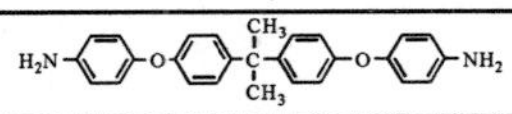 | 3.09 | 1kHz | 209 |
| 0.5 mol % 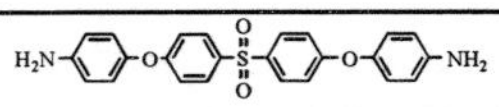 + 0.5 mol % 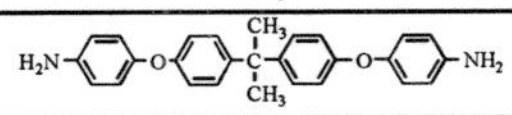 | 3.10 | 1kHz | 209 |
| 0.25 mol % 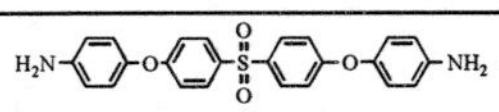 + 0.75 mol % 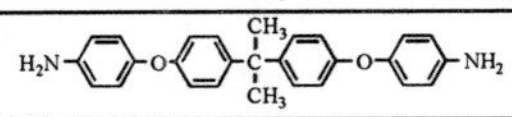 | 3.05 | 1kHz | 209 |
| 0.75 mol % 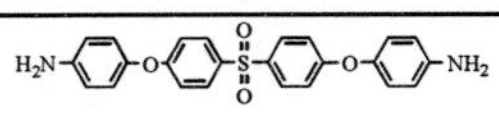 + 0.25 mol % 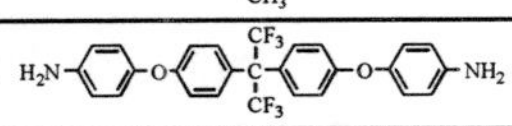 | 3.10 | 1kHz | 209 |
| 0.5 mol % 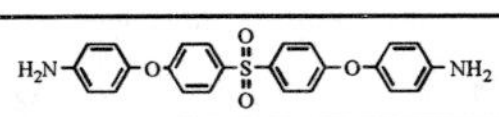 + 0.5 mol % 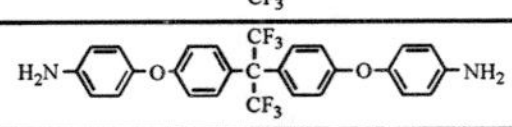 | 2.99 | 1kHz | 209 |
| 0.25 mol % 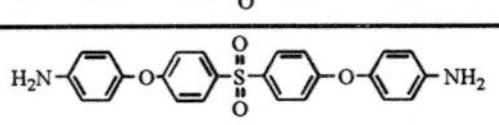 + 0.75 mol % 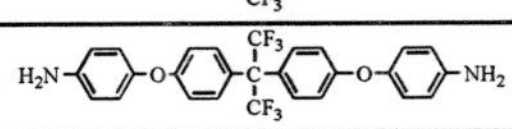 | 2.98 | 1kHz | 209 |

† dry sample
‡ wet sample

The creation of nanorpores in PIs is another way of reducing the dielectric constant. Here, a portion of the polymer is replaced with air, which has a dielectric constant of one. As a result, the dielectric constant is reduced. Nanoporous PI films were prepared in two steps. The first step is the preparation of poly(urethane-imide) films by casting method and in the second step these films were thermally treated above 300 °C to give nanoporous PI films. Solutions of PAA derived from 6FDA and pPDA and phenol blocked PU derived from 1,6-hexamethylene diisocyanate and poly(ethylene glycol) in 95:5 weight percent was mixed together and a film was casted on a glass plate. This film was later subjected to heat treatment to get nanoporous PI films. Dielectric constant of nanoporous PI film at 1kHz is 2.85 whereas the film without nanopores has dielectric constant 3.05 at 1kHz [140].

Metal containing PIs were developed from ruthenium-containing diamine [Ru(NH$_2$Phtpy)$_2$] [PF6]$_2$ (Figure 38) and 6FDA for optoelectronic applications.

$[Ru(NH_2Phtpy)_2][PF6]_2$ was synthesized by the direct complexation of ruthenium trichloride trihydrate with two equivalents of 4'-(p-aminophenyl)-2,2': 6',2" – terpyridine in moderate yield. Light emission was observed in single-layer light emitting diiodes, which was fabricated using the polymer film. [210].

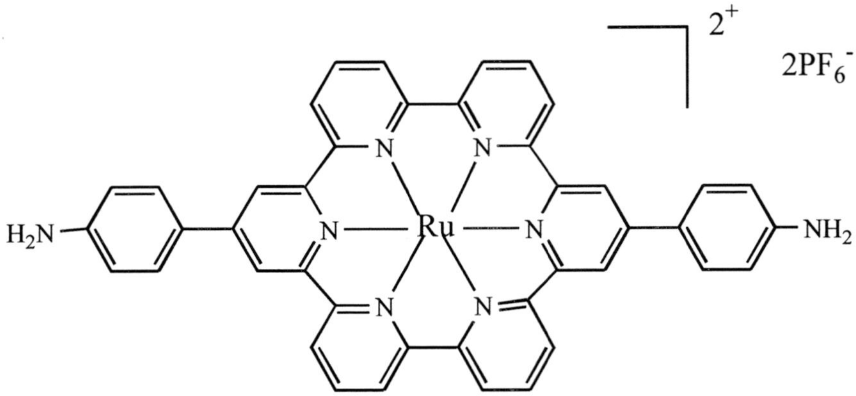


Fig. 38: Chemical structure of $[Ru(NH_2\ Phtpy)_2]\ [PF_6]_2$

PI containing hexylene spacer and a fluorostilbene unit in the side chains were prepared using 6FDA and 3,3'-bis[(4'-fluoro-4-stilbenyl)oxyhexyloxy]-4,4'-biphenyldiamine (FS6B) The possible use as a liquid crystal (LC) alignment layer in preimidized form of PI derived from 6FDA and 2,2'-bis(trifluromethyl)benzidine (TFMB) was investigated [211].

Conventional PIs used in semiconductor industry have poor characteristics that make them unsuitable for use as optical materials. Reuter et al [212] reported the optical properties of three kinds of 6FDA PIs (derived from 4,4-ODA, 3-6FpDA and PFMB) paying attention as an optical waveguide material for the first time. These PIs exhibit optical transmission losses of less than 1/10 of those of non-fluorinated PIs at the wavelength of 0.63 μm. In the past decade, Amoco introduced Ultradel 9020D and Amoco 4212. These commercial samples have low optical losses in the near infrared region, a broad range of refractive index control and excellent heat resistance. Franke et al [213] fabricated planar waveguides using 6FDA based PIs derived from 4-6FpDA and 3-6FpDA. They aimed to apply the waveguides to humidity sensor utilizing the good sensitivity of the propagation characteristics of PI waveguides to humidity. PI derived from 6FDA and 1,4-bis(4-amino-2-trifluoromethyl-phenoxy) benzene (ATPB), 1,4-bis(4-amino-2-trifluoromethyl-phenoxy) tetrafluorobenzene (ATPT), 1,3-bis(4-amino-2-triflurorömethyl-phenoxy) 4,6-dichlorobenznen(ATPD) were investigated for optical wave guide applications. These PIs were found to have low optical absorption loss in the optical communication wavelengths of 1.3 and 1.55 μm and low birefringence and water absorption and high thermal stability [214]. Optical polymers derived from 6FDA, PFMB and DCB [36] are suitable for optical waveguide applications because they wee found to have controllable refractive index and low optical loss in the optical communication wavelengths of 1.3 and 1.55 μm.

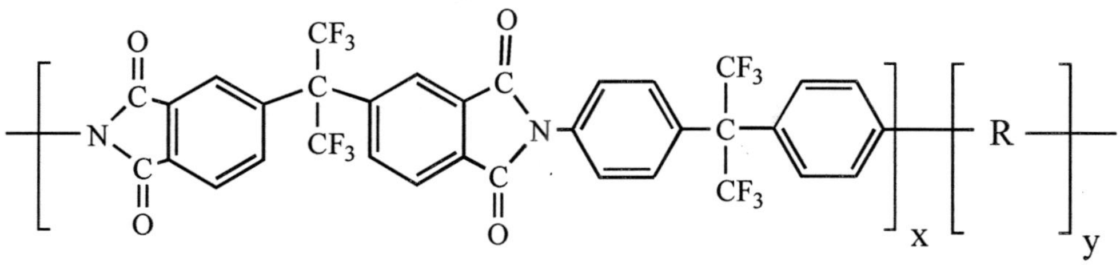

Ultradel 9020D

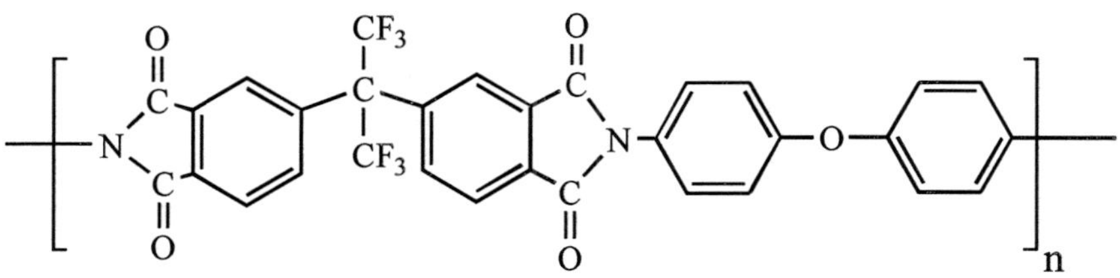

Amoco 4212

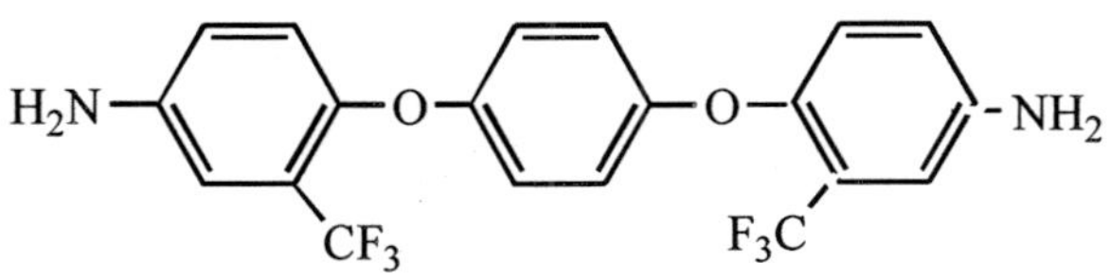

ATPB

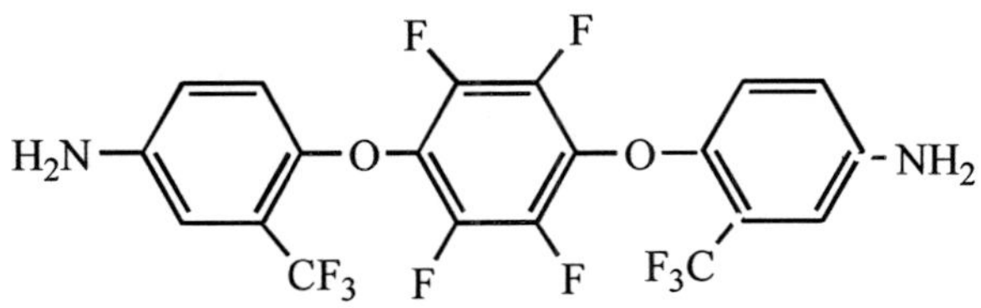

ATPT

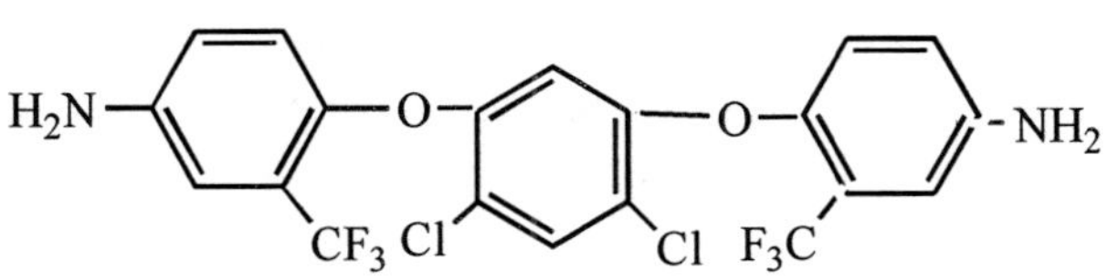

ATPD

Fig. 39: Chemical structure

## Gas Separation Applications

Compared to traditional separation processes namely cryogenic distillation, absorption and pressure swing adsorption, membrane based gas separations offer many significant advantages such as low energy consumption, low capital investment cost, simple and easy operation. Membranes for gas separations were first commercially used in the late 1970s for the separation of hydrogen from synthetic ammonia. [215]. Early membrane separation had low productivity because of large membrane thickness. Nowadays, the capacity of membrane technology for the generation of nitrogen from air has been increased to 19,100 $Nm^3h^{-1}$ [216]. Other applications of polymeric gas separation membranes are the removal of helium gas from natural gas, stripping carbon dioxide from natural gas, and oxygen enrichment from air [217].

Even though large number of polymeric materials have been studied for the above applications, 6FDA based PIs received special attention from both academia and industry because these membranes were found to have higher permeability and selectivity, which are very much essential to minimize capital and operating costs. However, the disadvantage is their tendency to plasticize even at a partial carbon dioxide pressure of 8–10 bar [218]. In membrane studies, plasticization is generally defined as an increase in the segmental motion of polymer chains, due to the presence of one or more sorbates, such that the permeability of both components increases and the selectivity decreases [219]. For many new potential applications such as high pressure carbon dioxide/ methane separations (such as natural gas upgrading, enhanced oil recovery and landfill gas cleaning), carbon dioxide separations for enhanced oil recovery, propylene/propane separations, butadiene/butane separations [220-222] plasticization resistance is required.

Cross-linking Chemistry is widely used to suppress the undesirable plasticization effects in these applications. Cross-linking modifications can be carried out by diverse methods such as thermal treatment [223], ion beam [224], UV irradiation [225] and chemical reaction [226]. Heating a mixture of PIs such as Thermid FA-700 with monomers or oligomers containing reactive acetylene end groups at 265 °C forms a semi-interpenetrating polymer networks, which was found to suppress the plasticization effect but permeability is reduced [227]. In another approach, 6FDA based PI containing ODA and NDA moieties were crosslinked at ambient temperature using p-xylene diamine [228]. In this chemical crosslinking, a significant reduction in carbon dioxide induced plasticization at least upto 550 psi was observed, whereas untreated hollow fiber membrane exhibited plasticization in the feed pressure is greater than 75 psi. However, carbon dioxide permeability is decreased significantly. Similarly, to suppress plasticization effect, hollow fiber membrane prepared from 6FDA and 2,6-DAT was chemically crosslinked using p-xylene diamine or m-xylene diamine at ambient temperature. It has been reported that both diamines have similar cross-linking effectiveness as illustrated from the values of $CO_2/CH_4$ selectivity and permeance [229]. In another study, $CO_2$ plasticization effect was suppressed by heating 6FDA-2,6-DAT hollow fiber membranes at 250 °C for 5 min. Thermal treatment resulted in more compact selective skin layer and substructure when compared to the untreated ones and this accounted for the anti-plasticization characteristics of hollow fiber membrane. Surprisingly, cross-linking was not observed after the heat treatment. [230].

6FDA - DDA has a higher permeability, but a relatively low selectivity for a specific gas pair, while 6FDA – pPDA has a higher selectivity with a relatively low permeability. Chung et al [231] synthesized various copolyimides to utilize the advantages of the above two homopolymers and studied the gas transport properties. 6FDA based copolyimides containing DABA and 6FpDA were synthesized and were crosslinked with ethylene glycol (EG) and aluminum acetylacetonate ($Al(AcAc)_3$) to suppress undesirable plasticization effects in carbon dioxide/methane separations [232]. 6FDA- TAPA based hyberbranched macromolecules were connected via chemical bonds through the chemical reaction between the terminal functional groups and a difunctional crosslinking agent such as ethylene glycol diglycidyl ether (EGDE), terephthaldehyde (TPL), 4,4'-ODA and DIDS. TPA crosslinked hyberbranched PI membranes displayed better separation performance than that of the linear analogues and many other linear polymeric membranes [119]. Chung et al investigated gas transport properties of 6FDA based PIs containing indan structure [46]. The potential of 6FDA based hollow fiber membrane for intravascular membrane oxygenation (IVOX) was studied with respect to oxygen transfer [233].

$$HO-CH_2-CH_2-OH \quad \text{EG}$$

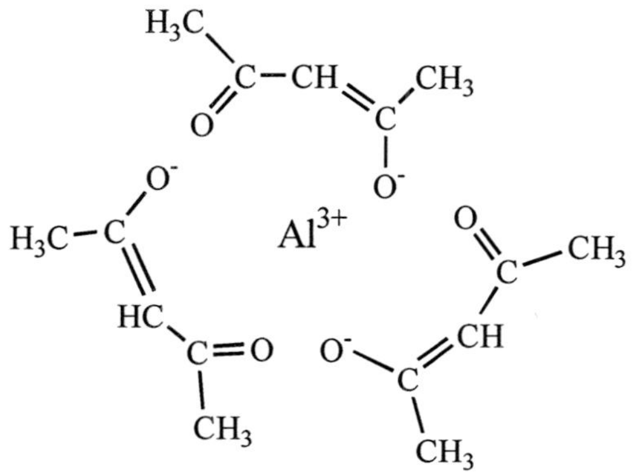

(Al(AcAc)₃)

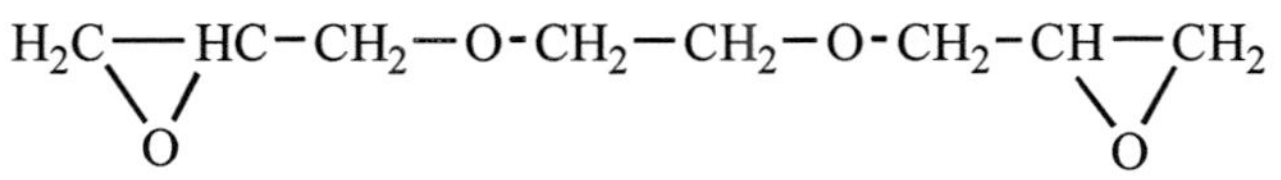

EGDE

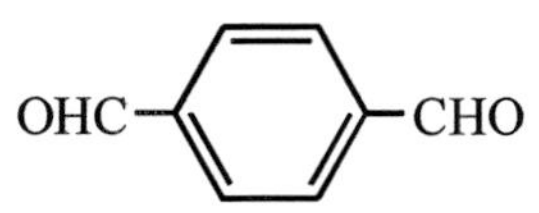

TPL

Fig. 40: Chemical structure

# Other Applications

6FDA PI prepared from mDA6FPPO was reported to have good atomic oxygen (AO) and UV resistance. These properties make it potentially useful for a variety of applications on space craft such as thin film membranes on antennae, second surface mirrors, themal/optical coatings and multi-layer thermal insulation (MLI) blanket materials [234].

Metal containing PI was prepared in a single step by polycondensing 6FDA with Bis(1,3-di-p-dimethylaminobenzylimidazolidinylidene)gold(I) (AuIm) (Figure 41) and has the potential for catalytic activity. Gold coordinated PI has increased solubility and thermal stability [235].

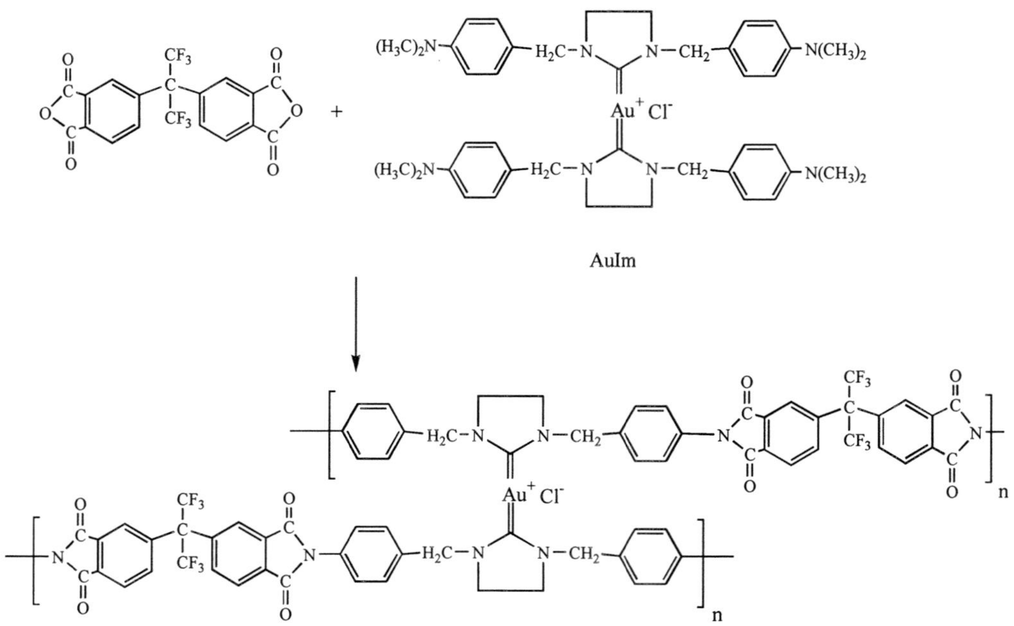


PI

Fig. 41: Reaction scheme of metal containing PI

Proton exchange membrane (PEM)/ polymer electrolyte fuel cells (PEFC) are known as efficient, non-polluting and low-noise electrical generators and are attracting increasing interest as a primary source for electrical vehicles and portable electric devices [236]. Department of Energy of United States estimates that oil imports could be reduced by 800,000 barrels per day if only 10 percent of domestic automobiles were powered by fuel cells. Therefore, the impetus to bring fuel cell technology into everyday use is gaining momentum. 6FDA based PI containing sodium sulfonate groups were synthesized from 4,4'-diaminodiphenyl sulfone (DDS) or 4,4'-(9-fluorenylidene dianiline) (FBPA) and sodium salt of 2,5-diaminobenzene sulfonic acid. Membranes prepared from these sulfonated PIs demonstrated moderate to high water absorption which is necessary for PEM fuel cells [237].

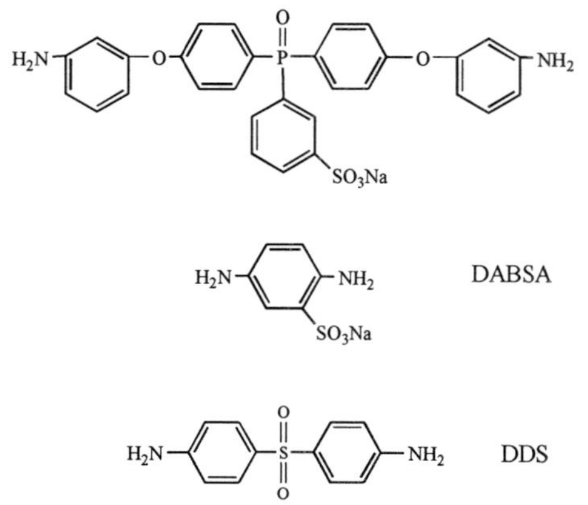

SBAPPO

DABSA

DDS

Fig. 42: Chemical structure

For more than a decade, polymers containing 2,5-diamino-1,4-benzoquinone functional group have been of interest, since Erhan's [238] observation that this class polymers have a strong affinity for the surface of iron. They have found that amine-quinone polymers could displace moisture from the surface of iron, making it hydrophobic. If these polymers would prevent moisture from adsorbing onto the iron surface, then they would inhibit corrosion. 6FDA based PIs derived from 2,5-bis(4,4'-methylene dianiline)-1,4-benzoquinone (AQMDA) and 2,5-bis(4,4'-oxydianiline)-1,4-benzoquinone (AQODA) were synthesized and evaluated for their ability to protect iron against corrosion by sodium chloride electrolyte [239, 240].

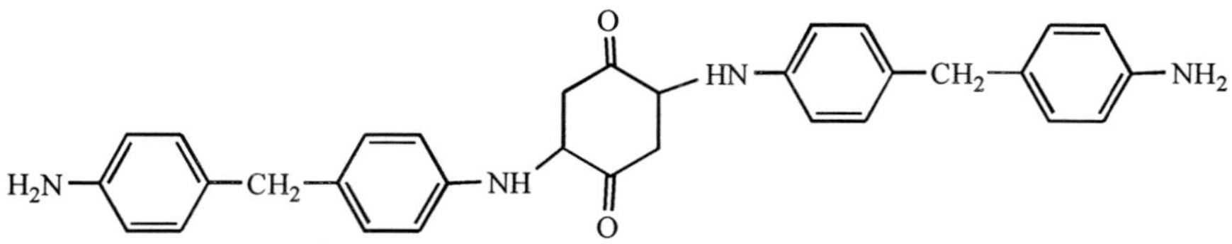


AQMDA

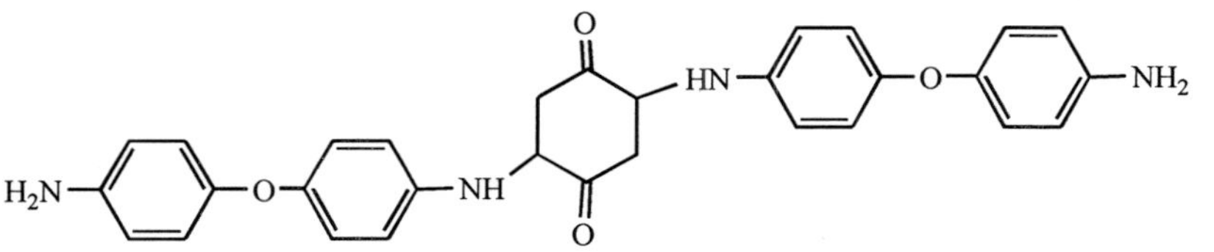


AQODA

Fig. 43: Chemical structure

# Conclusion

Even though 6FDA based PI is considered as first generation of fluorinated PIs, it is still extensively investigated for various applications because of its easy processability, good properties that it imparts to the resulting PI and easy availability as a commercial sample. Although a great deal of research and development on 6FDA PIs have been done in the past decades, and there have been many successful commercialized applications, this versatile dianhydride still has potential to form novel materials for new areas of applications in the future. Some of the good applications from past development still need to overcome cost barrier for real commercialization. Some of the applications that were not feasible in the past may become practical in future provided if we succeed in producing 6FDA at lower cost.

# References

[1] *Polyamic Acids and Polyimides Synthesis, Transformations and Structure*; Bessonov, M. I.; Zubkov, V. A. Eds.; CRC Press, Boca Raton, Ann Arbor, London, Tokyo, 1993.

[2] Dine-Hart, R. A.; Wright, W. W. *Makromol. Chem*. 1971, B143, 189-206.

[3] Kolegov, V. I.; Sklizkova, V. P.; Kudryavtsev, V. V.; Belen'kii, B. G.; Frenkel, S. Y.; Koton, M. M. *Dokl. Akad. Nauk*. USSR (Engl. Translation). 1977, 232, 848.

[4] Krishnan, P. S. G.; Vora, R. H.; Veeramani, S. *Int. J. Polym. Anal. Charc*. 2004, 9, 177-196.

[5] Krishnan, P. S. G.; Vora, R. H.; Chung, T-S.; Uchimura, S-I.; Sasaki, N. *J. Polym. Res.* 2004, 11, 299-308.
[6] Li, J.; Kato, J.; Kudo, K.; Shiraishi, S. *Macromol. Chem. Phys*. 2000, 201, 2289.
[7] Yin, J.; Ye, Y-F.; Wang, Z-G. *Eur.Polym. J.* 1998, 34, 1839-1843.
[8] St. Clair, A. K.; St. Clair, T. L.; Smith, E. N. *Polym. Prepr.* 1976, 17, 359.
[9] Yang, C. P.; Lin, J. H. *J. Polym. Chem.* Ed. 1994, 32, 423.
[10] Yang, J. Y.; Jung, B. T.; Suh, D. H. *Polymer*. 2001, 42, 8349-54.
[11] *Fluoropolymers* **I**. Synthesis; Hougham, G.; Cassidy, P. E.; Johns, K.; Davidson, T. Eds.; Kluwer Academic/Plenum Publishers, New York, 1999.
[12] *Fluoropolymers* **2**: Properties; Hougham, G.; Cassidy, P. E.; Johns, K.; Davidson, T. Eds.; Kluwer Academic/Plenum Publishers, New York, 1999.
[13] Coe, D. G. U. S. Patent 3,310, 573, 1967
[14] Alston, W. B.; Gratz, R. F. *U. S. Patent* **4**,758, 380, 1988
[15] Yang, S. Y.; Ge, Z. Y.; Yin, D. X.; Liu, J. G.; Li, Y. F.; Fan, L. *J. Polym. Sci. Part A Polym. Chem.* 2004, 42, 4243-52.
[16] Song, K. Y.; Moore, J. A. *Polym. Prepr.* 2000, 41, 192-3.
[17] Watson, K. A.; Connell, J. W. *NASA technical report*. 2000-45.
[18] Muraka, S. P. *Solid State Tech.* 1996, March, 83
[19] Trofimenko, S. In Advances in Polyimide Science and Technology; Feger, C.; Khojastech, M. M.; Htoo, S. Eds.; *Technomic*, Lancaster, PA, 1993, p 3-14.
[20] Myung, B. Y.; Kim, J. J.; Yoon, T. H. *J. Polym. Sci. Part A Polym. Chem.* 2002, 40, 4217-27.
[21] Matsuura, T.; Ishizawa, M.; Hasuda, Y.; Nishi, S. *Macromolecules*. 1992, 25, 3540.
[22] Liu, H.; Simone, C. D.; Scola D. A.; *J. Polym. Sci. Part A Polym. Chem.* 2003, 41, 2630-49.
[23] Savitski, E. P.; Li, F. M.; Lin, S. H.; McCreight, K. W.; Wu, W.; Hsieh, E.; Rapold, R. F.; Leland, M. E.; McIntyre, D. M.; Harris, F. W.; Cheng, S. Z. D.; Kwan, S. C. M.; Wu, C. *Int. J. Polym. Anal. Charc.* 1997, 4, 153-72.
[24] Chung, T-S.; Vora, R. H.; Jaffe, M. *J. Polym. Sci. Polym. Chem.* Ed. 1991, 29, 1207-12.
[25] Hedrick, J. L.; Charlier, Y.; DiPietro, R.; Jayaraman, S.; Mc Grath, J. E. *J. Polym. Sci. Part A Polym. Chem.* 1996, 34, 2867-2877.
[26] Gerber, M. K.; Pratt, J. R.; St. Clair, A. K.; St. Clair, T. L. *Polym. Prepr.* 1990, 31, 340-1.
[27] Jeong, K.U.; Jo, Y. J.; Yoon, T. H. *J. Polym. Sci. Part A Polym. Chem.* 2001, 39, 3335-47.
[28] Jeong, K.U.; Kim, J.J.; Yoon, T. H. *Polymer*. 2001, 42, 6019-30.
[29] Rogers, M.E.; Brink, M. H.; McGrath, J. E.; Brennan, A. *Polymer*. 1993, 34, 849-855.
[30] Wang, C-S.; Yang, R-W. *J. Appl. Polym. Sci.* 1997, 66, 609-17.
[31] Tamai, S.; Yamaguchi, A.; Ohta, M. *Polymer*. 1996, 37, 3683-92.
[32] Zoia, G.; Stern, S. A.; St. Clair, A. K.; Pratt, J. R. *J. Polym. Sci. Part B Polym. Phys.* 1994, 32, 53–58.
[33] Yang, C-P.; Su, Y-Y.; Wu, K-L. *J. Polym. Sci. Part A Polym. Chem.* 2004, 42, 5424-38.
[34] Hsiao, S.H.; Yang, C. P.; Chung, C. L. *J. Polym. Sci. Part A Polym. Chem.* 2003, 41, 2001-18.

[35] Yang, C. P.; Hsiao, S.H.; Yang, C.C. *J. Polym. Res.* 2004, 11, 23-9.
[36] Han, K.; Lee, J.; Rhee, T. H. *J. Appl. Polym. Sci.* 1999, 74, 107-12.
[37] Feiring, A. E.; Auman, B. C.; Wonchoba, E. R. *Macromolecules.* 1993, 26, 2779.
[38] Gibbs, H. H. *J. Appl. Polym. Sci.* 1979, 35, 207-22.
[39] Vannucci, R.D.; Malarik, D.; Papadopoulas, P.; Wafer, J. *22nd International Technical Conference SAMPE*, 1990, 22, 175-180.
[40] Serafini, T. T.; Cheng, P. G.; Ueda, K. K.; Wright, W. F. *22nd International Technical Conference, SAMPE*, 1990, 22, 94-107.
[41] Vora, R.H.; Wang, H.M.; Chung, T-S. *In Polyimides and Other High Temperature Polymers* Vol 1; Mittal, K.L.; Ed.; VSP International Science, NY, 2001; pp 33-53.
[42] Connolly, M. L. J. Mol. *Graphics.* 1993, 11, 139-141.
[43] Simone, C.D.; Scola, D. A. *Macromolecules.* 2003, 36, 6780-90.
[44] Rogers, F.E. *U. S. Patent* **3**,356,648, 1964.
[45] Yi, M. H.; Huang, W.; Lee, B. J.; Choi, K-Y. *J. Polym. Sci. Part A Polym. Chem.* 1999, 37, 3449.
[46] Liu, S. L.; Chng, M. L.; Chung, T. S.; Goto, K.; Tamai, S.; Pramoda, K. P.; Tong, Y. J. *J. Polym. Sci. Part B Polym. Phys.* 2004, 42, 2769-79.
[47] Farr, I. V.; Kratzner, D.; Glass, T. E.; Dunson, D.; Ji, Q.; McGrath, J. E. *J. Polym. Sci. Part A Polym. Chem.* 2000, 38, 2840-54.
[48] Volksen, W.; Cha, H. J.; Sanchez, M. I.; Yoon, D. Y. *React. Funct. Polym.* 1996, 30, 61-9.
[49] Takeichi, T.; Tanikawa, M. J. Polym. Sci. Part A Polym. Chem. 1996, 34, 2205 - 11.
[50] Takeichi, T.; Nakajima, K. *React. Funct. Polym.* 1996, 30, 75 - 83.
[51] Gajiwala, H. M.; Zand, R. *Macromolecules.* 1993, 26, 5976 – 83.
[52] Yang, C. P.; Lin, J.H. *J. Polym. Sci. Part A Polym. Chem.*1995, 33, 2183-91.
[53] Yagci, H.; Ostrowski, C.; Mathias, L. J. *J. Polym. Sci. Part A Polym. Chem.* 2000, 37, 1189-97.
[54] Tamami, B.; Yeganeh, H. *J. Polym. Sci. Part A Polym. Chem.*2001, 39, 3826–31.
[55] Chen, J. P.; Natansohn, A. *Macromolecules.* 1999, 32, 3171–77.
[56] Tanaka, K.; Osada, Y.; Kita, H.; Okamoto, K. I. *J. Polym. Sci. Part B Polym. Phys.* 1995, 33, 1907–15.
[57] Kreuz J.A.*; J. Polym. Sci. Part A Polym. Chem.* 1990, 28, 3787-93.
[58] Sroog, C.E. *Prog. Polym. Sci.*1991, 16, 561-694.
[59] McGrath, J. E.; Rogers, M. E.; Arnold, C. A.; Kim, Y. J.; Hedrick, J. C. *Makromol. Chem. Macromol. Symp.* 1991, 51, 103-25.
[60] Konas, M.; Moy, T. M.; Rogers, M. E.; Shultz, A. R.; Ward, T. C.; McGrath, J. E. *J. Polym. Sci. Part B Polym. Phys.* 1995, 33, 1429-39.
[61] Konas, M.; Moy, T. M.; Rogers, M. E.; Shultz, A. R.; Ward, T. C.; McGrath, J. E. *J. Polym. Sci. Part B Polym. Phys.* 1995, 33, 1441-48.
[62] Quaranta, A.; Carturan, S.; Maggioni, G.; Della Mea, G.; Ischia, M. Campostrini, R. *Appl. Phys. A – Mater. Sci. & Proc.* 2001, 72, 671-7.
[63] Quaranta, A.; Carturan, S.; Maggioni, G.; Milazzo, P. M.; Abbondanno, U.; Della Mea, G.; Gramegna, E.; Pieri, U. *IEEE Trans. Nucl. Sci.* 2001, 48, 219-224.
[64] Krishnan, P. S. G.; Vora, R. H.; Chung, T-S. *Polymer.* 2001, 42, 5165.
[65] Dorogy Jr, W. E.; St. Clair, A. K. *J. Appl. Polym. Sci.* 1993, 49, 501-10.
[66] Krishnan, P. S. G.; Vora, R. H.; Veeramani, S. *Plast. Rubb. Comp.* 2002, 31, 289-94.

[67] Krishnan, P. S. G.; Vora, R. H.; Veeramani, S.; Goh, S. H.; Chung, T-S. *Polym. Degrad. Stab.* 2002, 75, 273-85.
[68] Krishnan, P. S. G.; Veeramani, S. *Polym. Degrad. Stab.* 2003, 81, 225-32.
[69] Pithan, F.; Staudt-Bickel, C.; Hess, S.; Lichtenthaler. *Chem .Phys. Chem.* 2002, 3, 856-62.
[70] Hess, S.; Scharfenberger, G.; Staudt-Bickel, C.; Lichtenhaler, R. N. *Desalination.* **145**, 2002, 359-64.
[71] Xu, S.; Yang, M.; Wang, J.; Ye, H.; Liu, X. *Syn. Met.* 2003, 132, 145-9.
[72] Ren, L.; Zeng, F.; Ning, P.; Chen, Z.; Ko, T-M. *J. Appl. Polym. Sci.* 2000, 77, 3252-8.
[73] Iijima, T; Kaise, T.; Tomoi, M. *J. Appl. Polym. Sci.* 2003, 88, 1-11.
[74] Yoshino, M.; Nakamura, S.; Kita, H.; Okamoto, K.; Tanihara, N.; Kusuki, Y. *J. Membr. Sci.* 2003, 212, 13-27.
[75] Chang, K. Y.; Lee, Y. D. *J. Polym. Sci. Part A Polym. Chem.* 1993, 31, 2775-84.
[76] Chan, S. S.; Wang, R.; Chung, T. S.; Liu, Y.; *J. Membr. Sci.* 2002, 210, 55-64.
[77] Liu, S. L.; Wang, R.; Chung, T. S.; Chng, M. L.; Liu, Y.; Vora, R. H.; *J. Membr. Sci.* 2002, 202, 165-76.
[78] Ding, H.; Harris, F. W. *Pure Appl. Chem.* 1995, 67, 1997-2004.
[79] Rhee, S. B.; Park, W.; Moon, B. S.; Chang, J. Y. *Macromolecules.* 1993, 26, 404.
[80] Park, J. W.; Lee, M.; Lee, M-H.; Liu, J. W.; Kim, S. D.; Chang, J. Y.; Rhee, S. B. *Macromolecules.* 1994, 27, 3459-63.
[81] Maruo, Y. Y.; Andoh, Y.; Sasaki, S. J. Vac. Sci. Tech. A Vac. Surf. Films. 1993, 11, 2590-6.
[82] Yanagishita, H.; Nozoye, H.; Nakane, T. *Desalination.* 1993, 90, 55-63.
[83] Hsu, S. C.; Whang, W. T.; Chen, S. C. *J. Polym. Res.* 2003, 10, 7-12.
[84] Kumar, D.; Razdan, U.; Gupta, A. D. *J. Polym. Sci. Part A Polym. Chem.* 1992, 30, 1477-87.
[85] Liu, D. C. H.; Liaw, D. J.; Liaw, B. Y.; Tsai, C. H.; Ho, J. H.; Ho, T. I. *J. Chinese Chem. Soc.* 2000, 47, 583-88.
[86] Choi, W-S.; Harris, F. W. *Polymer.* 2000, 41, 6213-21.
[87] Wang, W. C.; Vora, R. H.; Kang, E.T.; Neoh, K. G.; Liaw, D. *J. Ind. Eng. Chem. Res.* 2003, 42, 784-94.
[88] Wang, W. C.; Ong, G. T.; Lim, S. L.; Vora, R. H.; Kang, E. T.; Neoh, K. G. *Ind. Eng. Chem. Res.* 2003, 42, 3740-9.
[89] Wang, W. C.; Vora, R. H.; Kang, E. T.; Neoh, K. G. *Polym. Eng. Sci.* 2004, 44, 362-75.
[90] Wang, W. C.; Vora, R.H.; Kang, E.T.; Neoh, K.G. *Macromol. Mater. Eng.* 2003, 288, 152-63.
[91] St. Clair, A. K.; St. Clair, T. L.; Slemp, W. S. *In Proceedings of the 2nd International Conference on Polyimides*; Weber, W.; Gupta, M.; Ed.; Society of Plastics Engineers: Poughkeepsie, NY, 1987; Chapter 2, pp 16-36.
[92] St. Clair, T. L.; "Structure Property relationship in Linear Aromatic Polyimides", In Polyimides; Wilson, D.; Stanzenberger, H. D.; Hergenrother, P. M. Eds.; Chapman and Hall: NY, 1990; pp 67.
[93] Yang, C. P.; Hung, K. S.; Chen, R. S. *J. Polym. Sci. Part A Polym. Chem.* 2000, 38, 3954–61.

[94] Vora, R. H.; Goh, S. H.; Chung, T-S. *Polym. Eng. Sci.* 2000, 40, 1318-29.
[95] Yang, C. P.; Yang, H-W. *J. Appl. Polym. Sci.* 2000, 75, 87-95.
[96] Bes, L.; Rousseau, A.; Boutevin, B.; Mercier, R.; Kerboua, R. *Macromol.Chem.Phys.* 2001, 202, 2954–61.
[97] Xu, J.; He, C.; Chung, T-S. *J. Polym. Sci. Part A Polym. Chem.* 2001, 39, 2998–3007.
[98] Zhou, H.; Chen, C.; Zhang, S.; Yang, S. *Polym. Prepr.* 2002, 43, 658-9.
[99] Kuo, W-J.; Hsiue, G-H.; Jeng, R-J. *Macromolecules.* 2001, 34, 2373–84.
[100] Yang, C. P.; Chen, J-A. *J. Polym. Sci. Part A Polym. Chem.* 1999, 37, 1681–91.
[101] Chan, S. H.; Wong, W. T.; Chan, W. K. *Chem. Mater.* 2001, 13, 4635–41.
[102] Marek, M.; Doskocilova, D.; Schmidt, P.; Schneider, B.; Kriz, J.; Labsky, J.; Puffr, R. *Polymer.* 1994, 35, 4881–8.
[103] Park, J. H.; Jung, J. C.; Sohn, B. H.; Lee, S. W. Ree, M. *J. Polym. Sci. Part A Polym. Chem.* 2001, 39, 3622–32.
[104] Banerjee, S.; Madhra, M. K.; Salunke, A. K.; Jaiswal, D. K. *Polymer.* 2003, 44, 613-622.
[105] Madhra, M. K.; Salunke, A. K.; Banerjee, S.; Prabha, S. *Macromol. Chem. Phys.* 2002, 203, 1238-48.
[106] Banerjee, S.; Madhra, M. K.; Salunke, A. K.; Maier, G. *J. Polym. Sci. Part A Polym. Chem.* 2002, 40, 1016-27.
[107] Ge, J. J.; Xue, G.; Li, F. M.; McCreight, K. W.; Wang, S. Y.; Harris, F. W.; Cheng, S. Z. D.; Zhuang, X. W.; Hong, S. C.; Shen, Y. R. *Macromol. Rapid. Commun.* 1998, 19, 619–23.
[108] Kwan, S. C. M.; Wu, C.; Li, F. M.; Harris, F. W.; Chang, S. Z. D. *Polym. Eng. Sci.* 1999, 39, 586-93.
[109] Yagei, H.; Ostrowski, C.; Mathias, L. J. 1999, *J. Polym. Sci. Part A Polym. Chem.* 1999, 37, 1189–97.
[110] Kumar, D.; Gupta, A. D. *Macromolecules.* 1995, 28, 6323–29.
[111] Omote, T.; Koseki, K.; Yamaoka, T. *Macromolecules.* 1990, 23, 4788-95.
[112] Feng, K.; Tsushima, M.; Matsumoto, T.; Kurosaki, *T. J. Polym. Sci. Part A Polym. Chem.*1998, 36, 685-93.
[113] Langlade, G. R.; Monjol, P.; Sekiguchi, H.; Mercier, R.; Sillion, B. *Polymer.* 1997, 38, 4965-72.
[114] Zhang, A.; Li, X.; Nah, C.; Hwang, K.; Lee, M-H. *J. Polym. Sci. Part A Polym. Chem.* 2003, 41, 22-9.
[115] Jikei, M.; Chon, S. H.; Kakimoto, M.; Kawauchi, S.; Imase, T.; Watanabe, J. *Macromolecules.* 1999, 32, 2061.
[116] Kricheldorf, H. R.; Bolender, O.; Wollheim, T. *High Perform. Polym.* 1998, 10, 217.
[117] Kim, Y.H. *J. Polym. Sci. Part A Polym. Chem.* 1998, 36, 1685-98.
[118] Bolton, D.H.; Wooley, K.L. *Macromolecules.* 1997, 30, 1890.
[119] Fang, J.H.; Kita, H.; Okamoto, K. *J. Membr. Sci.* 2001, 182, 245-56.
[120] Liu, Y.; Chung, T. S. *J. Polym. Sci. Part A Polym. Chem.* 2002, 40, 4563.
[121] Makita, S.; Kudo, H.; Nishikubo, T. *J. Polym. Sci. Part A Polym. Chem.* 2004, 42, 3697-3707.
[122] Suzuki, T.; Yamada, Y.; Tsujita, Y. *Polymer.* 2004, 45, 7167–71.
[123] Fang, J.; Kita, H.; Okamoto, K-I. *Macromolecules.* 2000, 30, 4639-46.

[124] Stenzenberger, H. D. *Adv. Polym. Sci.* 1994, 117, 165-220.
[125] Rossi, R. D. *Polyimides*, reprint from: Engineered Materials Handbook; 1992; Vol 3, Adhesives and Sealants, ASM International, Materials Park, Ohio pp 151-62.
[126] Baldwin, L. J.; Meador, M. A. B.; Meador, M. A. *Polym. Prepr.* 1988, 29, 236.
[127] Hirano, T.; Muramatsu, T.; Jnoue, H. *1st Japan Int. SAMPE Symp.* 1989, 1.
[128] Alam, S.; Varma, I. K. *J.Polym.Mater.* 1998, 15, 141-9.
[129] Alam, S.; Varma, I. K. *J. Thermal. Anal.* 1998, 51, 275-284.
[130] Mathur, A.; Varma, I. K. *Polymer.* 1992, 33, 4845-50.
[131] Alam, S.; Kandpal, L.D.; Varma, I. K. *J. Appl. Polym. Sci.* 1997, 861-9.
[132] Mathur, A.; Varma, I. K. *J. Appl. Polym. Sci.* 1992, 46, 1749.
[133] Alam, S.; Kandpal, L.D.; Varma, I. K. *J. Appl. Polym. Sci.* 1994, 53, 1073-81.
[134] Srivastava, A.; Varma, I. K. *J. Engg. Mat. Sci.* 2001, 8, 165-9.
[135] Chen Sr, P. N.; Vora, R.H.; Glick, M. M.; Jaffe, M. J. *21st International SAMPE Technical Conference* Sept 25-28, 1989 pp 755-66.
[136] Yang, C. P.; Chen, R. S.; Chen, S. H.; *J. Polym. Sci. Part A Polym. Chem.* 2001, 39, 93-104.
[137] Liao, D. C.; Hsieh, K. H. *J. Polym. Sci. Part A Polym. Chem.* 1994, 32, 1665.
[138] Jiang, B.; Hao, J.; Wang, W.; Jiang, L.; Cai, X. *J. Appl. Polym. Sci.* 2001, 81, 773.
[139] Zuo, M.; Takeichi, T. *Polymer.* 1999, 40, 5153-60.
[140] Krishnan, P. S. G.; Cheng, C. Z.; Cheng, Y. S.; Cheng, J. W. C. *Macromol. Mater. Engg.* 2003, 288, 730-6.
[141] Zuo, M.; Xiang, Q.; Takeichi, T. *Polymer.* 1998, 39, 6883-9.
[142] Takeichi, T.; Zuo, M.; Ito, A. *High Perform. Polym.* 1999, 11, 1.
[143] Mehdipour-Ataei, S.; Keshavarz, S. *J. Appl. Polym. Sci.* 2003, 89, 2567-72.
[144] Moon, Y. D.; Lee, Y.M. *J. Appl. Polym. Sci.* 1993, 50, 1461-73.
[145] Wang, G. P.; Chang, T. C.; Hong, Y. S.; Chin, Y. S. *Int. J. Polym. Anal. Char.* 2002, 7, 227-43.
[146] Akiyama, E. I.; Takamura, Y.; Nagase, Y. *Macromol. Chem. Phys.* 1992, 193, 1509-19.
[147] Yu, H. S.; Yamashita, T.; Horie, K. *Macromolecules.* 1996, 29, 1144-50.
[148] Yu, H. S.; Yamashita, T.; Horie, K. *Polym. J.* 1996, 28, 703-7.
[149] Pinniavaia T.; Beal, G. W. Eds.; Polymer-Clay *Nanocomposites*; John Wiley and Sons: New York, 2000.
[150] Alexandre, M.; Dubois, P. *Mater. Sci. Eng.* R. 2000, 28, 1-63.
[151] Ray, S. S.; Okamoto, M. *Prog. Polym. Sci.* 2003, 28, 1539-1641.
[152] Dagani, R. *Chem. Engg. News.* 1999, 77(23), 25.
[153] Miller, B. *Plast. Formulating Compound.* 1997, May/June, 30.
[154] Vora, R.H.; Pramoda, P.K.; Goh, S.H.; Chung, T.S.; Lim, Y.X.; Bang, T. K. *Macromol. Mater. Eng.* 2003, 288, 337-56.
[155] Sarathi, R.; Deepa, S.; Mishra, A.K.; Vora, R. H. *J. Metastable and Nanocrystalline Materials.* 2005, 23, 347-50.
[156] Jayaganthan, R.; Vora, R. H. *J. Metastable and Nanocrystalline Materials*, 2005, 23, 339-42.
[157] Jiang, L-Y.; Wei, K-H. *J. Appl. Polym. Sci.* 2004, 92, 1422-5.
[158] Southward, R. E.; Thompson, D.S.; Thornton, T.A.; Thompson, D. W.; St Clair, A. K. *Chem. Mater.* 1998, 10, 486-94.
[159] Thompson, D. S.; Thompson, D. W.; Southward, R. E. *Chem. Mater.* 1998, 14, 30-7.

[160] Qiu, F-X.; Zhou, Y-M.; Liu, J-Z. *Eur. Polym. J.* 2004, 40, 713-20.
[161] Yen, C-T.; Chen, W-C.; Liaw, D-J.; Lu, H-Y. *Polymer*. 2003, 7079-87.
[162] Chang, C-C. ; Chen, W-C. *Chem.Mater*. 2002, 14, 4242-8.
[163] Chang, C. C.; Wei, K. H.; Chen, W. C. J. *Electrochem*. Soc. 2003, 150, F147-50.
[164] Prasad, P. N.; Williams, D. J.; *Introduction to Nonlinear Optical Effects in Molecules and Polymers*, Wiley: New York, 1991.
[165] Wise, D. L.; Wnek, G. E.; Trantolo, D. J.; Copper, T. M.; Gresse, J. D. *Photonic Polymer Systems*, Marcel Dekker: New York, 1998.
[166] Leigh, W. B. *Devices of Optoelectronics*, Marcel Dekker: New York, 1996.
[167] Burland, D.; Miller, R.; Walsh, C. *Chem. Rev*. 1994, 94, 31.
[168] Li, F.; Kim, K. H.; Kulig, J. J.; Savitski, E. P.; Brittain, W. J.; Harris, F. W.; Cheng, S. Z. D.; Hubbard, S. F.; Singer, K. D. *J. Mater. Chem*. 1995, 5, 253.
[169] Lindsay, G. A.; Stenger-Smith, J. D.; Henry, R. A.; Hoover, J. M.; Nissan, R. A. *Macromolecules*. 1992, 25, 6075.
[170] Chen, L. T.; Tam, W.; Marder, S. R.; Stiegman, A. E.; Rikken, G.; Spangler, C. W. *J. Phys. Chem*. 1991, 95, 10643.
[171] Yamamoto, H.; Katogi, S.; Watanabe, T.; Sato, H.; Miyata, S.; Hosomi, T. *Appl. Phys. Lett*. 1992, 60, 935.
[172] Miyata, S.; Sasabe, H. *Poled Polymers and their Applications to SHG and EO devices*; Gordon and Breach Science: Australia, 1997.
[173] Chen, T-A.; Jen, A. K-Y.; Cai, Y. *J. Am. Chem. Soc*. 1995, 117, 7295-6.
[174] Lee, H-J.; Lee, M-H.; Han, S. G.; Kim, H-Y.; Ahn, J-H.; Lee, E-M.; Won, Y. H. *J. Polym. Sci. Part A Polym. Chem*. 1998, 36, 301-7.
[175] Leng, W. N.; Zhou, Y. M.; Xu, Q. H.; Liu, J. Z. *Polymer*. 2001, 42, 7749-54.
[176] Bes, L. Rousseau, A.; Boutevin, B.; Mercier, R.; Sillion, B.; Toussaere, E. *High Perform. Polym*. 2000, 12, 169-76.
[177] Marestin, C.; Mercier, R.; Sillion, B.; Chauvin, J.; Nakatani, K.; Delaire, J. A. *Synth. Metals*. 1996, 81, 143-6.
[178] Xu, B.; Peng, Z. *Polym. Prepr*. 2000, 41, 196-7.
[179] Gubbelmans, E.; Verbiest, T.; Beylen, M. V.; Persoons, A.; Samyn, C. *Polymer*. 2002, 43, 1581-5.
[180] Kim, T-D.; Lee, K-S.; Lee, G. U.; Kim, O-K. *Polymer*. 2000, 41, 5237-45.
[181] Yu, D.; Gharavi, A.; Yu, L. *Macromolecules*. 1995, 28, 784–6.
[182] Yu, D.; Gharavi, A.; Yu, L. *J. Am. Chem. Soc*. 1995, 117, 11680–86.
[183] Zhou, Y. M.; Leng, W. N.; Liu, X.; Xu, Q. H.; Feng, J.; Liu, J. Z. *J. Polym. Sci. Part A Polym. Chem*. 2002, 40, 2478–86.
[184] Leng, W. N.; Zhou, Y. M.; Xu, Q. H.; Liu, J. Z. *Polymer*. 2001, 42, 9253–9.
[185] Davey, M. H.; Lee, V. Y.; Wu, L-M.; Moylan, C. R.; Volksen, W.; Knoesen, Miler, R. D.; Marks, T. *J. Polym. Prepr*. 2000, 41, 853–4.
[186] Van den Broeck, K.; Verbiest, T.; Degryse, J.; Van Beylen, M.; Persoons, A.; Samyn, C. *Polymer*. 2001, 42, 3315–22.
[187] Sui, Y.; Liu, Y-G.; Yin, J.; Gao, J.; Zhu, Z-K.; Huang, D-Y.; Wang, Z-G. *J. Polym. Sci. Part A Polym. Chem.*1999, 37, 4330-6.
[188] Kuo, W-J.; Hsiue, G-H.; Jeng, R-J. *Macromolecules*. 2001, 34, 2373-84.
[189] Neher, D.; Kaltbeitzel, A.; Wolf, A.; Bubek, C.; Wegner, G. In Conjugated polymeric materials:Opportunities in electronics,optoelectronics and molecular electronics;

Bredas, J. L.; Chance, R. R.; Eds.; Kluwer Academic: Dordrecht, 1990; p 387.
[190] Hu, H. F.; Yao, K. L. *Phys. Lett.* 1997, 234, 310.
[191] Wada, T.; Matsuda, T.; Sasabe, H. *Mol. Cryst. Liq. Cryst.* 1994, 247, 139.
[192] Kaino, T.; Kurihara, T. *In Polymers for Lightwave and Integrated Optics*; Hornak, L. A.; Ed.; Marcel Dekker: New York, 1992; p. 647.
[193] Okawa, H.; Wada, T.; Sasabe, H. *Mater. Res. Symp. Proc.* 1992, 244, 263.
[194] Ho, Z. Z.; Ju, C. Y.; Htherington, W. M. *J. Appl. Phys.* 1987, 62, 716.
[195] Sakai, Y.; Ueda, M.; Yahagi, A.; Tanno, N. *Polymer*. 2002, 43, 3497-3504.
[196] St. Clair, A. K.; St. Clair, T. L.; Winfree, W. P. *Polym. Mater. Sci. Eng.* 1998, 59, 28.
[197] Hougham, G.; Tesoro, G.; Viehbeck, A.; Chapple-Sokol, J. D. *Macromolecules*. 1994, 27, 5964.
[198] Goto, K.; Akiike, T.; Inoue, Y.; Matsubara, M. *Macrmol. Symp.* 2003, 199, 321-31.
[199] Hougham, G.; Tesoro, G.; Shaw, J. *Macromolecules*. 1994, 27, 3642.
[200] Hougham, G.; Tesoro, G. Polym. *Mater. Sci. Eng.* 1989, 61, 369.
[201] Ichino, T.; Sasaki, S.; Matsuura, T.; Nishi, S. *J. Polym. Sci. Polym. Chem. Ed.* 1990, 28, 323.
[202] Yusa, M.; Takeda, S.; Miyadera, Y. *Polym. Prepr. Japan* (Engl Ed). 1990, 39, 897.
[203] Auman, B. C.; Highley, D. P.; Scherer, K. V. *Polym. Prepr.* 1993, 34, 389-90.
[204] Misra, A. C.; Tesoro, G.; Hougham, G.; Penkarkar, S. M. *Polymer*. 1992, 33, 1078.
[205] Matsuura, T.; Hasuda, Y.; Nishi, S.; Yamada, N. *Macromolecules*. 1991, 24, 5001.
[206] St. Clair, A. K.; St. Clair, T. L.; Winfree, W. P. *Polym. Mater. Sci. Eng.* 1988, 59, 28.
[207] Chung, T-S.; Pan, J.; Liu, S. L.; Mullick, S.; Vora, R. H. Polyimide: Syntheses, Characterization, *Blends and Applications In Advanced Functional Molecules and Polymers Nalwa*, H. S.; Ed.; Gordon and Breach Science Publishers: NV, 2001; pp157-198.
[208] Buchanan, R. A.; Mundhenke, R. F.; Lin, H. C. *Polym. Prepr.* 1991, 32, 193-4.
[209] Vora, R. H.; Krishnan, P. S. G.; Goh, S. H.; Chung, T-S. *Adv. Func. Mater.* 2001, 11, 361-73.
[210] Chan, W. K.; *Polyimides and Other High Temperature Polymers* Ed. Mittal, K. L. Vol. 1 pp. 243-9.
[211] Ha, K.; West, J. L.; Magyar, G. R.; *Mol. Cryst. Liq. Cryst.* 2001, 357, 263.
[212] Reuter, R.; Franke, H.; Feger, C. *Applied Optics*, 1988, 27, 4565-71.
[213] Franke, H.; Wagner, D.; Kleckers, T.; Reuter, R.; Vora, R. H.; Blech, B. A. *Applied Optics*, 1993, 32, 2927-35.
[214] Han, K. Korea. *Polym. J.* 2000, 8, 165-71.
[215] Stern, S. A. *J. Membr. Sci.* 1994, 94, 1-65.
[216] Maier, G. Angew. *Chem. Int. Ed.* 1998, 37, 2960-74.
[217] Prasad, R.; Shaner, R.L.; Doshi, K.J. *In Polymeric gas separation membranes*. Paul, D. R.; Yampol'skii, Y.; Eds.; CRC Press, Boca Raton, FL, 1994, p. 531.
[218] Koros, W. J. Walker, D. R. B. *Polym. J.* 1991, 5, 481.
[219] Koros, W.J.; Hellums, M. W. *In Encyclopedia of Polymer Science and Engineering*, John Wiley Sons: New York, 1989; Suppl. Vol., pp 724-802.
[220] Koros, W. J.; Fleming, G. K. *J. Membr. Sci.* 1993, 83, 1.
[221] Houde, A. Y.; Krishnakumar, B.; Charati, S. G.; Stern, S. A. *J. Appl. Polym. Sci.* 1996, 62, 2181.

[222] Okamoto, K.; Noborio, K.; Hao, J.; Tanaka, K.; Kita, H. *J. Membr. Sci.* 1997, 134, 171.
[223] Bos, A.; Punt, I. G. M.; Wessling, M.; Strathmann, H. *Sep. Purif. Tech.* 1998, 14, 27.
[224] Won, J.; Kim, M. H.; Kang, Y. S.; Park, H. C.; Kim, U. Y.; Choi, S. C.; Koh, S. K. *J. Appl. Polym. Sci.* 2000, 75, 1554.
[225] Hayes, R. A. U. S. *Patent* **4**, 717, 393 (1988).
[226] Liu, Y.; Chung, T. S.; Wang, R.; Li, D. F.; Chng, M. L. *Ind. Eng. Chem. Res.* 2003, 42, 1190.
[227] Bos, A.; Punt, I. G. M.; Wessling, M.; Strathmann, H. *J. Polym. Sci. Part B Polym. Phys.* 1998, 36, 1547.
[228] Ren, J.; Wang, R.; Chung, T-S.; Li, D-F.; Liu, Y. J. Membr. Sci. 2003, 222, 133-47.
[229] Cao, C.; Chung, T-S.; Liu, Y.; Wang, R. Pramoda, K. P. *J. Membr. Sci.* 2003, 216, 257- 68.
[230] Chung, T-S.; Ren, J.; Wang, R.; Li, D.; Liu, Y.; Pramoda. K. P.; Cao, C.; Loh, W. W. *J. Membr. Sci.* 2003, 214, 57-69.
[231] Lin, W-H.; Vora, R. H.; Chung, T-S. *J. Polym. Sci. Part B Polym. Phys.* 2000, 28, 2703-13.
[232] Wind, J. D.; Staudt-Bickel, C.; Paul, D. R.; Koros, W. J. Ind. Eng. Chem. Res. 2002, 41, 6139-48.
[233] Kanamori, T.; Niwa, M.; Kawakami, H.; Mori, Y.; Nagaoka, S.; Haraya, K.; Shinbo, T. *Asaio J.* 2000, 46, 612-9.
[234] Connell, J. W.; Watson, K. *A. High Perform. Polym.* 2001, 13, 23-34.
[235] Seckin, T.; Koytepe, S.; Ozdemir, I.; Cetinkaya, B. *J. Inorg. Organomet. Polym.* 2003, 13, 9-20.
[236] Lalande, G.; Guay, D.; Dodelet, J. P.; Majetich, S. A.; McHenry, M. E. *Chem. Mater.* 1997, 9, 784.
[237] Gunduz, N.; McGrath, J. E. *Polym. Prepr.* 2000, 41, 182-3.
[238] Kaleem, K.; Chertok, F.; Erhan, S. *Prog. Org. Coat.* 1987, 15, 63-71.
[239] Han, M. J.; Bie, H. M.; Nikles, D. E.; Warren, G. W. *Polym. Prepr.* 2000, 41, 1759-60.
[240] Han, M. J; Bie, H. M.; Nikles, D. E.; Warren, G. W. *Electroact. Polym. Corr. Cont.* 2003, 843, 175-81.

In: Frontiers in Polymer Research
Editor: Robert K. Bregg, pp. 107-132
ISBN: 1-59454-824-2


*Chapter 4*

# APPLICATIONS OF FUNCTIONAL POLYMERS FOR SEPARATIONS IN BIOCHEMICAL PRODUCTION

***Andrei A. Zagorodni***
Department of Materials Science and Engineering,
Royal Institute of Technology, Stockholm, Sweden
***Vladimir F. Selemenev***
Department of Analytical Chemistry, Voronezh State University, Voronezh, Russia

## Abstract

Chemical separation with functional polymers is a virtually important part of technologies producing different chemical and biochemical substances by cultivation of yeast, bacteria, or fungus. Each cultivation mixture is extremely complex and the product extraction/purification could be the most costly step. Chromatographic techniques (low-pressure liquid chromatography) can be considered as the main option to fulfil the extraction/purification task. The sorbents used in such processes differ from the materials used in analytical chromatography due to demands on the product quantity/purity rather than quality of the analytical signal. Functional polymers are highly advantageous for such separations. Even more, a careful selection of the polymer and operating conditions could allow replacing the costly chromatographic separation by more economically and environment-friendly processes based on selective sorption and stripping interactions.

The work describes applications of functional polymers for separations of bio-cultivated substances combining primary data with review of previously published works. An attention is paid to relationships between properties of the selected polymer, target product(s), and contaminants. Exploitation of these relationships for benefits of the separation efficiency is described. Specific phenomena and interactions taking place in phase of the polymer are discussed as well as effect of these phenomena on the separation processes. This includes specific interactions with functional groups and three-dimensional polymeric networks, transformations of substances in the polymer phase, dimerization, ion exchange isothermal supersaturation, etc.

A special section discusses changes taking place in phase of the functional polymers at continuous industrial use. This includes phenomena of semi-reversible and irreversible sorption, chemical and physical deterioration, aging, etc.

The systems selected to serve as major examples include amino acids and different ion exchange resins. This selection was done due to representativity of amino acids as an example of biochemical substances (labile charge, ability to form zwitterions that is almost specific for bio-products, tendency to form associates, etc.) while structures of these materials are well-known. Functional polymers and, particularly, ion exchange resins are materials of first choice for sorption-based separation of such substances.

## Introduction

Biochemical industry is a well-established branch continuing a rapid grows in different directions: quantity and diversity of products, development of new technologies, use of new biological species for industrial cultivation, etc. Most of the exploited biochemical processes consist of cultivation of yeast, bacteria, or fungus following with proper extraction procedures. Due to a high complexity of the cultivation mixtures, one or several extraction and purification steps are unavoidable for obtaining a reasonably pure target product. Besides the product, each mixture contains remains of the substrate, whole and parted biological cells, undesirable products of the microorganisms activity, products of chemical interactions between the mixture constitutes, substances initially introduced to provide the media desirable for microorganisms, etc. Chromatographic techniques and, particularly, low pressure liquid chromatography can be considered as the main option to fulfil the separation/purification task. The sorbents used in such processes differs from the materials used in analytical chromatography due to demands on the product quantity and purity rather than intensity and quality of the analytical signal. Functional polymers are highly advantages for such separations. Even more, a careful selection of the polymer and operating conditions could allow replacing the costly chromatographic separation by more economically and environmentally friendly processes based on selective sorption and stripping interactions.

## Interactions with Biochemically Produced Substances

All products of biochemical industry are organic substances of different complexity. Chemical structures of some of them (for example, amino acids) are well known, however, structures of many proteins, enzymes, nucleic acids, etc. are yet to be identified.[1] One of the common properties of these substances is their liability to chemical environment and operating conditions. For example, charge of many organic ions depends on *pH* of the solution. Even more, dependently on the conditions the same species can be cationic, anionic, or even zwitterionic because different groups of the same molecule can be oppositely charged (see Figure 1 for illustration).

All system of interactions in the separation system is affected by the recharging of target molecules. When ion exchange is discussed, the conventional reference is interactions between inorganic ions and functional polymers. Such reactions are usually described by the simple reaction

[1] One can argue that biochemists already know structures of proteins and nucleic acids. However, many biochemical approaches consider overall molecules or most important parts and, in many cases, do not take into account detailed chemical structures and groups that do not belong to the biologically active sites of the molecules.

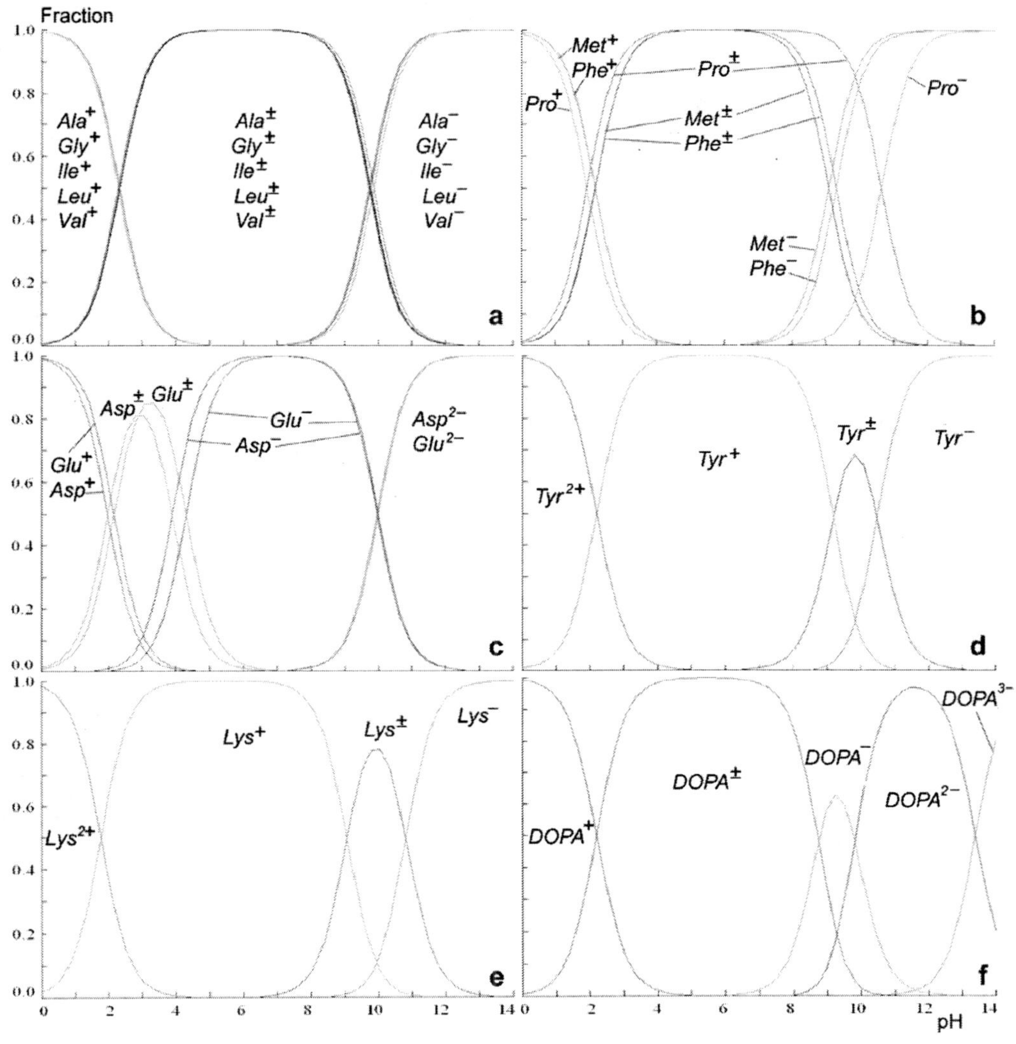


Figure 1 Ionization of amino acids in solutions. *pH* dependencies of ionic fractions: **a** - amino acids with aliphatic radicals (alanine, glycine, isoleucine, leucine, and valine); **b** - amino acids with aromatic radical (phenylalanine), sulfur-containing radical (methionine), and nitrogen in the cycle (proline); **c** - amino di-carboxylic acids (aspartic acid and glutamic acid); d - amino acid with phenol group (tyrosine); **e** - di-amino acid (lysine); **f** - dihydroxyphenylalanine (DOPA). One can see that there is no significant difference in ionization of amino acids that have aliphatic radicals while ionization of amino acids with more complicated radicals differs significantly

$$\overline{RH} + M^+ = \overline{RM} + H^+ \quad \text{or} \quad \overline{RH} + M^+ = \overline{R^{-+}M} + H^+ \qquad (1)^2$$

where $R$ represents a part of the cation exchanger bearing one functional group; $M^+$ is any single charged metal cation; bar denotes phase of the ion exchanger. In contrary, reactions in systems with bioorganic substances could be much more complicated. An example for

[2] The cation exchange reaction between hydrogen and a single charged cation is shown only as an example. Many similar examples can be written for cation exchange, anion exchange, chelating, and other interactions.

interactions between lysine ions and a strong cation exchanger is shown in Figure 2. The scheme is designed with suggestions that i) electrochemical behavior of the substance is similar in both phases: in aqueous solution and in internal solution of the functional polymer; ii) interactions of different ionic forms with the sorbent is independent on each other. The figure illustrates the possibility of recharging of the organic ion in both the solution (interactions **8-10**) and sorbent (interactions **1-3**) phases.[1]

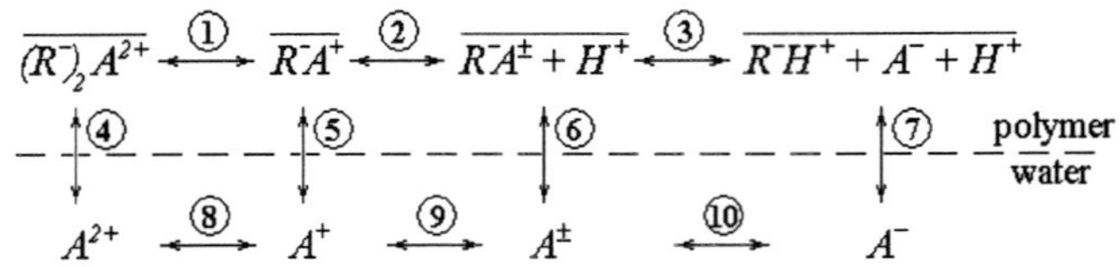


Figure 2 Interactions in the system including a strong cation exchanger and a di-amino acid (lysine can be named as an example)

Separation processes including biosynthesized substances are also affected by a complicated hydration of these objects.[2-8] This fact is defined by presence of both hydrophilic and hydrophobic parts in the same molecule. The hydrophilicity is provided by chargeable and polar groups such as *–COOH*, *-OH*, *-NH₃*, *=NH₂*, *≡N*, *=S*, etc. The hydrophobicity comes from hydrocarbon parts of molecules. Different ionic forms of the same substances are differently hydrated. As the result, even a recharging of the ion inside the polymer phase initiates a water transfer between the sorbent and solution.[9] Exchange of ions between solution of the targeted substance and functional polymer is inevitably followed by the water transfer

$$\overline{RH} + A^{+} + W \cdot \overline{H_2O} = \overline{RA} + H^{+} + W \cdot H_2O \qquad (2)$$[3]

where $A$ represents the organic ion. $W$ in equation (2) can be both positive and negative. Thermodynamic constant of such reaction includes a term reflecting the water transfer:

$$K_H^{A^+} = \frac{\overline{a}_{A^+} \cdot a_{H^+}}{\overline{a}_{H^+} \cdot a_{A^+}} \cdot \left( \frac{\overline{a}_W}{a_W} \right)^{\frac{dW}{dX_A}} = \frac{\overline{f}_{A^+}}{\overline{f}_{H^+}} \cdot \frac{\overline{C}_{A^+} \cdot a_{H^+}}{\overline{C}_{H^+} \cdot a_{A^+}} \cdot \left( \frac{\overline{a}_W}{a_W} \right)^{\frac{dW}{dX_A}} \qquad (3)$$

where $a$, $f$, and $C$ are activity, activity coefficient, and concentration, correspondingly. $W$ in equation (3) is the number of water molecules per one functional group of the ion exchanger; $a_W$ is activity of water molecules. Fraction of the ion $A^+$ in the sorbent is calculated as

$$X_A = \frac{\overline{C}_{A^+}}{\overline{C}_{A^+} + \overline{C}_{H^+}} \qquad (4)$$

Equations (2), (3), and (4) are written for the case of exchange between $H^+$ and a single changed organic cation. To obtain the thermodynamic constant (3), activities of all ions and water molecules in both phases shall be measured. However, so far, only first steps have been done to develop a direct measuring technique for activity determination in phase of swollen ion exchangers.[10] These methods are not ready yet to investigate systems that include complicated bio-produced molecules. Certain approaches and simplifications are used instead to estimate activities in the polymer phase. For example, if ions of the same charge present in the system (the case corresponding to equations (3) and (4)) equality of activity coefficients for both ions in the solid phase is suggested. The suggestion allows ignoring the ratio of activity coefficients in the right-hand instance of equation (3). This approach is widely used while being essentially incorrect. This incorrectness is because the activity coefficients of equally charged ions are equal only in diluted solutions while the total concentration of single charged ions in the swollen ion exchanger lays in the range between 0.5 M and 5 M (for most of commercial ion exchange resins). Even more assumptions shall be brought if the water transfer accomplishes the ion exchange ($dW \neq 0$)[11] that, as was said above, is the common case.

The term of equation (3) describing water transfer cannot be ignored. An illustration of its importance can be found in the difference between thermodynamic characteristics of cation exchange of α-amino acids bearing an additional polar or ionizable group. Figure 3[4] illustrates dependencies $\Delta H = f(\overline{X_A})$ and $\Delta S = f(\overline{X_A})$ for exchange of different amino acids on two polystyrene-divinylbenzene cation exchangers bearing sulfonic groups. The difference between these two materials is only in their macroscopic structure (KU-2 has a gel structure while KU-23 is macroporous) which affects the swelling and water transfer characteristics. The complicated shape of dependencies corresponding to KU-23 can be attributed to existing of two types of pores in this material. The ion exchange takes place on the surface of macropores as well as in the micropores between polymeric chains. While chemical properties of the differently located functional groups are the same, an absence of space restrictions in macropores allows formation of complicated hydration structures that is restricted in micropores (which are similar to pores of the gel exchanger). Figure 4 presents examples of aqua-complexes formed in the cation exchanger with and without steric restrictions. As shown by the right-hand scheme, water molecules in macroscopic pores form second hydration shells due to the hydrogen bond association that is not a common case in the restricted environment of gel sorbents (left-hand scheme). More examples can be presented for other amino acids. Table 1 summarizes data on hydration in such systems.

Ion exchange of the double-charged cation (interaction **4** of Figure 2) can be described by the reaction similar to reaction (2)

$$\overline{R^{-+}H} + \tfrac{1}{2}A^{2+} = \tfrac{1}{2}\overline{R_2A} + H^+ + W \cdot H_2O \qquad (5)^5$$

While this reaction represents a conventional ion exchange, its thermodynamic evaluation is even more difficult because of the power coefficients in the equation of the reaction constant

[3] Reaction (2) corresponds to the interaction **5** of Figure 2. It can be considered as the simplest case.
[4] These and following thermodynamic characteristics were calculated from equilibrium data.
[5] One of conventional ways to describe ion exchange reactions is to use equivalent scale instead of molar scale. This is the reason for use of fractional coefficients.

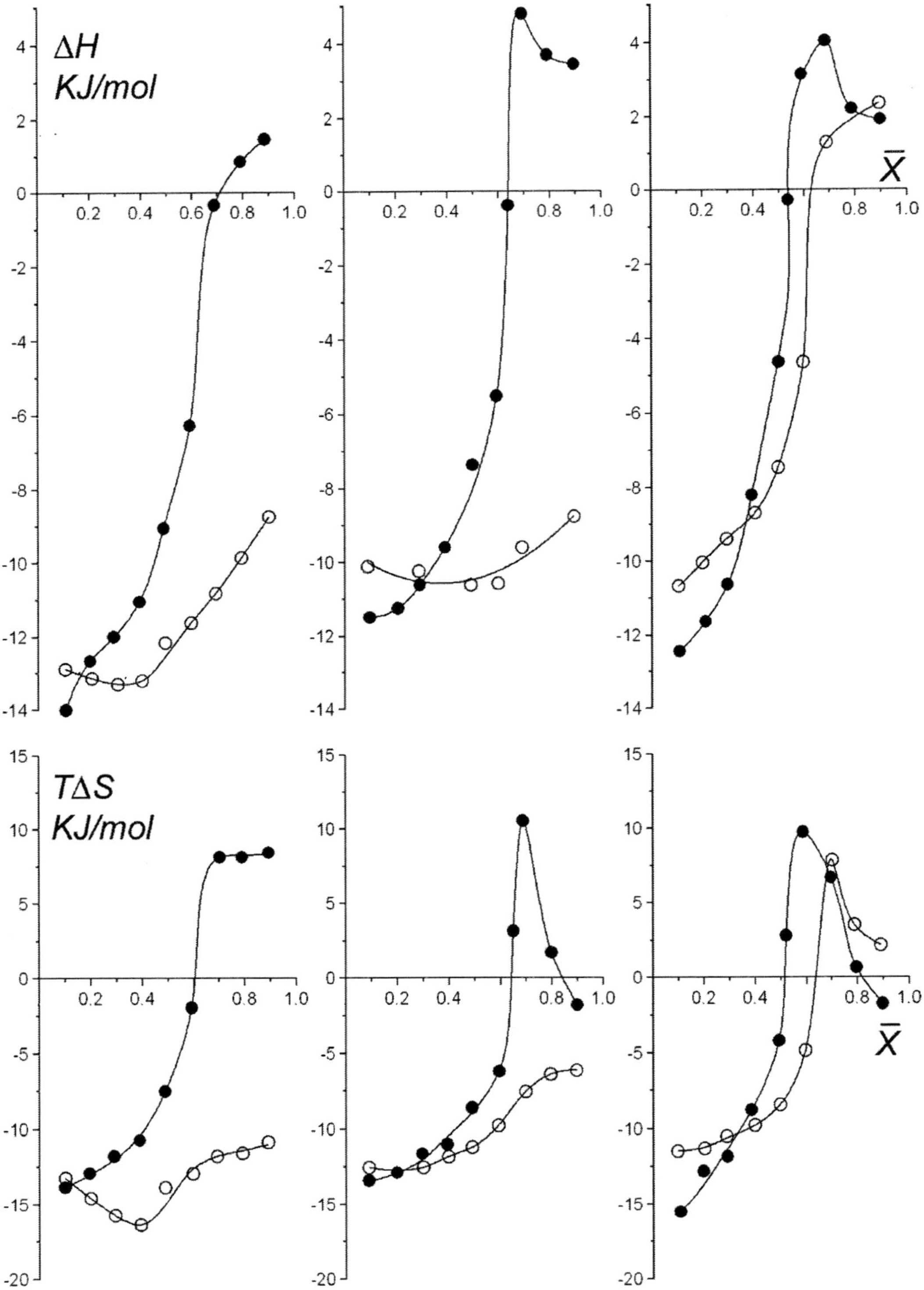


Figure 3 Thermodynamic functions of ion exchange between amino acids and hydrogen forms of gel cation exchanger KU-2-8 (open symbols) and macroporous cation exchanger KU-23 (30/100) (filled symbols): $Phe^+$ (left panels), $Tyr^+$ (central panels), and $Lys^{2+}$ (right panels)

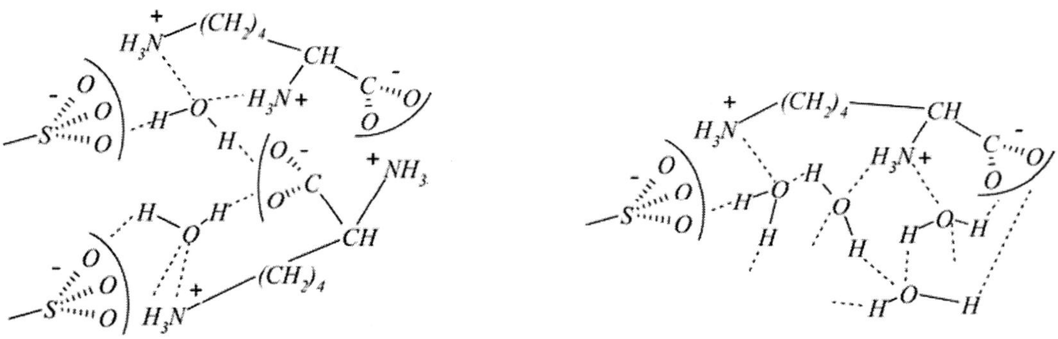

Figure 4 Hydration in the system single charged lysine ion - cation exchanger with (left) and without (right) steric restrictions. The schemes have been designed on base of IRS data for gel KU-2-8 and macroporous KU-23 (30/100) materials correspondingly. The angles between drown bonds are not representative

Table 1 Hydration numbers of amino acid cations loaded in strong cation exchangers. Two inorganic cations are presented for comparison. The table contains data obtained with different experimental methods. Reprinted from *Vestnik Voronezhskogo Gosudarstvennogo Universiteta*[12] © *V. F. Selemenev*

| **Ion** | **Isopiestic** | **DTA/DTG** | **NMR** | **IRS** | |
|---|---|---|---|---|---|
| | **Primary hydration shell** | | | | **Secondary hydration shell** |
| **Gel cation exchanger KU-2-8** | | | | | |
| $Glu^+$ | 2.8 | 4.4 | 3.8 | 4.3 | |
| $His^{2+}$ | 2.9 | | 4.1 | 3.9 | |
| $Lys^+$ | 2.4 | 3.6 | 3.4 | 3.6 | |
| $Lys^{2+}$ | 2.0 | 2.9 | 4.3 | 3.8 | |
| $Tyr^+$ | 3.8 | 4.0 | | 3.4 | |
| $H^+$ | 2.1 | 3.1 | 2.8 | 2.5 | |
| $Na^+$ | 1.8 | 2.3 | 3.1 | 2.0 | |
| **Macroporous cation exchanger KU-23** | | | | | |
| $Glu^+$ | 6.5 | 6.8 | | 8.4 | 21.0 |
| $Gly^+$ | | | | 5.6 | 16.0 |
| $His^+$ | | | | 6.0 | 19.0 |
| $His^{2+}$ | 8.0 | | | 6.6 | 13.5 |
| $Lys^{2+}$ | 5.8 | 4.2 | 6.9 | 6.2 | 18.6 |
| $Tyr^+$ | 6.5 | 6.9 | | 8.1 | 22.2 |
| $H^+$ | 4.2 | 4.6 | 6.2 | 5.8 | 14 |
| $Na^+$ | 3.0 | 3.4 | 4.9 | 5.0 | 12 |

$$K_{H^+}^{A^{2+}} = \frac{\left(\overline{f_{A^{2+}}}\right)^{1/2}}{\overline{f_{H^+}}} \cdot \frac{\overline{C}_{A^{2+}}^{1/2} \cdot a_{H^+}}{\overline{C}_{H^+} \cdot a_{A^{2+}}^{1/2}} \cdot \left(\frac{\overline{a}_W}{a_W}\right)^{\frac{dW}{dX_A}} \tag{6}$$

Thus, even the described above oversimplifying assumption of activity coefficients equality in the solid phase does not provide a ground for the straightforward procedure.

In contrary to conventional ion exchange, interactions between ion exchange groups and zwitterlites could be non-exchange while stoichiometric:

$$\overline{R^{-+}H} + H_3^+N - R'–COO^- = \overline{R^{-+}H_3N - R'–COO^{-+}H} \tag{7}$$

Reaction (7) corresponds to interaction **6** of Figure 2. Similar process takes place in anion exchange systems

$$\overline{R^{+-}OH} + {}^-OOC - R'–NH_3^+ = \overline{R^{+-}OOC - R'–NH_2} + H_2O \tag{8}$$

Please note that the absence of a bar over $H_2O$ in the right-hand side of equation (8) does not indicate a stoichiometric transfer of water molecules out of the exchanger phase.

Reactions (7) and (8) take place on hydrogen and hydroxide forms of ion exchangers correspondingly. Similar chemical equations can be written for salt forms (for example for $Na^+$ and $Cl^-$ forms) but these chemical reactions do not take place in any significant extent.[13]

The interaction **7** of Figure 2 is not a stoichiometric reaction. The co-ion $A^-$ can be sorbed by a cation exchanger only together with a neutralizing ion of the opposite charge. Such sorption is observed only in concentrated solutions (Donnan principle). However, due to the possibility for the ion recharging in the polymer phase (interaction **3**), some amount of the anion can be sorbed stoichiometrically together with the stoichiometric amount of the charge-compensating ion. The reaction combining interactions **7** and **3** of Figure 2 is

$$\overline{R^{-+}H} + A^- + H^+ = \overline{RAH} \qquad (9)^6$$

So far as the anionic forms of amino acids are more common for alkaline solutions, the following reaction shall be written instead of reaction (9)

$$\overline{R^{-+}H} + A^- + M^+ = \overline{RA^{-+}M} \tag{10}$$

The product of such interactions can be both non-ionized (reaction (9)) or ionized (reaction (10)) dependently on the associating ability of the charge-compensating ion. Accordingly to the experimental evidence, reaction (10) cannot convert the ion exchanger in

[6] The water transfer is not indicated in reaction (9) and later on for sake of simplicity.

the corresponding ionic form completely. Only a small fraction of groups participates due to the low thermodynamic benefit of the overall process.

The transfer of the co-ion is enhanced if this ion forms stable associates with ions of opposite charge that are present in the system. For example, accordingly to literature data,[14] *L*-lysine and *L*-ornithine form stable complexes with $Cu^{2+}$ under *pH* conditions where these amino acids exist as a mixture of the zwitterions and single charged anions. The most convenient representation of the complex formation is

$$Cu^{2+} + HA^{\pm} + A^{-} = CuA_2H^{+} \qquad (11)^7$$

The complexes formed by reaction (11) are rather strong with $\log\beta = 14.64$ for lysine and $\log\beta = 14.87$ for ornithine. A cation exchanger would readily sorb such complexes from alkaline solution

$$\overline{RNa} + CuA_2H^{+} = \overline{RCuA_2H} + Na^{+} \qquad (12)$$

rather than participate in reactions (9) or (10).

Formation of associates in the phase of polymer is another phenomenon specific for biochemical substances.[8,15-18] The following example shows formation of peptide-like associates between molecules of lysine in phase of a cation exchange polymer[15]

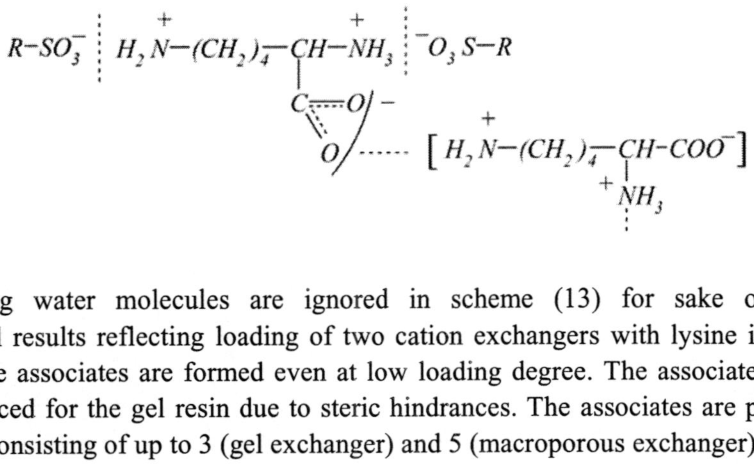

(13)

Hydrating water molecules are ignored in scheme (13) for sake of simplicity. Experimental results reflecting loading of two cation exchangers with lysine ion (Figure 5) show that the associates are formed even at low loading degree. The associate formation is less pronounced for the gel resin due to steric hindrances. The associates are polymolecular zwitterions consisting of up to 3 (gel exchanger) and 5 (macroporous exchanger) molecules of lysine. At low loading the lysine sorption is accompanied by the swelling reduction. However, at $\overline{X} \geq 0.6$ (gel) or $\overline{X} \geq 0.8$ (macroporous) the water content[8] is independent on the loading degree. It indicates that the exchanger phase contains only water molecules strongly associated in hydration shells. These shells can be assigned to $-SO_3^-$, $-COO^-$, and $-NH_3^+$ groups.

---

[7] The zwitterion is represented in equation (11) as $HA^{\pm}$ instead of the conventional $A^{\pm}$. It is done to emphasize the re-charging ability through the reaction $HA^{\pm} = H^{+} + A^{-}$.

[8] Water content in these and other experiments was determined by two methods: drying and DTG.

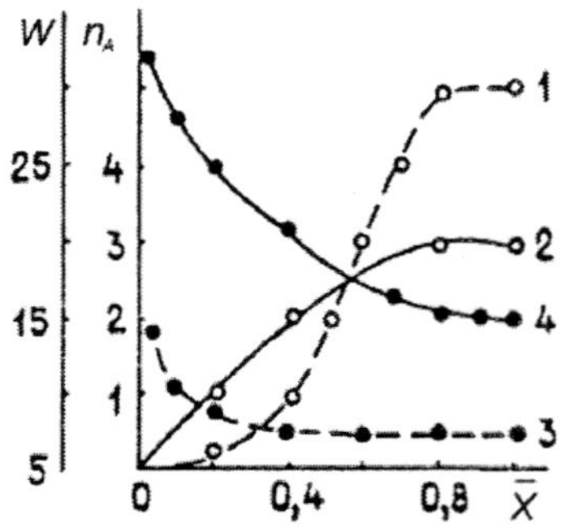


Figure 5 Composition of cation exchanger phase for gel KU-2-8 (curves 1 and 3) and macroporous KU-23 (30/100) (curves 2 and 4) cation exchangers differently loaded with lysine. Curves 1 and 2 show average number of lysine molecules in associates, $n_A$; curves 3 and 4 show number of water molecules per sulfuric group of the exchanger. Reprinted from *Vestnik Voronezhskogo Gosudarstvennogo Universiteta*[12] © *V. F. Selemenev*

In cases of low soluble substances (for example, tyrosine, tryptophan, or glutamic acid) the formed associates are so stable that they remain so in the solution after removal from phase of the polymer.[8,15,18-22] This phenomenon of ion exchange isothermal supersaturation takes place at *pH* equal or above the isoelectric point[9] for cation exchangers and equal or below the isoelectric point for anion exchangers. Breakthrough curves illustrating the column process are presented in Figure 6. The amino acid content at the column outlet achieves values few folds higher than the maximum solubility under these conditions: 3.5 folds for the anion exchanger and 2.1 folds for the cation exchanger. The obtained supersaturated solutions are stable up to several hours after losing the contact with the polymer. Examples of experimental results are presented in Table 2. One has to note that there is no any solid correlation between stability of the supersaturated solutions and achieved degree of supersaturation. The degree of supersaturation is related to the concentration achieved inside the resin under particular loading conditions and to rate of the amino acid elution while stability of the associates is attributed only to chemical properties of the particular amino acid. The *pH* range of the highest stability usually coincides with *pH* of the amino acid's side group ionization. For example, the most stable associates of tryptophan and tyrosine molecules can be expected in corresponding ranges $9.0 \le pH \le 13.0$ and $6.9 \le pH \le 9.5$. This is explained on base of IRS investigations[8,20] showing that stability of associates is defined by tunneling of proton between two groups of the same type as shown by the following schemes

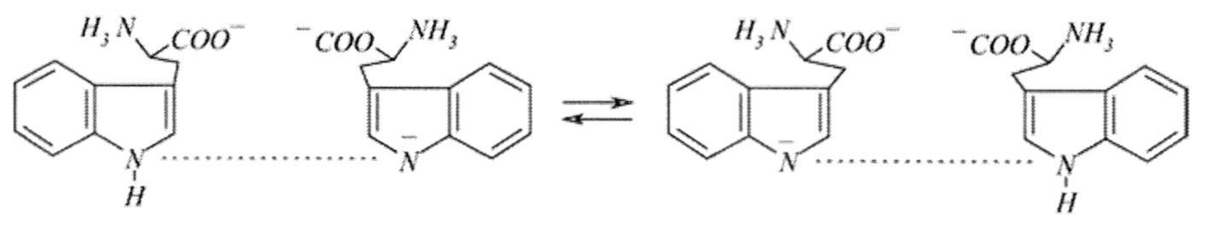


(14)

[9] Isoelectric point (*pI*) is the *pH* value of maximum concentration of the zwitterion.

Table 2 Results of experiments on ion exchange isothermal supersaturation. Initial forms of ion exchangers were $H^+$ and $OH^-$. Columns had diameter 1 cm. Initial height of the bed was 20 cm. Reprinted from *Vestnik Voronezhskogo Gosudarstvennogo Universiteta*[12] © *V. F. Selemenev*

| Amino acid | *pH* of initial solition | Overloading, mmol/g [a,b] | Water content, g/g [b] | $C/C_{max}$ in eluate | $\overline{C}/C_{max}$ [c] | Stability, min [d] |
|---|---|---|---|---|---|---|
| **KU-23 (30/100)** | | | | | | |
| *Glu* | 3.1 | 0.20 | 1.020 | 0.6 | 1.0 | - |
| | 4.3 | 0.32 | 0.826 | 2.1 | 6.3 | 10 |
| | 5.8 | 0.50 | 0.795 | 3.0 | 10.2 | 18 |
| | 12.2 | 0.54 | 0.734 | 4.2 | 20.2 | 25 |
| *Tyr* | 5.6 | 0.41 | 0.856 | 1.0 | 8.7 | - |
| | 8.1 | 1.03 | 0.820 | 1.8 | 114 | 35 |
| | 9.9 | 1.60 | 0.758 | 3.6 | 380 | 90 |
| | 11.0 | 1.06 | 0.859 | 2.1 | 19 | 50 |
| Trp | 5.9 | 0.40 | 0.902 | 0.4 | 3.0 | - |
| | 10.5 | 0.86 | 0.800 | 1.8 | 7.4 | 20 |
| | 11.3 | 2.02 | 0.706 | 3.0 | 19.5 | 32 |
| | 12.0 | 1.70 | 0.720 | 2.2 | 15.3 | 32 |
| *His* | 3.1 | 0.82 | 1.060 | 0.7 | 1.0 | - |
| | 5.9 | 1.86 | 0.830 | 1.7 | 7.0 | 40 |
| | 8.3 | 2.00 | 0.763 | 2.0 | 9.8 | 86 |
| | 9.9 | 2.76 | 0.732 | 2.3 | 14.0 | 90 |
| **AV-17-2P** | | | | | | |
| Glu | 0.9 | 0.65 | 0.630 | 2.1 | 11.6 | 40 |
| | 2.1 | 0.76 | 0.606 | 4.4 | 14.1 | 62 |
| | 3.1 | 0.26 | 0.814 | 1.6 | 3.2 | 30 |
| | 4.3 | 0.14 | 0.920 | 0.8 | 1.4 | - |
| *Tyr* | 0.9 | 1.59 | 0.526 | 3.5 | 106.0 | 80 |
| | 3.6 | 1.78 | 0.512 | 5.2 | 126.0 | 186 |
| | 5.3 | 0.76 | 0.690 | 1.4 | 9.6 | 65 |
| | 8.2 | 0.18 | 0.803 | 0.2 | 0.6 | - |
| *Trp* | 0.9 | 0.96 | 0.690 | 2.6 | 14.2 | 20 |
| | 1.9 | 1.14 | 0.668 | 2.3 | 12.0 | 35 |
| | 2.8 | 1.89 | 0.610 | 4.1 | 20.5 | 66 |
| | 5.8 | 0.52 | 0.724 | 0.3 | 4.5 | - |
| His | 1.9 | 1.35 | 0.560 | 1.8 | 5.9 | 100 |
| | 2.8 | 1.60 | 0.574 | 2.1 | 7.2 | 95 |
| | 3.6 | 2.06 | 0.592 | 2.8 | 10.8 | 118 |
| | 5.9 | 2.00 | 0.760 | 0.9 | 1.5 | - |

[a] Overloading was calculated as the total loading minus the total exchange capacity.

[b] The overloading and water content were calculated per dry weight of resins determined for $H^+$ or $Cl^-$ form.

[c] Concentration of the amino acid in the polymer phase ($\overline{C}$) was calculated per water content.

[d] Stability of the supersaturated solution was taken as the time between the elution moment and the moment when crystals of the amino acid first appeared.

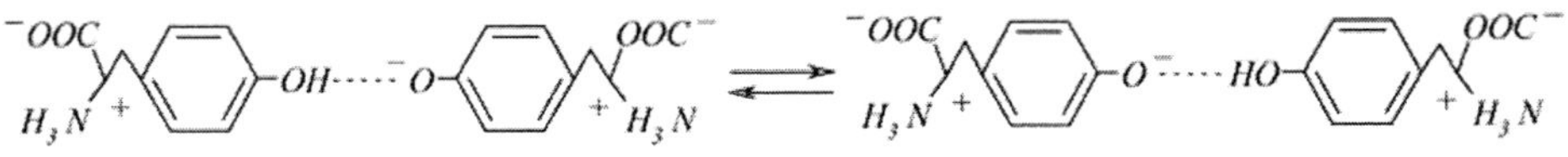
(15)[10]

Probability of such interactions correlates with reduction of the water amount in the exchanger phase (see Table 2). This can be explained by the increase of the internal solution concentration and, as the result, shorter average distances between the dissolved molecules.[11] Such hydrophobization can be related to an "extra crosslinking" of the polymer chains by the formed tunneling bonds.

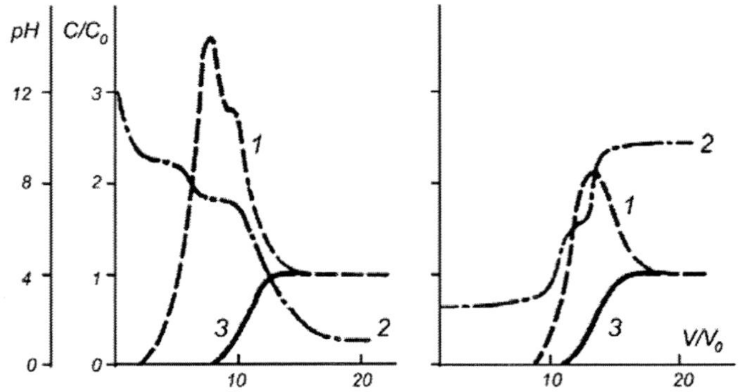


Figure 6 Breakthrough curve of $Tyr^{\pm}$ sorption on $OH^{-}$ form of anion exchanger AV-17-2P (left panel) and $H^{+}$ form of cation exchanger KU-23 (30/100) (right panel). 1 - concentration of tyrosine, 2 - *pH*, 3 - concentration of $Cl^{-}$ (left) or $Na^{+}$ (right). $C/C_0$ is the ratio between the concentration at the column outlet and the concentration of the solution supplied in the column; $V/V_0$ is the ratio between the volume of pumped solution and initial volume of the ion exchanger loaded in the column (column diameter was 1 cm, initial height of the bed was 20 cm). Reprinted from *Vestnik Voronezhskogo Gosudarstvennogo Universiteta*[12] © *V. F. Selemenev*

The phenomenon of ion exchange isothermal supersaturation can be successfully used for purification of corresponding substances. The discussed above Figure 6 shows breakthrough curves for treatment of a tyrosine solution with two different ion exchangers. These plots obviously resemble typical curves of frontal chromatography processes. The amino acid is displaced by $Cl^{-}$ (left panel) or $Na^{+}$ (right panel). The difference from conventional frontal chromatography can be found in chemical peculiarity of the process. First, the anion exchanger retains non-anionic species of the amino acid and the cation exchanger retains non-cationic species. The obtained product concentration is much higher than the maximum solubility. The solution is stabile for a time sufficient for its removal from the column, i.e. the column is not clogged by crystals of the amino acid that cold be expected for a sharp elution of a low soluble substance. The supersaturated solution is transferred in an appropriate

[10] The association mechanism represented by schemes (14) and (15) differs from scheme (13). In practice both mechanisms of the association take place. A domination of one of them is determined by properties of the amino acid and by the overall chemical system.

[11] One has to note that mobility of large organic molecules in internal solution of the polymer is significantly reduced in comparison with mobility of these substances in aqueous phase and even with mobility of inorganic ions inside the same polymer.

container where the precipitation process takes place without an additional effort. The technique is applicable for treatment of industrial solutions and allows an one step operation for the purification and extraction of crystals.

Table 3 Thermodynamic functions characterizing sorption of amino acids under conditions favourable for formation of supersaturated solutions. Reprinted from *Vestnik Voronezhskogo Gosudarstvennogo Universiteta*[12] © *V. F. Selemenev*

| **Ion exchanger** | **Ion** | ***pH*** | ***-ΔG*** | ***-ΔH*** | **TΔS** |
|---|---|---|---|---|---|
| KU-23 (30/100) | $Glu^{\pm}$ | $\geq pI$ | 17.3 | 0 | 17.3 |
| | $Tyr^{\pm}$ | $\geq pI$ | 17.8 | 0 | 17.8 |
| | $Trp^{-}$ | $\geq pI$ | 17.6 | 0 | 17.6 |
| | $His^{\pm}$ | $\geq pI$ | 16.0 | 0.3 | 15.7 |
| AV-17-2P | $Glu^{\pm}$ | $\leq pI$ | 81.8 | 49.6 | 32.2 |
| | $Tyr^{\pm}$ | $\leq pI$ | 79.1 | 47.0 | 32.1 |
| | $Trp^{-}$ | $\leq pI$ | 74.6 | 47.6 | 27.0 |
| | $His^{\pm}$ | $\leq pI$ | 72.2 | 44.1 | 28.1 |

Loading of an anion exchanger with zwitterions of an amino acid is characterized by significant values of $\Delta H$ (Table 3). Similar $\Delta H$ variations were not observed for cation exchangers. The high temperature sensitivity of interactions inside the anion exchanger allowed to develop reagent-free techniques for separation of amino acids from their mixtures. The driving force of the separation is altering of the temperature. The high value of $\Delta H$ indicated the high temperature dependency of the equilibrium constant

$$\left.\frac{d}{dT}\ln K\right|_P = \frac{\Delta H}{RT^2} \tag{16}$$

where $T$ is absolute temperature; subscript $P$ indicates a constant pressure. The difference between the conventional and dual-temperature separation approaches is illustrated by Figure 7. At the first step of the dual-temperature cycle the system is held at a first temperature. After loading of the column the temperature is changed without interrupting of the raw solution flow (second step). The solute s accumulated in the column is now expelled into the fluid due to the altering of the equilibrium constant (see equation (16)). As the result, the effluent has higher solute concentration for some time. After completion of the second step the temperature is changed back to the first value starting the new separation cycle.[12] Such dual-temperature ion exchange separations can be highly efficient if one or several involved elementary interactions are characterized by a high $\Delta H$.[23,24] Of course, the dual-temperature system cannot provide a complete separation of ions, however it allows

[12] The dual-temperature separation can be carried out by applying several techniques with both fixed and moving bed modes of operation. These techniques are not discussed within this publication. A reader can be sent to available literature on the described here straightforward separation (often referred as the temperature-swing mode),[23-25] parametric pumping,[26-30] moving bed dual-temperature separation,[31,32] and rocking wave technique.[33]

significant enrichment/depletion of the target ion concentration. The published data report on up to five order of magnitude difference between the concentrated and depleted solutions obtained with some dual-temperature techniques. An example of the straight-forward dual-temperature separation of two amino acids is presented in Figure 8. The figure illustrates pumping of a *Tyr/Phe* mixture through a column initially loaded with $OH^-$ form of AV-17-2P anion exchanger. The temperature-dependent concentration variations increase during first couple of cycles and remain repeatable during the following operation.

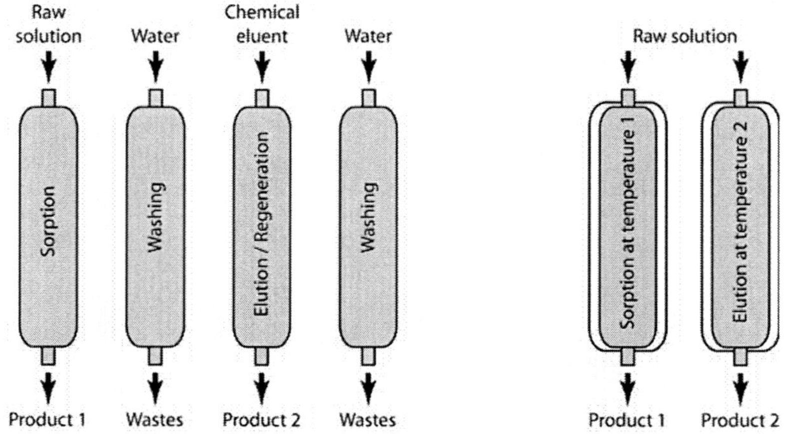


Figure 7 Schemes of conventional four-step (left) and dual-temperature two-step (right) column operational cycles. Simplest cases are presented. If two dissolved substances are separated with the dual-temperature technique, the stages could be assigned to: i) sorption of first substance with desorption of second substance; ii) sorption of second substance with desorption of first substance. As one can see, the dual-temperature scheme does not involve an addition of an external chemical that is inevitable in the conventional scheme (elution/regeneration step).

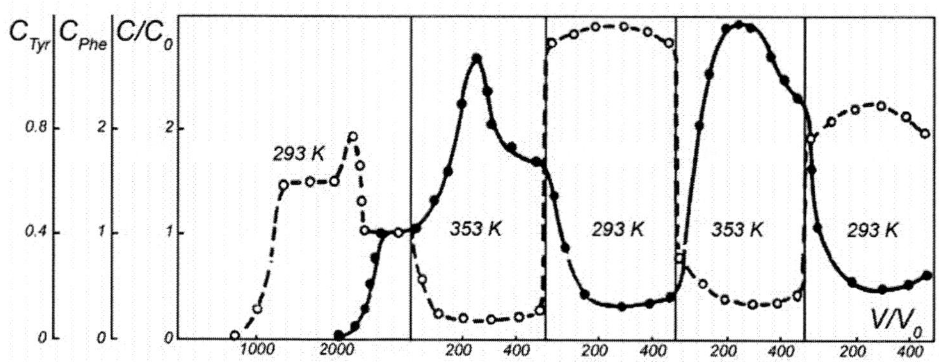


Figure 8 Separation of tyrosine (open symbols) and phenylalanine (filled symbols) by dual-temperature column operation. The column (diameter 1 cm) was initially loaded with $OH^-$ form of AV-17-2P strong anion exchanger (20 cm height of the bed), *pH* of the initial solution was 7. First (left) panel shows the initial loading. Other panels show two separation cycles. The following cyclic operations results in curves similar to curves presented for last two half-cycles of the figure. Reprinted from *Vestnik Voronezhskogo Gosudarstvennogo Universiteta*[12] © *V. F. Selemenev*

## Industrial Solutions

Independently on the particular biosynthesis procedure, the result is a solution containing a complicated mixture of the product with other inorganic and organic substances. The mixture can also contain insoluble particles: cells and their parts, particulate substrate, etc. A presence of useful byproducts justifies practical separation systems including several separation steps and allowing extraction of many components. A good example of such system is extraction of amino acids contained in molasses of sugar production.[13] The separation scheme is presented in Figure 9. The raw mixture contains sugar, all or almost all main amino acids, inorganic anions, dyes, colloids, and suspended particles. The treatment is performed after removal of a major part of solids and colloidal substances.[14] The efficient separation is achieved through a careful selection of polymeric sorbents. An important issue is *pH* levels specifically controlled for each ion exchange step. The scheme presented in Figure 9 does not include any *pH* adjustment steps, however, other similar technologies involve corresponding additions of chemicals.

The solution is pumped through the column ***1*** containing a non-exchange polymeric sorbent *Styrosorb* that is highly efficient for removal of organic dyes.[34,35] This step allows decolorizing of the solution and removal of high molecular weight organic impurities.

The following treatment with $Cl^-$ form of a strong anion exchange material (column ***2***) is performed at *pH* near isoelectric point of neutral amino acids. Glutamic acid and aspartic acid are anions while alkaline amino acids are cations. As has been shown above (see discussion following equation (10)), sorption of zwitterions and co-ions of amino acids by salt forms of ion exchangers does not take place due to low thermodynamic benefit of such processes. As the result the column retains *Glu* and *Asp* anions allowing almost free pass to neutral (zwitterions) and alkaline (cations) amino acids.

The next column contains $H^+$ form of a weak cation exchanger (column ***3***). This column retains alkaline amino acids and phenylalanine. Neutral amino acids are not sorbed through reaction (7) due to the high affinity of the weak cation exchanger to hydrogen ion.

Mixture of *Glu* and *Asp* is eluted from the column 2 by an acidic solution (conventionally *HCl*) and sent to the forth column containing $OH^-$ form of a strong anion exchanger. This step exploits the phenomenon of ion exchange isothermal supersaturation. The column concentrates these two amino acids simultaneously separating them from impurities of inorganic anions. The concentrated solution at the column output contains a mixture of both amino acids. However, concentration of glutamic acid two-three folds exceeds its solubility; this product precipitates as pure crystals.

---

[13] The sugar production cannot be named as belonging to biochemical industry. Nevertheless this example was selected due to its high representativity for separation of bio-related substances.

[14] The removal of solid constitutes is performed to avoid clogging of the packet bed columns. Expanded bed or fluidized bed techniques can be successfully applied instead to treat solid-containing mixtures and, hence, skip the removal of solids. However, these options mean reduction of the column efficiency. Actual column technique shall be selected considering peculiarity of each particular separation step.

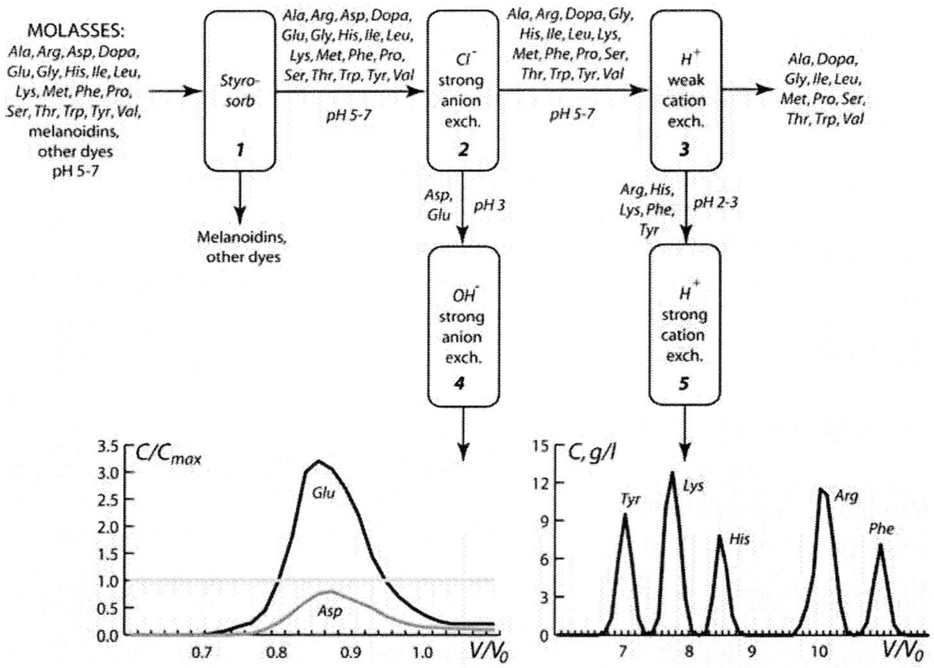


Figure 9 Scheme of amino acids extraction from sugar beet molasses (upper panel). The bottom panels show breakthrough curves for elution. The bottom-left panel shows the breakthrough curves of *Asp* and *Glu* elution from the column ***4*** accomplished with the supersaturation of glutamic acid. The bottom-right panel shows chromatographic elution from the column ***5***. $V/V_0$ is the ratio between the volume of pumped eluent and volume of resin bed in the column

The last column with $H^+$ form of a strong cation exchanger (column ***5***) gets an acidic mixture of five amino acids eluted from column ***3***: *Arg*, *His*, *Lys*, *Phe*, and *Tyr*. The process is performed in the way to achieve a chromatographic separation of these products. One has to note that the sequence of chromatographic elution does not reflect ionizing properties of these amino acids. Different retention of different substances could be explained by additional non-ionic interactions between these products and polystyrene-divinylbenzene matrix of the polymer.

In contrary to the described above process, microbiological production of amino acids usually targets obtaining of one concentrated product in the cultivation mixture. However, other valuable substances are also present in each broth. Even more, some amino acids can be obtained only as solutions with higher concentration of another substance of the same type. This requires application of separation schemes similar to one in Figure 9. Practical technologies for extraction of different amino acids from microbiological broth (with melanoidins are the main dye contaminants) are summarized in the following list.

**Lysine production.** The raw solution contains 30 g/l *Lys*, 5-6 g/l *Ala*, 1-2 g/l *Met*, and 1-2 g/l homoserine at *pH* 1-2. The treatment can be performed in one or two stages dependently on desired quality of the product.

The solution is pumped through a column containing a strong cation exchange resin in $NH_4^+$ form (for example, KU-2-8). Amount of the solution treated in one cycle , $V/V_0$, is 10-12 volumes per volume of the sorbent bed. The column traps lysine and major part of melanoidins. Due to a frontal chromatography process, a major part of alanine can be collected at the column outlet in form of a concentrated solution (up to 30 g/l in 1.5 $V/V_0$). Methionine and homoserine are not collected and disposed in mixture with other contaminants. 2 M $NH_4OH$ is used to elute the lysine and regenerate the sorbent. The eluate contains $Lys^-$ and $Lys^{\pm}$. Content of melanoidins is approximately 80 % of the initial. Lysine of this quality is used after crystallization and drying as an additive for cattle and poultry feed.

The second purification step produces lysine of quality required in pharmaceutics and food industry. The step conventionally consists of decolorizing with a strong macroporous anion exchanger, for example AV-17-2P, in $OH^-$ form. The sorbent treats 4-5 $V/V_0$ of the lysine solution obtained in the first purification step (*pH* 9-10). The frontal chromatography process at this step allows concentrating the lysine up to 340 g/l. The content of dyes is 20 % in comparison with the solution before the treatment with anion exchanger. The sorbent is regenerated with 1 M *NaOH*. A modern non-exchange sorbent Styrosorb can be used instead with even better efficiency of the dyes removal. Such column treats up to 50 $V/V_0$ of the solution removing up to 95 % of the dyes, however the concentrating of lysine does not happen.

**Glutamic acid production.** 30-35 g/l *Glu* raw solution contains additional amino acids: *Ala, Ser,* and *Thr* (5-7 g/l of each). The treatment is performed in one or two stages. An additional step can be performed to obtain serine as a secondary product.

Decolorizing of the raw solution (*pH* 5-6) with a Styrosorb column (50 $V/V_0$) is performed. The sorbent is regenerated with a solution containing 2 M *NaCl* and 0.1 M *NaOH.*

The decolorized solution is treated with $OH^-$ form of a strong macroporous anion exchanger. Concentration of amino acids is continuously monitored outside the column. The process of ion exchange isothermal supersaturation allows obtaining pure crystals of glutamic acid. The column is regenerated with 0.5 M *NaOH.*

Solution from the previous step containing a mixture of *Ala, Ser,* and *Thr* is treated with a weak anion exchanger in $OH^-$ form. The process allows obtaining supersaturated solution of *Ser* while two other amino acids are disposed.

**Serine production.** The raw solution contains 15 g/l of serine and 30 g/l of other smino acids (*Ala, Glu, Thr*, and *Val*).

Decolorizing of the raw solution (*pH* 5-6) with a Styrosorb column is performed. If 50 $V/V_0$ is treated, the removal of dyes is 80 %. The decolorizing efficiency can be increased up to 95 % reducing the treated volume down to 15 $V/V_0$. The regeneration of the sorbent is performed with mixed solution containing 2 M *NaCl* and 0.1 M *NaOH.*

The solution passing through the Styrosorb column is treated with $OH^-$ form of a strong macroporous anion exchanger after *pH* adjustment to 9-10 with *NaOH* (3-5 $V/V_0$). The process allows removal of *Glu* as a supersaturated (up to 52 g/l) solution. The following crystallization allows obtaining pure glutamic acid as the secondary product. The column is regenerated with 0.5 M *NaOH.*

The mixture of *Ser, Ala, Thr*, and *Val* that passes through the previous column is adjusted to *pH* 5-6 with *HCl* and pumped through a column containing a weak anion exchanger in $OH^-$ form (5-6 $V/V_0$). The supersaturated solution of *Ser* (72-75 g/l) is used to obtain pure crystals. Other two amino acids are disposed. The regenerating solution is 1 M *NaOH.*

**Proline production.** The raw solution contains 15-25 g/l *Pro*, 15-25 g/l *Leu* and *Val*, 2-5 g/l *Ala.*

The solution is decolorized with Styrosorb (50 $V/V_0$). The regenerating solution is 0.5 M *NaOH*.

The following treatment is performed with $OH^-$ form of a strong macroporous anion exchanger after *pH* adjustment to 6-7 with *NaOH*. Up to 8 $V/V_0$ is treated. The $OH^-$ form of the exchanger increases the solution *pH* up to 10. $Pro^{\pm}$ passes trough the column being concentrated up to 40 g/l while anions (at that *pH*) of *Ala*, *Val*, and partially *Leu* are trapped. Last fractions of the effluent contain purified leucine which can be extracted as the secondary product. The elution of leucine trapped by the column is performed with 0.5 M *HCl*. The obtained *Leu* solution has concentration 32 g/l. Other two amino acids are disposed at the regeneration. If need, the anion exchange treatment stage can be repeated twice.

**Phenylalanine production.** The raw solution contains 15-20 g/l *Phe* and 15-20 g/l total concentration of other amino acids: *Ala*, *Lys*, *Pro*, *Tyr* (5-7 g/l), and *Val*.

The solution is decolorized with Styrosorb (50 $V/V_0$). The regeneration is performed with 0.5 M *NaOH*.

3 $V/V_0$ of the solution with *pH* 5-6 is treated with $OH^-$ form of a strong macroporous anion exchanger after *pH* adjustment to 6-7 with *NaOH*. *Phe* and *Tyr* are trapped due to combined ion exchange and molecular sorption. Other three amino acids pass through the column and disposed. The mixture of *Phe* and *Tyr* is eluted from the column with 0.5 M *HCl*.

The mixture of two amino acids is adjusted to *pH* 6-7 and sent to another column containing the strong macroporous anion exchanger[15] for a dual-temperature separation. Two fractions collected at different temperatures are enriched with two different amino acids.

## Changes in Continuously Used Materials

So far as ion exchange materials are continuously used in the sorption/elution cycles, deterioration of these materials is of primary concern. Such deterioration can reduce efficiency of the process and contaminate the final product with outcomes of the polymer decomposition.

Physical deterioration is a mechanical destruction of the material beads causing a discharge of the exchanger from the reactor. The reason for physical deterioration is movements of the beads and swelling/shrinking of the material in each production cycle. This causes a need in regular additions of the fresh material in amounts up to 25 % per year (an estimation for packed bed columns). Chemical deterioration is more complicated. It comprises of several processes:[36]

Breaking of covalent bonds between the resin matrix and functional groups results in losses of the groups and reduction (or even total loss) of the ion exchange capacity.[37] Some processes alter ion exchange groups degrading them to non-exchange structures. This affects selectivity and other properties of the material.[36]

Some irreversible or semi-reversible sorbed species serve as extra cross-linking of the polymer. [38-40] It results in a decrease of the matrix flexibility, fragility of the beads, and reduction of permeability for ions. The physical deterioration is increased and kinetic properties are impaired.[41]

---

[15] When the process started first time $OH^-$ form of the anion exchanger is used.

Irreversible interactions between the sorbent and the sorbed organic species also result in the third kind of the chemical deterioration. Large organic species can physically block the functional groups altering the capacity.[38,39]

Among a wide diversity of applicable investigation methods vibration spectroscopy is the most convenient tool to reveal changes taking place inside the polymer phase.[1,36] Few discussed below examples are based on results of IR spectroscopic investigations.

**Alkaline degradation of amino-groups** can take place when anion exchangers are regenerated with hot alkali.[16] Such degradation of quaternary amines, $-N^+(CH_3)_3$, functional groups (strong anion exchangers) is described by the following reactions:[36]

$$R-CH_2-\overset{+}{N}(CH_3)_3OH^- \underset{NaOH}{\overset{T^o}{=}} R-CH_2-N(CH_3)_2 + CH_3OH \quad (17)$$

$$R-CH_2-N(CH_3)_2 + H_2O \underset{NaOH}{\overset{T^o}{=}} R-CH_2-NH-CH_3 + CH_3OH \quad (18)$$

$$R-CH_2-NH-CH_3 + H_2O \underset{NaOH}{\overset{T^o}{=}} R-CH_2-NH_2 + CH_3OH \quad (19)$$

The degraded material contains different amines: quaternary, ternary, secondary, and primary. This degradation can be seen from comparison of spectra presented in Figure 10. The spectral bands revealing this degradation are 885-900 $cm^{-1}$ ($-N^+(CH_3)_3$ group), 2538-2620 $cm^{-1}$ ($O$-$H$ stretching vibrations in $-N^+(CH_3)_3\cdots OH^-$), 975, 1668 $cm^{-1}$ (deformation vibrations of the same structure), and 1640 $cm^{-1}$ (bending vibrations of water associated to these ionic pairs). Newly appearing primary, secondary and ternary amines are reflected by new responses at 835, 1146, 1300, 1350, 3290, and 3300-3325 $cm^{-1}$.[36] Reactions (17)-(19) alter the selectivity while the total ion exchange capacity is not affected.

The alkaline regeneration affects polyethylene-polyamine ion exchangers even more (Figure 11). The alkali causes a sequence if reactions transforming original pyridine groups of this ion exchanger to pyridone:

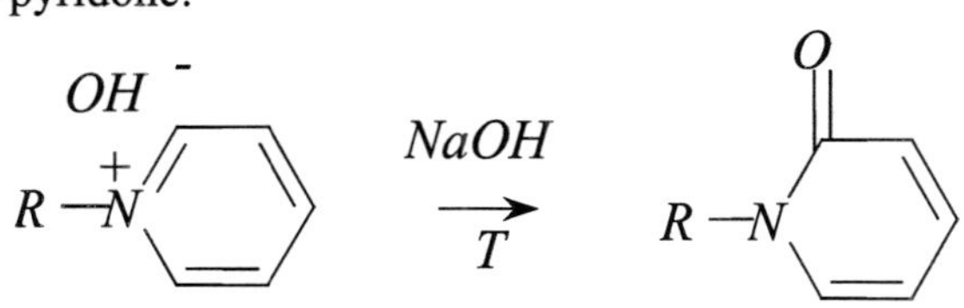


(20).

One of the most representative evidence of the degradation is the appearance of new spectral bands corresponding to the $C=O$ vibrations at 1702 $cm^{-1}$ and 906 $cm^{-1}$. Bands reflecting different vibrations of the pyridine ring disappear (857-876, 970-990, and 1479-1490 $cm^{-1}$), reduce the intensity (760-768 $cm^{-1}$), or change the frequency (1580-1586 $cm^{-1}$).[36] Few more spectral responses indicate these changes. For example, when

[16] The "hot" regeneration is applied to improve efficiency of the procedure and to benefit from the simultaneous removal of non-exchangeable sorbed large organic molecules (for example dyes).

alkali is removed from the material, pyridone formed by reaction (20) can associate with a proton. The charged group exists as two forms staying in equilibrium:[42,43]

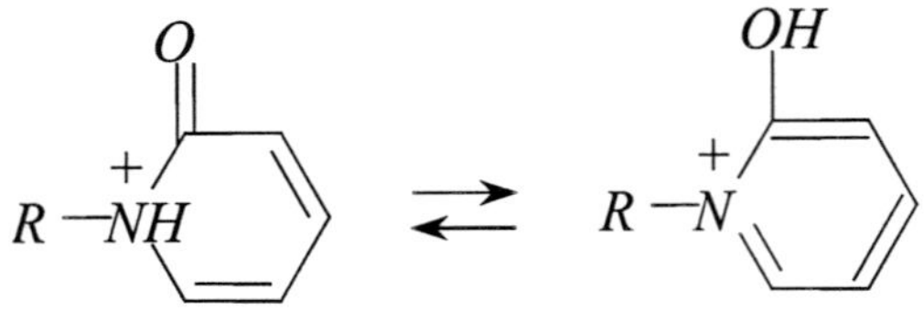

(21).

The absorption at 1310-1314 and 1345-1365 cm$^{-1}$ increases that reflects the presence of *OH*-groups.[36]

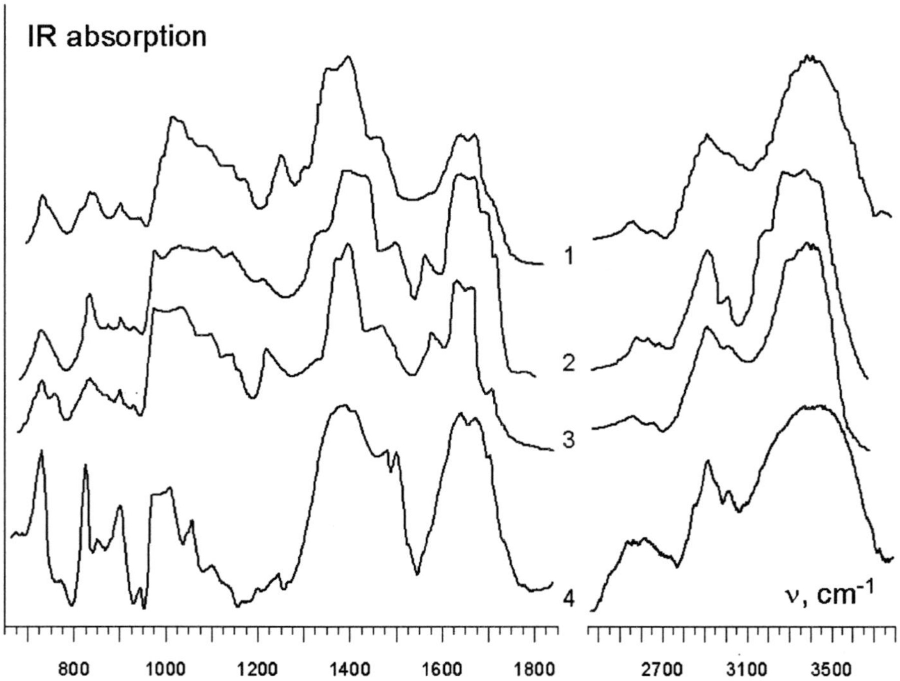


Figure 10 IR spectra of strong anion exchanger AV-17-2P: (1) exposed in 3 M *NaOH* during 900 h at 80 °C; (2) exposed in 60 % sucrose solution during 450 h at 70 °C; (3) used in sugar production (900 h of operation); (4) initial *OH*⁻ form. Reprinted from *Reactive and Functional Polymers*[36] with permission from *Elsevier* ©

Degradation of the anion exchanger functional groups was confirmed by chemical experiments. The results are presented in Figure 12 illustrating the altering of functional groups and of the ion exchange capacity.

Functional polymers exploited in biochemical productions are subjects for a dramatic chemical deterioration due to an irreversible sorption of organic dyes. The dyes bear different functional groups (*-COOH*, *-NH*$_2$, *-OH*, *-O-C-O-*, etc.) [44,45] which provide an ability for different types of interactions inside the sorbent phase: ion exchange, hydrogen binding, dipole-dipole and van der Waals interactions, etc. Hydrocarbon radicals of dyes participate in hydrophobic interactions with polymeric matrix. The chemical interactions are accomplished with physical effects: chain twisting, physical trapping of large organic molecules inside microcavities, diffusion difficulty, etc. Figure 13 illustrates degradation of a strong cation exchanger continuously exploited for extraction of lysine from microbiologically produced broth. The incomplete coinciding of $\overline{C}_D$ and $A_{UV}$ curves could indicate different accumulation rate for different melanoidins. Figure 14 shows IR spectra of this cation exchanger before and after the contact with dyes produced at microbiological synthesis of lysine. Sorption of dyes is indicated by new spectral bands at 1454, 1494, 1620-1630, 3416 cm$^{-1}$ (amino groups), 1407, 1727 cm$^{-1}$ (carboxylic groups), 1601 cm$^{-1}$ (amino and carboxylic groups), 3007 cm$^{-1}$ (water associated with $COOH$), 2840 and 2907 cm$^{-1}$ (charged amines and -$CH_2$-).[36] Amino- and carboxylic groups are not structure members of the fresh sulfuric cation exchanger, but they are an integral part of the melanoidins. [44,45]

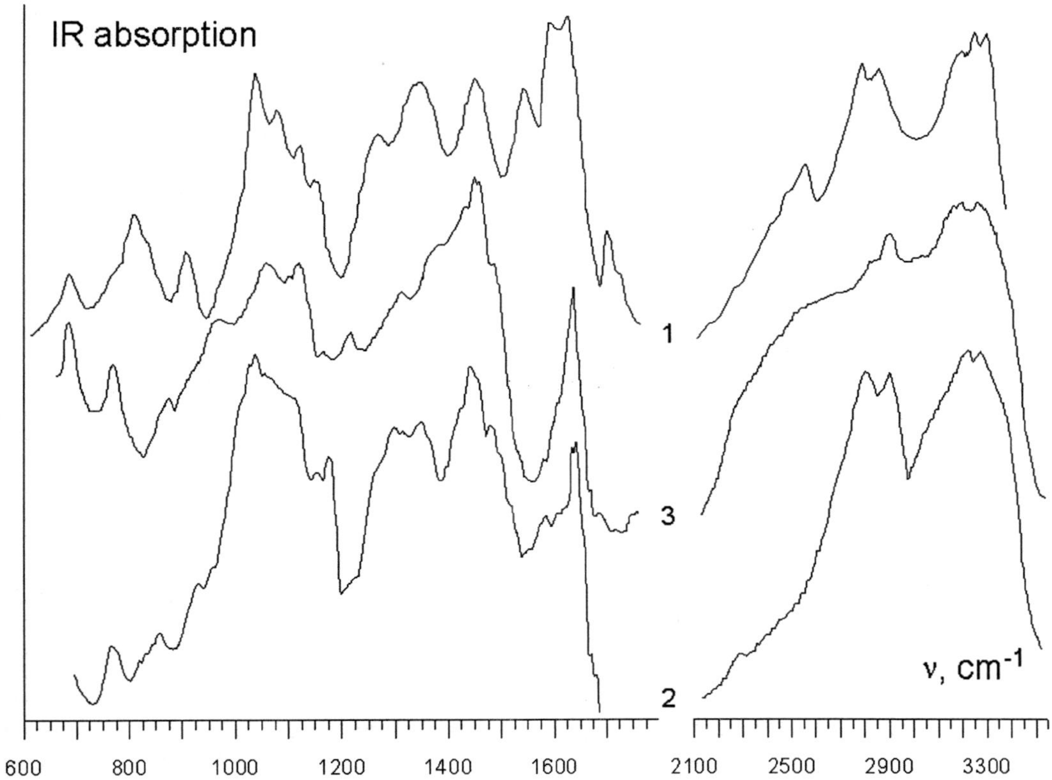


Figure 11 IR spectrum of strong polyethylene-polyamine ion exchanger AV-16G after an alkali treatment (1). The spectra of fresh $OH^-$ (2) and $Cl^-$ (3) forms are presented for comparison. Reprinted from *Reactive and Functional Polymers*[36] with permission from *Elsevier* ©

Treatment with $HCl$ partially regenerates the ion exchanger allowing a repeated use of this material. A continuous use of the exchanger results in the following alteration of the spectrum. The exchange capacity is reduced due to reduction in number of acting sulfuric groups. This allows for an increase of carboxylic groups dissociation. The accumulation is reflected in corresponding spectra by a dramatic increase of the maximum at 1405-1407 cm$^{-1}$ and by an appearance of a new strong maximum at 1372 cm$^{-1}$ (charged $COO^-$ groups). A reduction in the number of the protonated $NH_3^+$ (1447-1454 cm$^{-1}$, 1485-1494 cm$^{-1}$) and an increase in the number of the non-protonated $NH_2$ (1620-1633 $cm^{-1}$) amino groups can also be observed. Accumulation of dyes in the resin phase is followed by intra- and intermolecular interactions, namely by formation of internal salts ($-COO^- \; H_2\overset{+}{N}-$). This is reflected by a split of the maximum at 1601 $cm^{-1}$ into two bands at 1584-1587 and 1610-1614 $cm^{-1}$. The IR spectra do not contain any evidence of covalent bonds formation. Hereby, the dyes are fixed in the polymer through physical and weak chemical interactions. This conclusion was confirmed by regeneration experiments. No solid limit was found for the regeneration degree, however the 100 % regeneration was never achieved.[36]

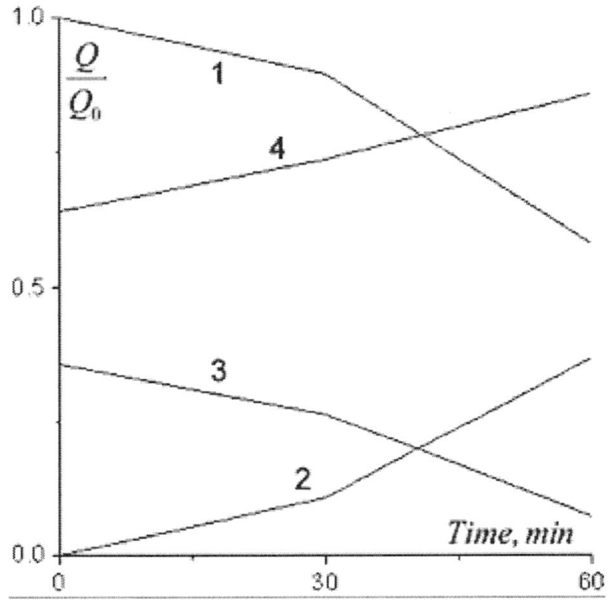


Figure 12 Ion exchange capacity of anion exchangers treated with 1.5 M *NaOH* at 60 °C. The values are expressed as part of the total ion exchange capacity of fresh materials ($Q_0$). **1**, **2** - AV-17-2P; **3**, **4** - AV-16G. **1**, **3** - strong alkaline groups; **3**, **4** - weak alkaline groups

The internal chemical environment of the polymer phase can be harsh enough to alter some organic substances. Even sucrose, for example, can be converted inside the phase of a strong anion exchanger ($OH^-$ form) at elevated temperature.[36] The hydroxyl form of the polymeric ion exchanger has the same effect on sucrose as soluble alkali which convert the sugar to dyes. Sucrose exposed to an alkali environment participates in a number of degrading reactions.[46,47] Compounds containing carboxyl, aldehyde, ketone, and hydroxyl groups are formed[48] as well as carboxylic acids, oxymethylfurfural, and polymeric dyes. A following exposure of these substances to the alkali medium results in formation of macromolecular dyes[48] that are irreversible sorbed through the described above mechanism.

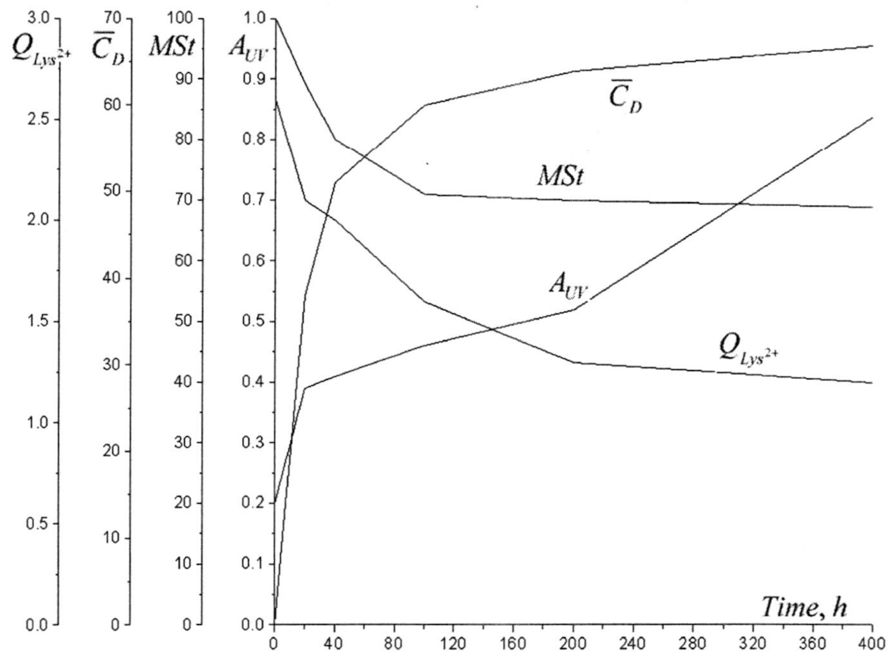


Figure 13 Degradation of a strong cation exchanger KU-2-8 exploited in lysine production. $Q_{Lys^{2+}}$ is total ion exchange capacity toward $Lys^{2+}$, mmol per gram of dry exchanger; $\overline{C}_D$ is amount of irreversibly sorbed dyes, mg per gram of dry exchanger; *MSt* is mechanical stability of beads expressed as a fraction of mechanical stability of the fresh ion exchanger; $A_{UV}$ characterizes coloring of the exchanger (UV absorption of 1 cm bed of the beads at 285 nm)

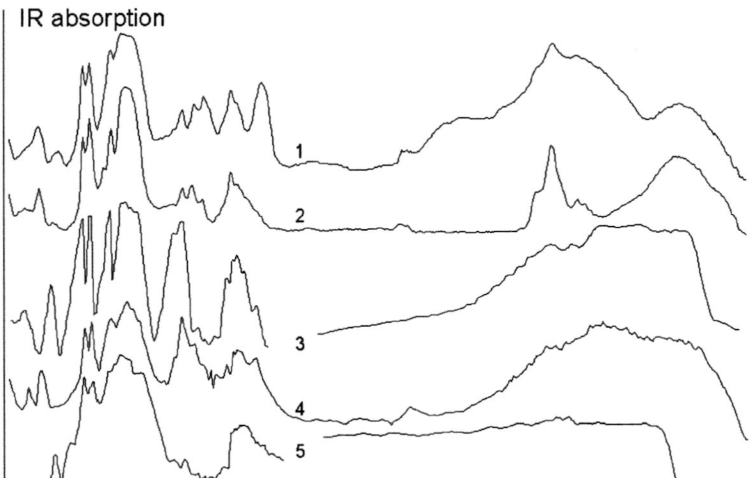


Figure 14 IR spectra of KU-2-8 interacted with lysine production dyes: (1) saturated with dyes (model system); (2) after regeneration stage (model system); (3) exposed to dye solution for 400 h; (4) used in 100 cycles in lysine purification; (5) initial $H^+$ form. Reprinted from *Reactive and Functional Polymers*[36] with permission from *Elsevier* ©

Ion exchangers exploited in biochemical separations suffer from combined influence of deteriorating factors. As the result, a periodical replacement of the sorbent is inevitable. However, deteriorated materials are not necessary subjects for disposal. Secondary reuse of them could be considered due to their complexing and weak ion exchange abilities.[49,50]

## Symbols Used

$A$ - organic ion;
$A_{UV}$ - UV absorption;
$a$ - activity of ion;
$a_W$ - activity of water molecules;
$C$ - concentration;
$C_0$ - concentration of solution before treatment;
$C_{max}$ - solubility;
$f$ - activity coefficient;
$K$ - thermodynamic constant of chemical reaction;
$M^+$ - any single charged metal cation;
$MSt$ - characteristic of mechanical stability;
$n_A$ - number of associated amino acid molecules;
$P$ - pressure;
$pI$ - isoelectric point;
$Q$ - ion exchange capacity;
$R$ - part of the exchanger bearing one functional group; gas constant;
$T$ - absolute temperature;
$V$ - volume;
$V_0$ - volume of ion exchanger loaded in column;
$W$ - number of water molecules;
$X$ - molar fraction;
$\Delta H$ - enthalpy change;
$\Delta S$ - entropy change.

## References


[1] Selemenev, V. F.; Zagorodni A. A. *Reactive and Functional Polymers* 1999, 39, 53.
[2] Selemenev, V.F.; Kotova, D. L.; Uglyanskaya, V. A.; Oros, G. Yu. *Russian Journal of Physical Chemistry* 1986, 60, 2269.
[3] Krysanova, T. A.; Kotova, D. L.; Selemenev, V. F.; Davydova, E. G. *Russian Journal of Physical Chemistry* 2004, 78, 1156.
[4] Zyablov, A. N.; Eliseeva, T. V.; Selemenev, V. F. *Russian Journal of Physical Chemistry* 2003, 77, 1959.
[5] Kotova, D. L.; Krysanova, T. A.; Selemenev, V. F.; Beilina, D. S. *Russian Journal of Physical Chemistry* 2001, 75, 1860.
[6] Zyablov, A. N.; Eliseeva, T. V.; Selemenev, V. F.; Samoilova, N. N. *Russian Journal of Physical Chemistry* 2001, 75, 473.

[7] Kotova, D. L.; Rozhnova, O. I.; Selemenev, V. F.; Peregudov, Y. S. *Russian Journal of Physical Chemistry* 2000, 74, 2071.

[8] Selemenev, V. F.; Zagorodni, A. A.; Uglyanskaya, V. A.; Zavyalova, T. A.; Chikin, G. A. *Russian Journal of Physical Chemistry* 1992, 66, 1555.

[9] Kotova, D. L.; Selemenev, V.F.; Kravchenko T.A. *Russian Journal of Physical Chemistry* 1997, 71, 1854.

[10] Molochnikov, L. S.; Kovalyova, E. G.; Grigorev, I. A.; Zagorodni, A. A. *Journal of Physical Chemistry B* 2004, 108, 1302.

[11] Biesuz, R.; Zagorodni, A. A.; Muhammed, M. *Journal of Physical Chemistry B* 2001, 105, 4721.

[12] Selemenev, V. F. *Vestnik Voronezhskogo Gosudarstvennogo Universiteta. Seriya* 2 *Estestvennye Nauki* 1996, #2, 151.

[13] Samsonov, G. V.; Kuznetsova, N. P. *Doklady Akademii Nauk SSSR* 1957, 115, 351.

[14] Conato, C.; Contino, A.; Maccarroneb, G.; Magri, A.; Remelli, M.; Tabbi, G. *Thermochimica Acta* 2000, 362, 13.

[15] Selemenev, V. F.; Stroiteleva, N. V.; Zagorodni, A. A.; Oros, G. Yu.; Chilin, G. A. *Izvestiya Vysshikh Uchebnykh Zavedenii, Pishchevaya Tekhnologiya* 1983, #5, 39.

[16] Selemenev, V. F.; Chikanov, V. N.; Frelikh, P.; Zagorodni, A. A.; Sholin, A. F. *Russian Journal of Physical Chemistry* 1990, 64, 3330.

[17] Selemenev, V. F.; Kotova, D. L.; Amelin, A. N.; Zagorodnii, A. A. *Russian Journal of Physical Chemistry* 1991, 65, 522.

[18] Muraviev, D. *Langmuir* 1998, 14, 4169.

[19] Muravev, D. N. *Russian Journal of Physical Chemistry* 1979, 53, 438.

[20] Selemenev, V. F.; Zagorodnii, A. A.; Polupanov N. V.; Ogneva, L. A. *Russian Journal of Physical Chemistry* 1986, 60, 871.

[21] Muraviev, D.; Khamizov, R.; Tikhonov, N. A. *Solvent Extraction and Ion Exchange* 1998, 16, 151.

[22] Selemenev, V. F.; Chikin, G.A.; Khokhlov, V. Ju. *Solvent Extraction and Ion Exchange* 1999, 17, 851.

[23] Zagorodni, A. A.; Muhammed, M. In: *Ion exchange developments and applications*; Greig, J. A.; Ed., Royal Chemical Society: Cambridge, UK, 1996, 446.

[24] Zagorodni, A. A.; Muraviev, D. N.; Muhammed, M. *Separation Science and Technology* 1997, 32, 413.

[25] Ivanov, V. A.; Timofeevskaja, V. D.; Gavlina, O. T.; Gorshkov, V. I. *Microporous and Mesoporous Materials* 2003, 65, 257.

[26] Wilhelm, R. H.; Rice, A. W.; Bendelius, R. A. *Industrial & Engineering Chemistry Fundamentals* 1966, 5, 141.

[27] Camero, A. A.; Sweed, N. H. *AIChE Journal* 1976, 22, 369.

[28] Rice R. *Separation and Purification Methods* 1976, 5, 139.

[29] Simon G.; Hanak L.; Grevillot G.; Szanya T.; Marton G. *Journal of Chromatography B-Biomedical Applications* 1995, 664, 17.

[30] Simon G.; Grevillot G.; Hanak L.; Szanya T.; Marton G. *Chemical Engineering Science* 1997, 52, 467.

[31] Gorshkov, V. I.; Kurbanov, A. M.; Apolonnik, N. B. *Russian Journal of Physical Chemistry* 1971, 45, 1686.

[32] Nikolaev, N. P.; Muraviev, D. N.; Muhammed, M. *Separation Science and Technology* 1997, 32, 849.
[33] Tikhonov, N. A.; Fokina, O.V.; Sokol'skii, D.A.; Khamizov, R. K. *Russian Chemical Bulletin* 1997, 46, 2053.
[34] Tsyurupa, M. P.; Davankov, V. A. Reactive & Functional Polymers 2002, 53, 193.
[35] Davankov, V. A.; Tsyurupa, M. P.; Tarabaeva, O. G.; Shacaeva, A. S. In: *Ion exchange developments and applications*; Greig, J. A.; Ed., Royal Chemical Society: Cambridge, UK, 1996, 209.
[36] Zagorodni, A. A.; Kotova, D. L.; Selemenev, V. F. *Reactive and Functional Polymers* 2002, 53, 157.
[37] Tulupov, P.E. *Stability of Ion Exchange Materials*, Khimia: Moscow, 1984.
[38] Bodamer, G.; Kunin, R. *Journal of Industrial and Engineering Chemistry* 1951, 43, 1082.
[39] Hale, D. K.; Packham, D. I.; Pepper, K. W. *Journal of the Chemical Society* 1953, 75, 844.
[40] Polanskiy, N. G. *Catalysis by Ion Exchangers*, Khimia: Moscow, 1973.
[41] Meleshko, V. P.; Shamritskaja, I. P.; Chikin, G. A.; Selemenev, V. F.; Zavjalova, T. A. *Teoriya i Praktika Sorbtsionnyh Protsessov* 1971, 5, 30.
[42] Dneprovski, A. C.; Temnikova, T. I. *Theoretical Base of organic transformations*, Khimia: Leningrad, 1979.
[43] Matsuda, Y.; Ebata, T.; Mikami, N. *Journal of Physical Chemistry* 1999, 110, 8397 (and references therein).
[44] Meleshko, V. P.; Shamritskaja, I. P.; Selemenev, V. F.; Chikin, G. A. *Sakharnaya Promyshlennost* 1975, #1, 23.
[45] Sapronov, A. P.; Kolcheva, P. A. *Dyes and Their Influence on Sugar Quality*, Pischevaja Promyshlennost': Moscow, 1975.
[46] Lange, S.; Pieck, R.; Rens, G.; *Londres, Compte-rendu de la X Assemble Commision Internationale Technique de sucrerie*, 1957, 10, 89.
[47] Prey, V.; Andres, H. *Zeitschrift fuer die Zuckerindustrie* 1971, #6, 267.
[48] Kochetkov, N. K.; Bochkov, A. F.; Dmitriev, E. A.; Usov, A. I.; Chizhov, O. S.; Shibaev, V. N. *Chemistry of Carbohydrates*, Khimiya: Moscow, 1967.
[49] Selemenev, V. F.; Chikin, G. A.; Halabuzar, V. G.; Kovaleva, T. A.; Belogortsev, Yu. A.; Zagorodni, A. A.; Luzhkov, A. M.; Grigorev, E. F. Pat. USSR 1163639, 1985.
[50] Selemenev, V. F.; Chikin, G. A.; Vikulina, G. L.; Obraztsov, A. A.; Sidorova, E. V. Pat. USSR 1627539, 1991.

In: Frontiers in Polymer Research
Editor: Robert K. Bregg, pp. 133-153
ISBN: 1-59454-824-2


*Chapter 5*

# NEW DEVELOPMENTS IN CATIONIC PHOTOPOLYMERIZATION: PROCESS AND PROPERTIES

***Marco Sangermano[1], Roberta Bongiovanni, Giulio Malucelli and Aldo Priola***
Dipartimento di Scienza dei Materiali e Ingegneria Chimica Politecnico di Torino, C.so Duca degli Abruzzi 24, 10129 Torino Italy

## Abstract

Cationic photopolymerization of vinyl ether, epoxy and oxetane systems has been reported. A structure-properties relationship by varying different additives in the cationic photocurable formulations has been investigated. The presence of hydroxyl containing compounds in the photocurable formulation induced an increase on rate of polymerization and final conversion. The use of fluorine containing monomers as additives to UV curable systems allowed the modification of the surface properties of the UV-cured films obtaining a high hydrophobic surface while the bulk properties remained unchanged. A wide range of hyperbranched polymers (HBP) was also investigated as additives in cationic photopolymerization of epoxy systems. The HBP were inserted into the polymeric network either by a copolymerization or through a chain transfer reaction involving the hydroxyl groups. Notwithstanding the low commercial availability of photocurable monomers, it has been shown that is possible to modulate and tailor the final properties of the UV-cured films by varying properly the additive in the system controlling the mechanism of the curing process.

## 1 Introduction

The UV-induced polymerization of multifunctional monomers has found a large number of applications in various industrial fields [1], mainly in the production of films, inks and coatings on a variety of substrate, including paper, metal and wood. Moreover varieties of

[1] E-mail address: marco.sangermano@polito.it

high-tech, such as the coating of optical fibers and the fabrication of printed circuit boards, have been developed.

This technology allows obtaining a quick transformation of a liquid monomer into a solid polymer having tailored physical-chemical and mechanical properties. The main advantage on using UV-radiation, to initiate the chain reaction, lies in the very high polymerization rates that can be reached under intense illumination, so that the liquid to solid change takes place within a fraction of a second [2]. Besides, the solvent free formulation makes UV-curing an environmental friendly technique. Another distinct feature of light induced reaction is that polymerization will occur only in the illuminated areas, thus allowing complex relief patterns to be produced after solvent developments [3].

In the UV curing process, radical or cationic species are generated by the interaction of the UV light with a suitable photoinitiator. Concerning cationic photopolymerization process, onium salt (iodonium I, or sulfonium II) represents the most widely used class of photoinitiators [4]. Such properties like thermal stability and inactivity towards polymerizable monomers at ambient temperature, made these salts very attractive in photocurable formulations [5].

| $Ph_2I^+X^-$ | $PH_3S^+X^-$ |
|---|---|
| iodonium salt | sulfonium salt |
| **(I)** | **(II)** |

The cationic portion of the salt is the light-absorbing component. For this reason the structure of the cationic controls the UV-adsorption characteristics: the photosensitivity, quantum yield, whether the compound can be photosensitized and the ultimate thermal stability of the salt.

However is the nature of the anion that determines the strength of the acid formed during photolysis and its corresponding initation efficiency. The nature of the anion also determines the character of the propagating ion pair. This has a direct impact on the kinetics of polymerization and whether terminations can occur [6].

Thus, onium salts may be viewed as photoacid generators. Anion such as $BF_4^-$, $PF_6^-$, $AsF_6^-$ and $SbF_6^-$ are most useful and, under UV radiation, they generate "superacids" with Hammet acidities ranging respectively from – 15 to – 30 [7].

The larger the negatively charged anion the more loosely it is bound and the more active the propagating cationic species is in the polymerization. The order of reactivity is the following: $SbF_6^- > AsF_6^- > PF_6^- > BF_4^-$ [4].

The mechanism of photodecomposition is quite complex and a simplified schematic representation is here reported for a diaryliodonium salt.

$$Ar_2I^+ MtX_n^- \xrightarrow{h\nu} \left[ Ar_2I^+ MtX_n \right]^1 \longrightarrow \left\{ \begin{array}{c} ArI^{\overset{+}{\bullet}} MtX_n^- + Ar^{\bullet} \\ \\ ArI + Ar^+ MtX_n^- \end{array} \right\} \Longrightarrow HMtX_n$$

The mechanism involves first the photoexcitation of the diaryliodonium salt and then the decay of the resulting excited singlet state with both heterolytic and homolytic cleavages;

cations and radicals fragments are formed simultaneously. The aryl cations and aryliodine cation radicals generated are very reactive species and can react further with monomers to give the protonic acid, which is the actual initiator of cationic polymerization [8,9].

The cationic photoinduce process present some advantages compared to the radical one [10] in particular: lack of inhibition by oxygen, low shrinkage and good adhesion and mechanical properties of the UV-cured materials. Moreover, the monomers employed are generally characterized by being less toxic and irritant with respect to acrylates and methacrylates, largely employed in radical photopolymerization.

Different types of monomers and oligomers have been proposed and reported in literature for cationic process mainly: epoxides [11,12], vinyl ethers [13-15] and propenyl ethers [16-18]. More recently attention has been devoted to the synthesis and cationic photopolymerization of oxetane monomers [19-21]. The high ring strain, very similar to that of epoxides, and the higher basicity of the heterocyclic oxygen in oxetane than that for oxirane oxygen in epoxides, make oxetanes interesting alternative monomers in cationic UV-curing applications [22]. In general, cationic photopolymerization allow obtaining a large number of polymers containing heteroatoms in their backbone.

In this work an overview on the research studies accomplished by our group, in the field of cationic photopolymerization, is reported.

Our recent work in this area can be divided into three main lines:

1. Investigation of the effect of the hydroxyl containing additives on rate of polymerization and final properties of UV cured films.
2. Investigation of the effect of fluorine-containing additives on surface properties of UV cured films.
3. Investigation of the presence of hyperbranched polymers on the photocure process and on the properties of UV cured films.

# 2 Experimentals

## 2.1 Materials

The following resins were employed and their structures are reported in Table 1: Triethylenglycol divinyl ether (from ISP, London, DVE3) Diethylenglycoldivinyl ether (from Aldrich, DVE2) and Bis-(4-vinyl oxy butyl) isopthalate (Vectomer® 4010, from Allied Signal Inc., USA IVE), were employed as vinyl ether monomers. 3,4-Epoxycyclohexylmethyl-3',4'-epoxycyclohexyl carboxylate, (*UVR 6110* from Dow, CE) was used as dicycloaliphatic epoxy resin. Bis[1-ethyl(3-oxetanyl)]methyl ether (OXT-221, from Toagosei, Japan, DOX) was used as an oxetane resin.

Table 1: chemical structure of photocurable resins

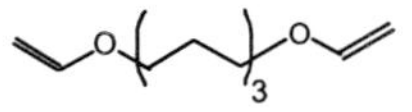
DVE3

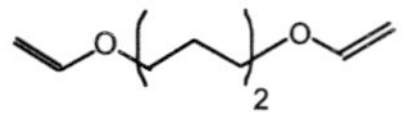
DVE2

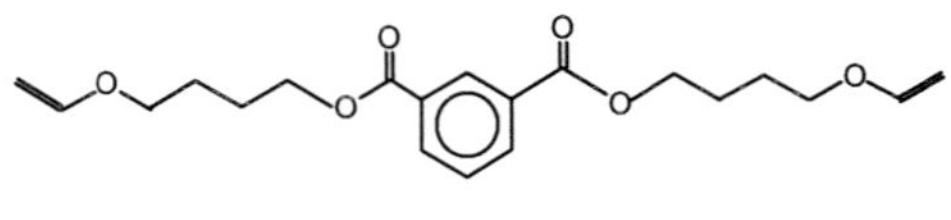
IVE

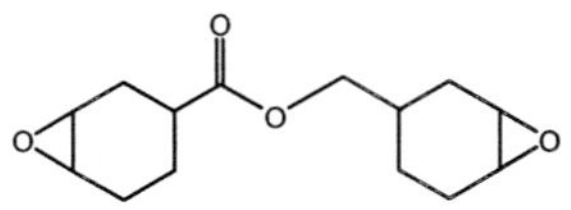
CE

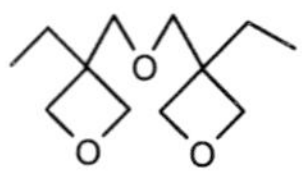
DOX

## 2.2 Additives

The different additives were purchased from Aldrich: 1-Butanol, 2-Phenyl-2-propanol, Diethylenglycol mono vinyl ether (HDVE2), Hydroxyl functionalized polybutadiene ($M_n$ = 1200; 55% 1,4 trans; 20% 1,4 cis; 25% 1,2; hydroxyl number 1.7 meq/g, PBOH).

## 2.3 Film Preparation and Characterization

The photopolymerizable formulations were prepared by adding to the monomers, or mixtures containing different additives, 2wt% of cationic photoinitiator (Triphenylsulfonium hexafluoroantimonate, Cyracure UVI6974, as 50% solution in propylene carbonate, from Dow). The films were obtained by coating the mixtures on different substrate and the curing reaction was performed, in air, by using a Fusion lamp (H-bulb), with radiation intensity on the surface of the sample of 280 mW/cm$^2$, and a belt speed of 6 m/min.

The kinetics of the photopolymerization was determined by FT-IR spectroscopy. A KBr disk was coated with the photocurable mixtures and a Genesis Series ATI Mattson (USA) Spectrometer was used for recording of FT-IR spectra. The decrease in the absorbance due to epoxy groups in the region 760-780 cm$^{-1}$ and vinyl groups (1620 cm$^{-1}$) or oxetane groups (980 cm$^{-1}$) was monitored as a function of irradiation time. Three to five runs were performed for each experiment and the results averaged. Variation in the experimental conditions (light intensity, humidity, temperature) caused slight differences in the kinetic curves. For this reasons all the kinetic curves contained in a Figure were performed on the same day and under the same conditions, thus good reproducibility was obtained.

The Gel Content of the films was determined by measuring the weight loss after 24 hours extraction at room temperature with chloroform.

The density of the photocured films was measured by means of a pycnometer. The solvent employed was water; 3 to 5 measurements were performed on small pieces of polymers at room temperature and the measured values were averaged.

Impact resistance measurements were performed on a dog-bone shape specimen, with a ATS FAAR Charpy pendulum, according to the ASTM D 256 method.

DSC measurements were performed using a Mettler DSC30 (Switzerland) instrument, equipped with a low temperature probe.

Dynamic mechanical thermal analyses (DMTA) were performed with a Rheometric Scientific MKIII (UK) instrument, at a frequency of 1 Hz in the tensile configuration.

The thermal stability of the photocured films was assessed by thermogravimetric analysis (TGA Leco TGA/601). Experiments were performed in air on samples of about 10 mg at a heating rate of 2 °C/min from 25 to 600 °C.

All the film characterizations were performed after storage the films for 24 hours and than a treatment in ammonia saturated atmosphere, in order to stop the acidic species.

# 3 Results and Discussion

## 3.1 Hydroxyl Additives

In a network forming polymerization, the growing chain end becomes fixed to the immobile three-dimensional network. When chain transfer occurs, new chain carrying species are formed which are not bound to the network and are, therefore, highly mobile. Higher polymerization rates result from removal of the restriction of impeded diffusion as well as from a decrease in the average crosslink density due to the chain transfer reaction. The use of alcohols as a chain transfer agents was first indicated by Penczek and Kubisa [23] then confirmed by Crivello [24] in cationic photopolymerization process.

We started our investigations studying the effect of hydrogen donor additives on the curing kinetics and on the properties of cure vinyl ether systems [25]; for this purpose different alcoholic additives were added to DVE3 and DVE2.

In Figure 1 the kinetic curves of pure DVE3 and in the presence of 5 wt% of n-butanol or 5 wt% of 2-phenyl-2-propanol are reported.

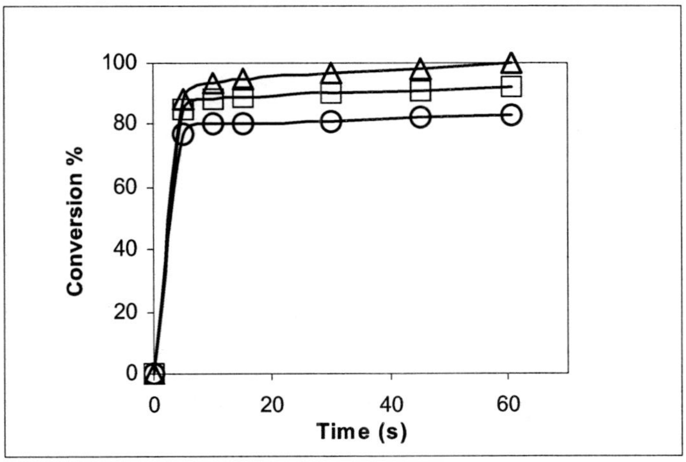


Figure 1: FT-IR conversion curves for DVE3 (O) and in the presence of 5 wt% of n-butanol (□), 2-phenyl-2-propanol(△)

It is possible to observe an increase of rate of polymerization and final double bond conversion in the presence of the alcoholic additives. A complete conversion is reached when the tertiary alcohol is added.

It is possible to explain these results taking into account that OH groups can interact with the growing carbocationic chain through a chain transfer mechanism, as reported in Scheme 1.

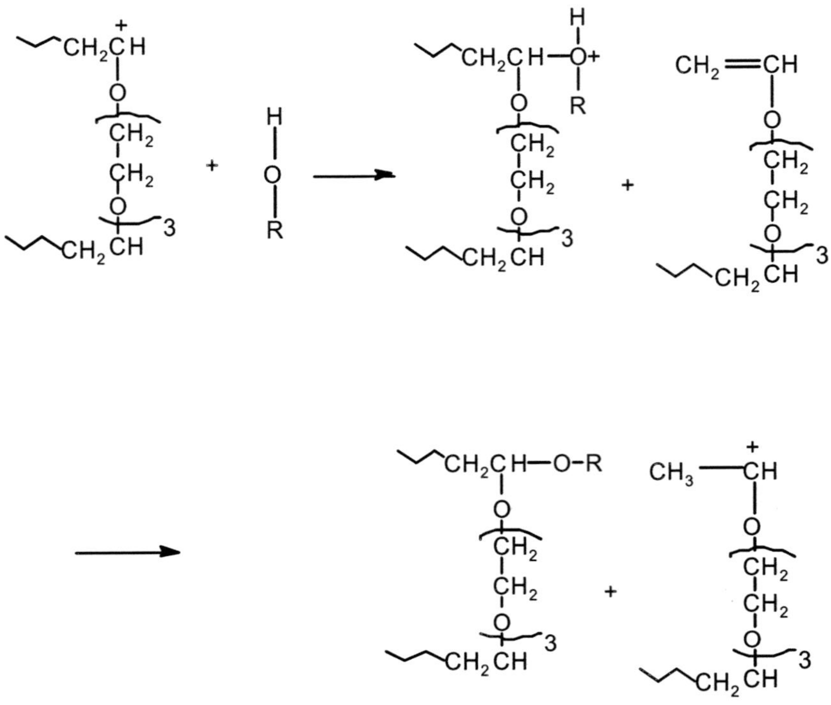

Scheme 1

The schematized reaction shows that during the chain transfer mechanism the growing chain is terminated by reaction of the cation with the hydroxyl group and ether linkage is formed. The occurrence of this mechanism was evidenced by FT-IR analysis showing a decrease of the OH band at 3600 $cm^{-1}$ and a strong increase of the ether band at 1100 $cm^{-1}$ after UV irradiation [26].These results are confirmed by NMR analysis [27].

The addition of a *monofer* as HDVE2, containing both a hydroxyl group and a vinyl ether copolymerizable double bond, was also investigated and similar results were obtained, due to the chain-transfer reaction.

As a result of the chain transfer mechanism the network structure results more flexible. This flexibilization is reflected by the lowering of the glass transition temperature of the cured films, as reported in Table 2. It is particularly interesting to observe that, at the same OH group concentration, the Tg values are higher in the presence of HDVE2 compared to the systems containing alcohols. These results can be attributed to the photopolymerization of the HDVE2 double bonds, which increases the crosslinking density of the network [28].

After vinly ether systems,we pursuing our research investigating the effect of hydroxyl containing additives on photopolymerization of epoxy systems.

The kinetic of photopolymerization of a typical epoxy monomer (CE) is reported in Figure 2. Notwithstanding its high reactivity CE resin gives rise to photocured films containing around 55% of unreacted epoxy groups after one minute of irradiation.

Polymerization proceeds quite rapidly in the first step, then slows down markedly because of the vitrification effect (Tg of cured film= 190 °C): as a consequence a large amount of epoxy groups remain trapped into the glassy polymer network because of the diffusion prevention.

The addition in the mixture of PBOH in different amount ranging between 10 to 50 wt%, induces a clear increase of the curing rate and on the epoxy group final conversion [29].

Table 2: Properties of UV cured films of vinyl ether systems

| **Sample** | **Conversion % FT-IR** | **Gel Content %** | **Tg [°C] DSC** |
|---|---|---|---|
| DVE3 | 82 | 98 | 37 |
| DVE2 | 83 | 97 | 42 |
| DVE3/n-Butanol | | | |
| 95:5 mol/mol | 95 | 95 | 22 |
| 90:10 mol/mol | 97 | 95 | 14 |
| DVE3/2-Phenyl-2-propanol | | | |
| 95:5 mol/mol | 97 | 95 | 19 |
| 90:10 mol/mol | 100 | 95 | 15 |
| DVE3/HDVE2 | | | |
| 95:5 mol/mol | 96 | 97 | 30 |
| 90:10 mol/mol | 100 | 98 | 18 |

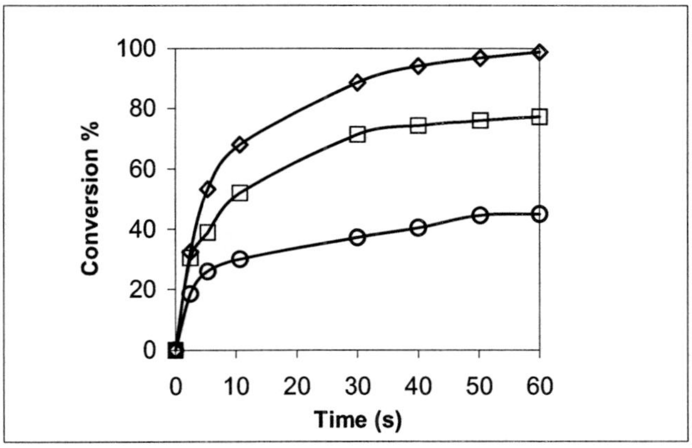


Figure 2: FT-IR conversion curves for CE (O) and in the presence of 20 (□) and 50 (△) wt% of PBOH

A complete conversion is achieved in the pèresence of 50 wt% of PBOH. Moreover the gel content is always high (>97%, see Table 2) indicating that the polybutadiene additive is tightly crosslinked into the polymer network. These results indicate that the PBOH interact with the growing chain through a chain-transfer reaction involving its terminal hydroxyl groups.

The photocured films were characterized by DSC and DMTA analyses. The data, reported in Table 2, show that by increasing the amount of PBOH in the photocurable mixture a decrease of Tg is induced. The flexibilization of the network is obtained as a consequence of the chain transfer reaction and to the introduction in the polymeric network of the very flexible PBOH structure.

From data reported in table 3 it can be observed that the Tg values obtained by DMTA are always higher than those obtained by DSC, similar results were reported previously and attributed to a frequency effect [30,31].

Table 3: properties of UV cured films of epoxy systems

| Sample | Conv. % FT-IR | Gel content % | Tg [°C] DSC | Tg [°C] DMTA | Impact Resistance $J/cm^2$ |
|---|---|---|---|---|---|
| CE | 45 | 95 | 195 | 214 | 0,45 |
| CE/PBOH | | | | | |
| 90:10 wt/wt | 56 | 97 | 187 | 203 | 0,70 |
| 80:20 wt/wt | 77 | 98 | 170 | 184 | 1,18 |
| 50:50 wt/wt | 98 | 98 | 153 | 171 | 1,27 |

Izod impact tests were performed on photocured films. From data reported in Table 3 an increase of the impact resistance is evident by increasing the amount of PBOH in the photocurable mixture. This result can be attributed both to the flexibilization of the polymeric network in the presence of PBOH, as well as to its flexible and tough properties.

This result is particularly interesting for epoxy thermosets, which are characterized by good mechanical properties, but they are brittle and fragile. By increasing the toughness properties of these photocured resins it might be possible to broad their applications.

## 3.2 Fluorinated Additives

The use of fluorinated structures in polymeric networks is very attractive due to their peculiar characteristics, connected to the presence of fluorine, such as chemical and thermal stability, weathering resistance, low surface tension, hydrophobicity and oleophobicity, optical and electrical behaviour [32]. It seemed interesting to employ fluorinated structures in UV-curable systems in order to combine the properties of these molecules and the advantages of the UV-curing technology, giving rise to cured products with outstanding properties.

The fluorine monomers employed were:

- monofunctional fluorinated vinyl ethers having structure

$$H_2C{=}CHOCH_2CH_2Rf$$

where $R_f$ is $C_6F_{13}$ (FVE1) or $C_8F_{17}$(FVE2), whose synthesis is reported elsewhere [33]

- monofunctional epoxides commercially available: their structures is

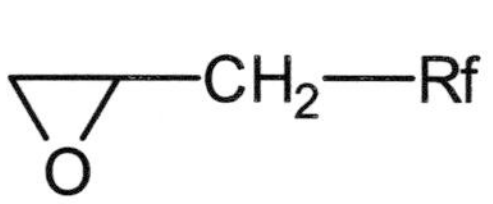


where Rf is $-(CF_2)_8F$ (FEP1) or $-O-CH_2(CF_2)_8H$ (FEP2)

- A monofuncitonal fluorinated epoxide synthesized on pourpose (FOX) [34]

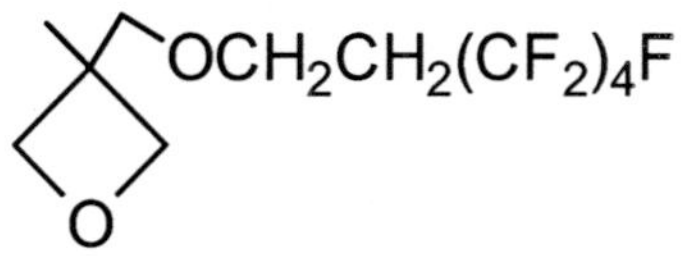


Because they are monofunctional monomers, they do not form network polymers, so these fluorinated monomers have been copolymerised with the hydrogenated resins bearing the same reactive functionality.

The fluorinated vinyl ethers ($CH_2$=CH-O-Rf), prepared by trans-etherification of ethyl vinyl ether and fluorinated alcohols have been added in low concentration (between 01.-5 wt%) to IVE, which is a typical cationic resin: its kinetics is not affected by the presence of the comonomers. The cured films obtained are transparent and show a deep change in the surface properties.

In Figure 3 the wettability of the UV cured films containing the different fluorinated comonomers is compared. The higher surface activity of the higher M.W. monomer is evident and due to the higher fluorine content.

The surface modification is selective in the same way as discussed in previous works from our group for the radical systems [35-37].

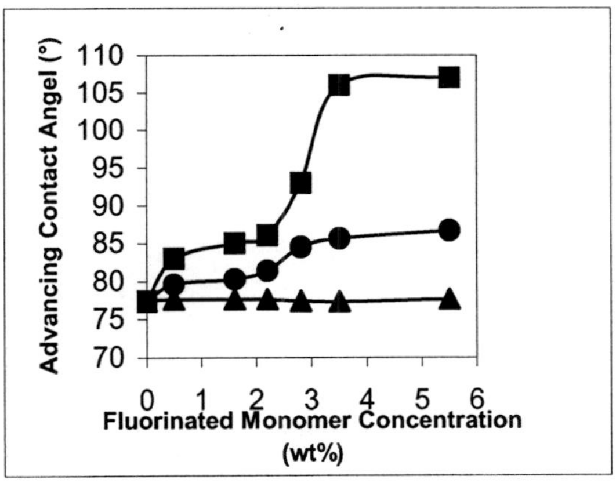


Figure 3: Contact Angle of a photocured vinyl ether system in the presence of $C_6F_{13}$(●) $C_8F_{17}$(■) vinyl ethers; and (▲) glass side.

In agreement with this behaviour, the adhesion of these films on polar substrates, as glass and wood, is good and not influenced by the presence of the fluorinated additives [33].

Coming to the epoxy systems, their surface modification is important due to the polar characteristics of these monomers.

The fluorinated epoxy comonomers have been mixed, in the range 0-1 wt%, with CE. The mixtures, coated on glass substrates and UV cured, give rise to transparent films with interesting surface properties, as reported in Figure 4 [38]. It can be seen that although the fluorinated chain length is the same for both comonomers, there is a difference in their surface activity.

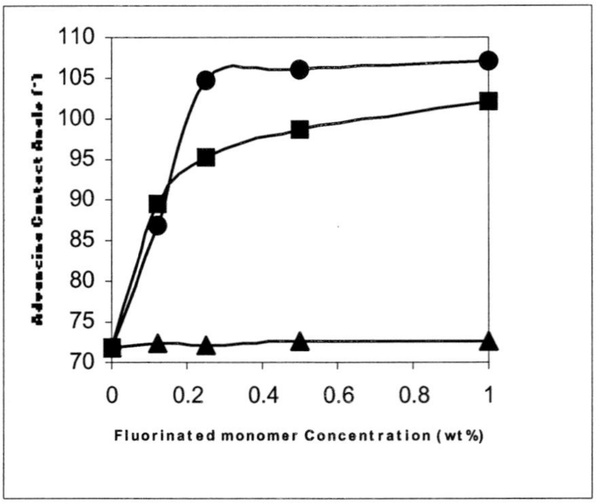


Figure 4: Contact Angle of a photocured epoxy system in the presence of $-(CF_2)_8F$ (●) -O-$CH_2(CF_2)_8H$ (■) monofunctional epoxy; and (▲) glass side.

This can be attributed to the effect of the different fluorinated end groups: $CF_2H$ and $CF_3$ respectively. In particular, the $CF_3$ end group when aligned at the surface guarantees a lower surface free energy.

The XPS data (Figure 5) confirm different surface enrichments of the two fluorine monomers at the air-side of the films, together with a different concentration profile. In the external layer (t.o.a. 25°) the fluorine concentration reaches that of the pure monomer [38].

Fluroinated oxetane monomer (FOX) was added to DOX in the range between 5 to 15 wt%. In this range of composition homogeneous and transparent formulations were obtained. The highest concentration of the fluorinated monomer in the reference resin, DOX, as assessed by visual inspection, corresponds to the solubility limit.

By increasing the amount of FOX monomer in the photocurable formulation, it is possible to observe an increase of oxetane final conversion together with an increase on gel content and decrease of glass transition temperature ( see Table 4).

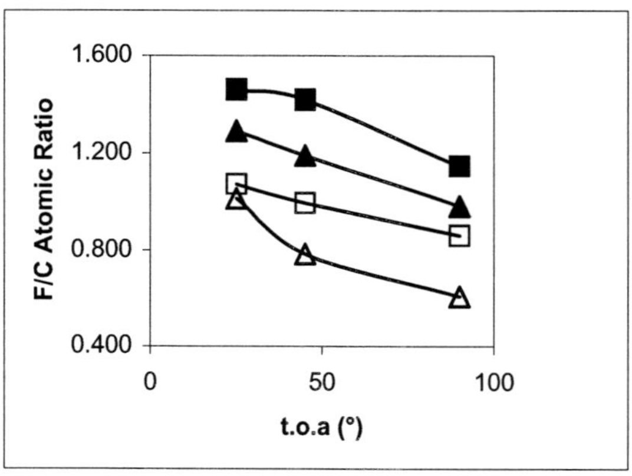


Figure 5: F/C Atomic ratio, at different depth of XPS analysis, for the epoxy photocured system in the presence of different fluorinated co-monomers: 0.2 (△) and 1 wt% (▲) -O-$CH_2(CF_2)_8H$; 0.2 (□) and 1 wt% (■) $-(CF_2)_8F$ epoxy co-monomer.

Table 4: Properties of DOX/FOX photocured copolymer

| Sample | Oxetane conversion % | Tg [°C] DSC | Gel content % |
|---|---|---|---|
| DOX | 86 | 65 | 90 |
| DOX/FOX | | | |
| 95:5 wt/wt | 90 | 58 | 95 |
| 90:10 wt/wt | 96 | 46 | 98 |
| 85:15 wt/wt | 100 | 38 | 97 |

A complete oxetane conversion is obtained in the presence of 15 wt% of FOX. In Figure 6 a portion of the FT-IR spectra for the DOX/FOX 85:15 wt/wt mixture before (a) and after (b) UV-curing is reported, showing the complete disappearance of the oxetane IR peak centered at 830 $cm^{-1}$.

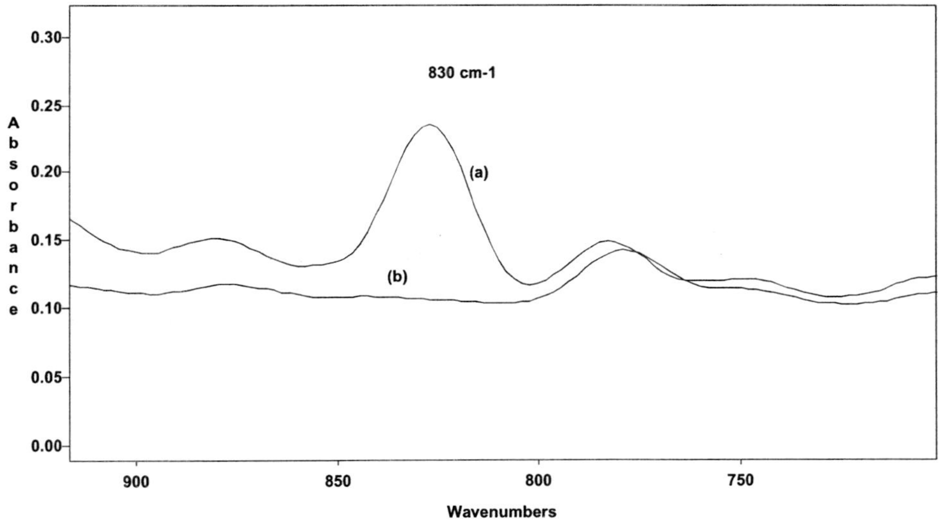


Figure 6: FT-IR spectra of the DOX/FOX 85:15 wt/wt mixture before (a) and after (b) UV-curing

It is possible to explain these results by taking into account that the FOX monomer, being a monofunctional additive, induces a decrease of the crosslinking density of the network. As a consequence, the network structure is more flexible and the mobility of the reactive species is retained so that the polymerization can be completed. In other words, the decrease of the functionality of the reactive system delays the vitrification effect.

The FOX monomer induces also an important modification of the surface properties: when the fluorinated monomer is added to the photocurable formulation and coated on a glass substrate, the wettability of the air-side of the film changes, as evidenced in Figure 7.

The high hydrophobic surface is a consequence of the segregation effect of the fluorinated oxetane which, due to its low surface energy, migrates towards the air interface and concentrates in the outermost layer.

The wettability of the films with hexadecane was also measured and reported in Table 5. An increase in the oil repellence, in the presence of the fluoroadditive, is evident because of the increasing hexadecane contact angle with increasing FOX monomer concentration.

From the contact angle measurements with water and hexadecane, by means of the harmonic method [38], the surface tension of the photocured films were calculated and a decrease on its value by increasing the amount of fluorinated additive was evidenced.

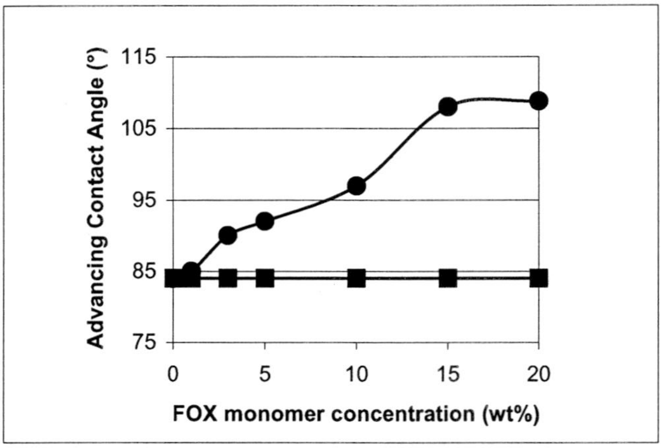


Figure 7: Advancing contact angle measurements with water for photocured films at the air side (●) and glass side (■)

Table 5: Surface properties of DOX/FOX photocured copolymer

| Sample | Hexadecane ϑadv. (air side) | Water ϑadv. (air side) | Surface Tension (mN/m) |
|---|---|---|---|
| DOX | 0 | 85 | 36 |
| DOX/FOX | | | |
| 95:5 wt/wt | 15 | 92 | 32 |
| 90:10 wt/wt | 27 | 97 | 29 |
| 85:15 wt/wt | 35 | 108 | 24 |

Particularly interesting is to observe an increase on thermal stability of photocured films in the presence of FOX monomers as reported by TGA analysis in Figure 8.

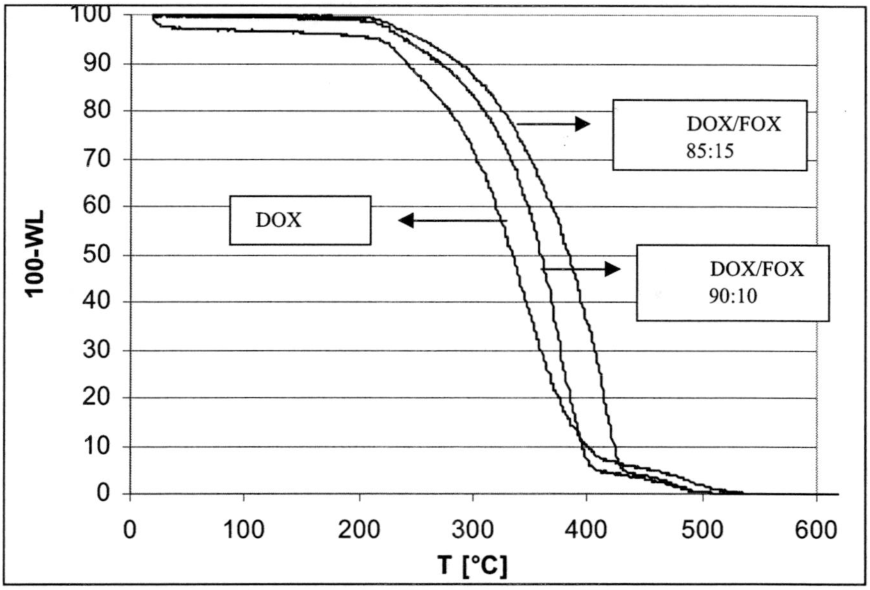


Figure 8: Termogravimetrical analyses of the photocured DOX and DOX/FOX films

All the films start to degrade at around 225 °C, but while the DOX photocured film yields a 10 wt% loss around 235 °C, the same loss is achieved only around 270 °C and around 300 °C for the films containing 10 and 15 wt% respectively of the FOX comonomer. Comparing the T50 data, the results have the same trend: while a DOX homopolymer 50% weight loss is reached at around 320 °C, for the DOX/FOX 90:10 sample it is reached at around 360 °C and for the DOX/FOX 85:15 specimen at around 400 °C.

## 3.3 Hyperbranched Additives

Hyperbranched polymers (HBP) belong to a new group of molecules called dendritic polymers, which have peculiar and often unique properties. They are characterized by a highly branched backbone, which give access to a large number of reactive groups. Their structure give them excellent flow and processing properties, which have attracted great attention [39,40]. Hypebranched polymers are attractive because they resemble dendrimers, but they can be produced more easily on a large scale at a reasonable cost.

HBP with acrylate, vinyl ether, allyl ether or epoxy functions were studied as multifunctional crosslinker [41-43] in coatings and thermosets [44,45].

Considering their characteristics we were interested to employ HBP in UV curable formulations in order to improve the final properties of cured films. The schematic structures of HBP employed as additives in this investigation are reported in Table 6.

Table 6: Schematic structures of HBP

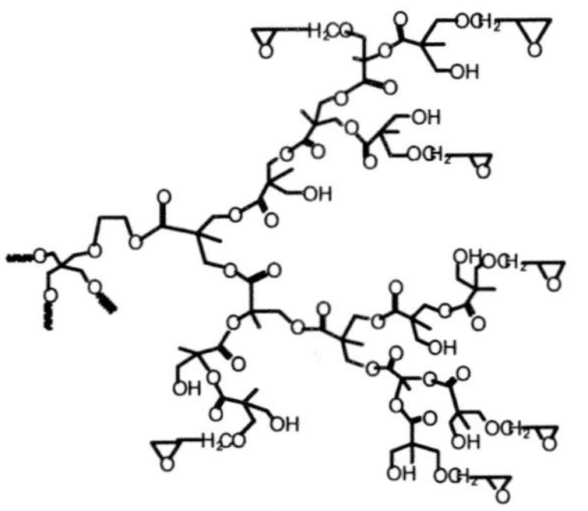

Boltorn E1

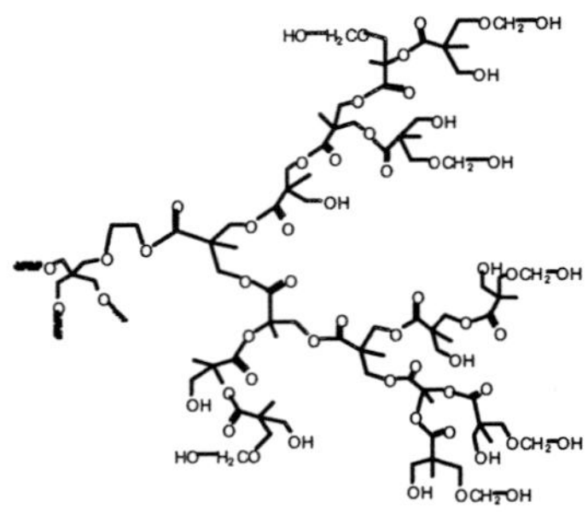

Boltorn, H20, H30, H40

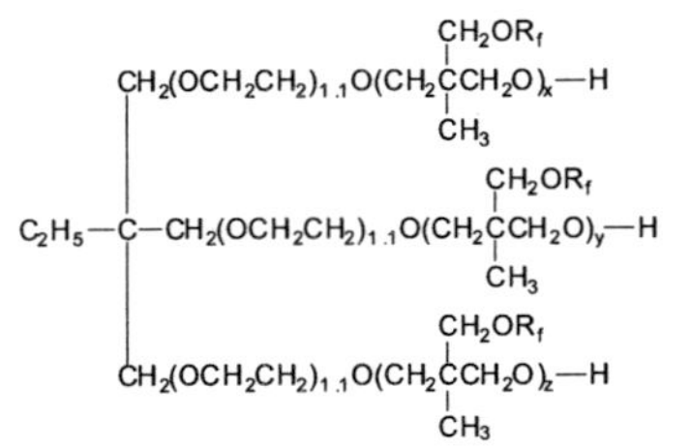

AT-1023 $R_f = CH_2CF_3$ x + y + z = 10.5
AT-1024 $R_f = CH_2CF_2CF_3$ x + y + z = 12.0
AT-1025 $R_f = CH_2CH_2(CF_2)_4F$ x + y + z = 8.0

Fluorinated-HBP

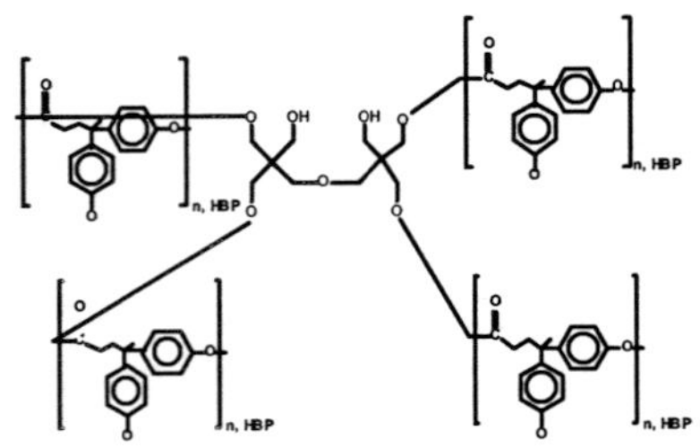

AZ129

The first type of HBP investigated involved aliphatic polyester Boltorn®, starting with an epoxy functionalized (Boltorn® E1) [46]. The conversion curves as a function of irradiation time showed a high conversion for the E1 resin, compare with that of cycloaliphatic monomer (Figure 9). This effect was attributed to the high number of epoxy groups per molecule, to its high mobility and to the presence of residual OH groups.

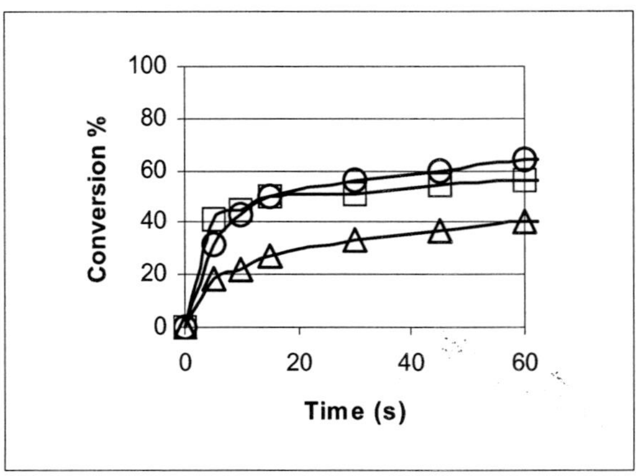


Figure 9: Conversion vs. time curves of the pure HBP(O), CE(□) and DGE(△) resin.

The use of HBP as additive in the epoxy systems did not significantly change the viscosity of the mixture, in agreement with the characteristics of the hyperbranched products. Than it did not affect the rate of polymerization, but on the other hand deeply modified the properties of UV cured films by inducing a strong flexibilization effect.

Subsequently the use of hydroxyl functionalized HBP (Boltorn H20, H30 and H40) was investigated. Their presence induced an increase of the final epoxy conversion, which was interpreted on the basis of a chain transfer mechanism. A decrease of the Tg values by increasing the amount of HBP additive in the photocurable formulation was observed. Moreover a clear increase in toughness was obtained and attributed to the plasticization effect by the presence of HBP.

The density of the photocured films, in the presence of different amounts of the different HBP in the photocurable formulations, was measured by means of a picnometer and the data are reported in Table 7 together with the Tg and resilience values.

It is possible to observe a quite significant increase of the density by increasing the amount of HBP in the photocurable formulation, while it is substantially independent on the type of HBP employed; this is because the density of the starting HBP is practically independent on the molecular weight, due to the peculiar characteristics of the HBP.

The density increase in the photocured films can be connected to a free-volume decrease, and it indicates that the molecules are more densely packed together determining a decrease of the macroscopic volume. This culd be an interesting property for applications in gas barrier coatings.

Following, a phenolic polyester HBP (AZ129) was synthesized [47] and employed as a chain transfer agent for the CE resin. The CE/AZ129 ratio was varied in the range between 5-30 wt%. The presence of such additive induced a chain transfer reaction involving the OH phenolic groups, which results in a flexibilization of the network and an increase on the curing rate. The kinetic curves for CE and CE/AZ129 mixture are reported in Figure 10.

Table 7: properties of Boltorn containing systems

| **Sample** | **Tg [°C] DSC** | **Tg [°C] DMTA** | **Resilience J/cm$^2$** | **Density g/cm$^3$** |
|---|---|---|---|---|
| CE | 195 | 214 | 0,17 | 1,10 |
| CE/H20 | | | | |
| 90:10 wt/wt | 98 | 144 | 0,32 | 1,20 |
| 80:20 wt/wt | 70 | 111 | 0,66 | 1,38 |
| 50:50 wt/wt | 21 | 79 | 0,51 | 1,40 |
| CE/H30 | | | | |
| 90:10 wt/wt | 104 | 180 | 0,41 | 1,28 |
| 80:20 wt/wt | 97 | 162 | 0,57 | 1,37 |
| 50:50 wt/wt | 26 | 86 | 0,68 | 1,42 |
| CE/H40 | | | | |
| 90:10 wt/wt | 82 | 142 | 0,30 | 1,27 |
| 80:20 wt/wt | 72 | 113 | 0,51 | 1,40 |
| 50:50 wt/wt | 22 | 64 | 0,71 | 1,42 |

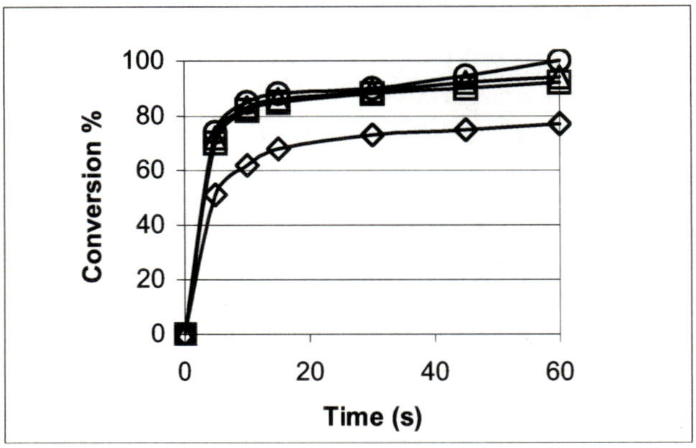


Figure 10: FT-IR conversion curves as a function of irradiation time fo CE (◇) and CE/AZ129 95:5 (□), 90:10 (△) and 80:20 (O).

A complete epoxy group conversion is obtained already in the presence of 20 wt% of AZ129. The Tg data are reported in Table 8 together with the impact resistance and E' values. It is particular interesting to observe a peculiar results due to the multicrosslinker behaviour of HBP: the increase on toughness properties is accompanied with with an increase on E' values notwithstanding the flexibilization of the network due to the presence of the HBP flexible structure.

The AFM analysis in tapping mode of photocured films showed the formation of nanophasic HBP separate domains with very good interphasic adhesion with epoxy matrix.

Table 8: Properties of UV cured films in the presence of AZ129

| Sample | Tg [°C] DMTA | Resilience J/cm$^2$ | E' @ 50 °C |
|---|---|---|---|
| CE | 214 | 0,27 | 633 |
| CE/AZ129 | | | |
| 95:5 | 180 | 0,28 | 1010 |
| 90:10 | 183 | 0,30 | 1393 |
| 80:20 | 180 | 0,33 | 1290 |
| 70:30 | 178 | 0,40 | 1300 |

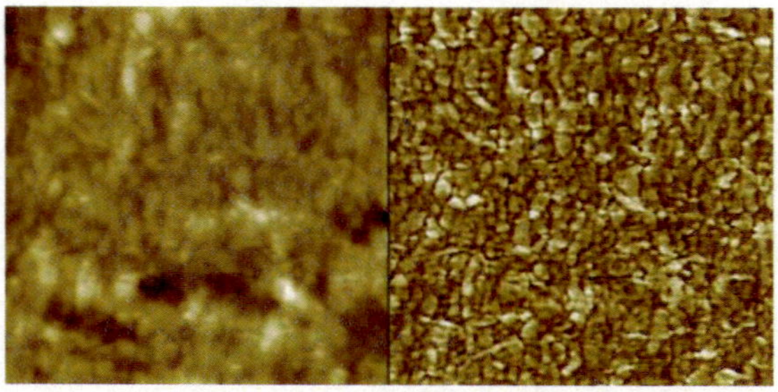

Figure 11: AFM topographic images (left topography, right phase) of photocured CE/AZ129 70:30

The AFM measurements were done on the films coated on glass substrates in the tapping mode. The scan conditions were chosen according to Maganov [48] (free amplitude >100 nm, set-point amplitude ratio 0.5) in order to get stiffness contrast in the phase image that means bright features in the phase image are stiffer than dark.

Finally different fluorinated containing HBP were synthesized starting from different fluorinated oxetanes [49] and employed as additives. As observed for the other hydroxyl containing HBP, they interact with the polymeric carbocation through a chain transfer mechanism inducing an increase in the final epoxy conversion [50]. High gel content values (>96%) for the photocured films confirm that the HBP additive is tightly crosslinked to the polymeric network.

Concerning the bulk properties different phenomena were observed by using different epoxy resins. For the CE and DGE monomers, by increasing the amount of HBP in the photocurable resin, a decrease of the Tg value is evidenced. This flexibilization can be explained by the occurance of the chain transfer. The more flexible network structure, characterized by a lower glass transition temperature, is a consequence of the decreasing polyether chain length.

The HBP molecule can act as a multifunctional crosslinker because it is characterized by a large amount of hydroxyl group on its surface. As a matter of fact, by using resins characterized by lower Tg and by higher chain mobility after curing, the multicrosslinker effect of the HBP becomes evident. By increasing the amount of the HBP in photocurable resins as HDGE and SIOG, an increase of the glass transition temperature is induced (see data on Table 9).

Table 9: Thermal properties of epoxy photocured films in the presence of fluorinated HBP.

| Formulation | Tg [°C] DSC | Tg [°C] DMTA |
|---|---|---|
| CE | 188 | 224 |
| CE/AT-1023 95:5 | 170 | 207 |
| CE/AT-1023 90:10 | 162 | 192 |
| CE/AT-1023 85:15 | 155 | 183 |
| DGE | 20 | -- |
| DGE/AT-1023 95:5 | 15 | -- |
| DGE/AT-1023 90:10 | 12 | -- |
| DGE/AT-1023 85:15 | 10 | -- |
| HDGE | -25 | -- |
| HDGE/AT-1023 95:5 | -20 | -- |
| HDGE/AT-1023 90:10 | -15 | -- |
| HDGE/AT-1023 85:15 | -15 | -- |
| SIOG | -70 | -70 |
| SIOG/AT-1023 95:5 | -65 | -68 |
| SIOG/AT-1023 90;10 | -60 | -62 |
| SIOG/AT-1023 85:15 | -60 | -60 |

In conclusion, when the photocured polymer is characterized by low mobility of the network (high Tg) the chain transfer effect, which induces a decrease of the crosslinking density (with a decrease of the glass transition temperature) is the main evidenced effect due to the presence of the hydroxyl functionalized HBP. When the mobility of the network is high (low Tg) the multicrosslinker effect, which induces an increase of the crosslink density (with an increase of the glass transition temperature), becomes more important with respect to the flexibilization effect induced by the chain transfer.

Because of the presence of fluorinated free-dandling chains in the HBP molecule, we were interested in investigating the surface effect of the presence of this additive in the photocurable formulations. Different epoxy formulations, in the presence of different amount of HBP, were coated on a glass substrate and photocured. Contact angle measurements were performed on UV cured films. The data reported in table 10 show an increase on hydrophobicity in the presence of fluorinated HBP.

Table 10: Contact angle measurements of photocured films in the presence of poly-FOX

| Formulation | θadv.[°] Air side | θadv.[°] Glass side |
|---|---|---|
| CE | 75 | 75 |
| CE/AT-1023 95:5 | 93 | 75 |
| CE/AT-1023 90:10 | 95 | 75 |
| CE/AT-1023 85:15 | 94 | 75 |
| CE//AT-1024 85:15 | 102 | 75 |
| CE/AT-1025 85:15 | 107 | 75 |
| DGE | 72 | 72 |
| DGE/AT-1023 95:5 | 95 | 72 |
| DGE/AT-1023 90:10 | 97 | 72 |
| DGE/AT10123 85:15 | 97 | 72 |
| HDGE | 78 | 78 |
| HDGE/AT-1023 95:5 | 94 | 78 |
| HDGE/AT-1023 90:10 | 93 | 78 |
| HDGE/AT-1023 85:15 | 95 | 78 |

# 4 Conclusions

Cationic photopolymerization of vinyl ether, epoxy and oxetane systems has been reported.

The presence of hydroxyl containing compounds in the photocurable formulation induced an increase on rate of polymerization and final conversion. These results can be interpreted on the basis of a chain transfer reaction with a flexibilization of the network. In the presence of epoxy-hydroxy-funcitonalized polybutadiene the Tg values of the cured films decreased sharply confirming the strong flexibilization effect together with an increase on toughness properties.

The use of fluorine containing monomers as additives to UV curable systems allowed the modification of the surface properties of the UV-cured films obtaining a high hydrophobic surface while the bulk properties remained unchanged. The hydrophobicity is due to the selective enrichment of the fluorinated monomers at the film surface, as confirmed by XPS analysis.

A wide range of hyperbranched polymers (HBP) was investigated as additives in cationic photopolymerization of epoxy systems. The HBP were inserted into the polymeric network either by a copolymerization or through a chain transfer reaction involving the hydroxyl groups. By varying the type and concentration of HBP a modification of the bulk properties of photocured films was induced. An increase of toughness properties together with a flexibilization was obtained without affecting the processability and the viscosity of the photocurable mixture. In the presence of fluorine-containing HBP, a surface modification was also induced, with an increase in hydrophobicity depending on the length of the fluorine pendant chain.

In conclusion: the great versatility of cationic photopolymerization process has been demonstrated. Notwithstanding the low commercial availability of photocurable monomers, it has been shown that is possible to modulate and tailor the final properties of the UV-cured films by varying properly the additive in the system controlling the mechanism of the curing process.

### Acknowledgements

Many thanks are due to Prof. B. Voit from IPF in Dresden, Germany, for the synthesis of AZ129 and for her constant supervision and valuable advices. Perstorp AB, Sweden and Omnova Solution INC. USA are acknowledge for supplying different HBP. Andreas Janke (IPF) and Hartmut Komber (IPF) are gratefully acknowledged for the AFM measurements and NMR analysis, as well Andrea Quaglia from Floramo Corporation (Italy) for TGA measurements.

## References

[1] Dufour P., in "Radiation Curing in Polymer science and Technology", J.P. Fouassier and J.F. Rabek Ed., Elsevier London, 1993, Vol.I, p.1.

[2] Decker C., *Prog. Polym. Sc.*, 1996, 21, 593.

[3] Schlegel L., Schabel W., in Dufour, P. in "Radiation Curing in Polymer science and Technology", J.P.. Fouassier and J.F. Rabek Ed., Elsevier London, 1993, Vol.I, p. 119.

[4] Crivello J.V., in "Photoinitiators for free-radicals cationic and Anionic photopolymerization", G. Bradley Ed., Wiley, New York, 1998, p. 329.

[5] Crivello J.V., *J. Polym. Sc. Polym. Chem.*, 1999, 37, 4241.

[6] Crivello J.V., *Makromol. Chem. Macromol. Symp.*, 1987, 188, 145.

[7] Olah G.A., Surya Prakash G.K., Sommer J., in Superacid, Wiley, New York, 1985, pp. 7-8.

[8] Derktar J.L., Hacker N.P. *J. Org. Chem.*, 1989, 55,639.

[9] Fouassier J.P., Burr D., Crivello J.V. ,*J. Macrom. Sc.,* 1994, A31, 677.

[10] Takimoto Y., In "Radiation Curing in Polymer science and Technology", J.P. Fouassier and J.F. Rabek Ed., Elsevier London, 1993, Vol.III, p. 269.

[11] Decker C., Thi Viet T.N., Thi H.P., *Polym. Int.*, 2001, 50, 986.

[12] Crivello J.V., Ortiz R.A., *J. Polym. Sc. Polym. Chem.*, 2001, 39, 2385.

[13] Roffey C.G., in “photopolymerization of surface coatings”, Wiley, New York, 1982, p. 74.
[14] Decker C., Morel F., *Polym. Mat. Sc. and Eng.*, 1997, 76, 10.
[15] Sangermano M., Malucelli G., Bongiovanni R., Priola A., Annby U., Rehnberg N., *Europ. Polym. J.*, 2002, 38, 655.
[16] Crivello J.V., Di Jo K., *J. Polym. Sc. Polym. Chem.*,1993, 31, 1437.
[17] Crivello J.V., Di Jo K., *J. Polym. Sc. Polym. Chem.*, 1993, 31, 1483.
[18] Sangemrnao M., Malucelli G., Bongiovanni R., Priola A., Annby U., Rehnberg N., *Polym. Int.*, 2001, 50, 998.
[19] Sangermano M., Malucelli G., Bongiovanni R., Priola A., *Europ. Polym. J.,* 2004, 40, 353.
[20] Hato H., Sasaki H., *Polym. Preprint,* 2001, 42(2), 729.
[21] Suzuki H., Sasaki H., *Polymer Preprint,* 2001, 42(2), 733.
[22] Sasaki H., Crivello J.V., *J. Macr. Sc. Pure & Appl. Chem.*, 1992, A29(10), 915.
[23] Penczek S., Kubisa P., Szamauski R., *Macromol. Chem. Macromol. Symp.*, 1986, 3, 203.
[24] Crivello J.V., Conlon D.A., Olson D.R., Webb K.K., *J. Rad. Tech.*, 1986, 13, 3.
[25] Sangermano M., Malucelli G., Morel F., Decker C., Priola A: *Europ. Polym. J.,* 1999, 35, 639.
[26] Sangermano M., Bongiovanni, R., Malucelli, G., Priola, A. *Polym. Bull.*, 1999, 42, 641.
[27] Sangermano M., Spera S., Bongiovanni R., Priola A., Busetto C., *Macrom. Chem. & Phys.*, 2000, 201(17), 2441.
[28] Bongiovanni R., Malucelli G., Sangermano M., Priola A., *Macromol. Symp.*, 2002, 187, 481.
[29] Sangermano M., Malucelli G., Bongiovanni R., Gozzelino G., Peditto F., Priola A., *J. Mat. Sc.*, 2002, 37, 4753.
[30] Malucelli G., Gozzelino G., Bongiovanni R., Priola A., *Polymer*, 1996, 37, 2565.
[31] Nielsen L.E., in Mechanical Properties of Polymers and Composites, Marcell Dekker, New York, 1994.
[32] Tomas R., in “Fluoropolymers 2: properties”, G. Houghman, K. Johns, P.E. Cassidy, T. Davidson ed., Plenum Press, New York, 1999, Chap. 4.
[33] Bongiovanni R., Sangermano M., Priola A., Malucelli G., Leonardi A., Ameduri B., Pollicino A., Recca A., *J. Polym. Sc. Polym. Chem.*, 2003, 41, 2890.
[34] Sangermano M., Bongiovanni R., Malucelli G., Priola A., Thomas R.R., Medsker R.E., Kim Y., Kausch C.M., *Polymer*, 2004, 45, 2133.
[35] Ameduri B., Bongiovanni R., Lombardi V., Pollicino A., Priola A., Recca A., *J. Polym. Sc. Polym. Chem.*, 2001, 39, 4227.
[36] Bongiovanni R., Malucelli G., Priola A., Tonelli C., Simeone G., Pollicnio A., *Macromol. Chem. Phys.*, 1998, 199, 1099.
[37] Bongiovanni R., Malucelli G., Sangermano M., Priola A., *Prog. Org. Coat.,* 1999, 36, 70.
[38] Wu S., in Polymer Interface and adhesion, M. Dekker ed., New York, 1982, Ch. 4.
[39] Kim Y., *J. Polym. Sc. Polym. Chem.*, 1998, 36, 1685.
[40] Voit B., *J. Polym. Sc. Polym. Chem.*, 2000, 38, 2505.
[41] Johansson M., Malmstrom E., Hult A., *J. Polym. Sc. Polym. Chem.*, 1993, 31, 619.

[42] Schmaljohann D., Voit B., Jansen J., Hedriks P., Loontjiens J.A., *Macromol. Mat. Eng.*, 2000, 275, 31.
[43] Johansson M., Hult A., *J. Coat. Tech.*, 1995, 67, 35.
[44] Hult A., Johansson M., Malmstrom E., *Macr. Symp.*, 1995, 98, 1159.
[45] Gopala A., Wu H., Xu J., Heiden P., *J. Appl. Polym. Sc.*, 1999, 71, 1809.
[46] Sangermano M., Malucelli G., Bongiovanni R., Priola A., Harden A., Renhberg N., *Polymer Eng. & Sc.*, 2003, 43(8), 1460.
[47] Sangermano M., Priola A., Malucelli G., Bongiovanni R., Quaglia A., Voit B., Ziemer A., *Macromol. Mat. & Eng.*, 2004, 289, 442.
[48] Maganov S.N., Elings V., Whangbo M.H., *Surface Sc*ience, 1997, 375, L385-L391.
[49] Kausch C.M., Leising J.E., Medsker R.E., Russel V.M., Thomas R.R., *Langmuir*, 2002, 18, 5933.
[50] Sangermano M., Malucelli G., Bongiovanni R., Vescovo L., Priola A., Thomas R.R., Kim Y., Kausch C.M., *Macromol. Mat. Eng.*, 2004, 289, 722.

In: Frontiers in Polymer Research
Editor: Robert K. Bregg, pp. 155-169

ISBN: 1-59454-824-2


*Chapter 6*

# DETERMINATION OF PHENOLIC COMPOUNDS IN WINES WITH TYROSINASE MODIFIED ELECTRODES

***Isıl Narlı, Senem Kıralp and Levent Toppare****
Department of Chemistry, Middle East Technical University, 06531 Ankara, Turkey

## Abstract

Wines, particularly red wines contain numerous biologically active compounds, the most important of which are polyphenols, whose nutritional importance is attributed to their antioxidant power.

This research was carried out to evaluate the phenolic capacity of two red wines produced in Turkey. Analysis was performed by using enzyme electrodes constructed by the immobilization of tyrosinase in conducting copolymers.

Immobilization matrices were synthesized by copolymerization of terephthalic acid bis-(2-thiophen-3-yl ethyl) ester (TATE) with pyrrole. Immobilization of enzyme was performed via entrapment in conducting copolymers using electrochemical polymerization of pyrrole.

Measurements were performed by using Besthorn's Hydrazone method which includes spectrophotometric analysis of quinones produced by enzyme. Enzyme electrodes were characterized in terms of maximum reaction rate ($V_{max}$) and Michaelis-Menten constant ($K_m$). In addition to kinetic parameters, stability of enzyme electrodes towards environmental conditions such as pH and temperature was investigated. Usage stability and shelf-life analysis were also examined.

It is known from previous studies that free enzyme could not be used in phenolic determination studies in wines because of inhibitory effects of various substances naturally found in red wines. To understand the behavior of immobilized tyrosinase toward the inhibition, benzoic acid was used as the inhibitor and inhibition constant $K_i$ was determined.

* E-mail address: toppare@metu.edu.tr, Corresponding author. Tel: +90-312-2103251; fax: +90-312-2101280

# Introduction

One of the most "versatile" enzymes in nature is tyrosinase (EC. No. 1.14.18.1). Tyrosinase was discovered by Bertnard and Bourquelot about 100 years ago. In analyzing certain varieties of mushrooms, they observed that as oxidation progressed, the mushrooms changed color and finally become dark brown or black. Subsequently studies showed that this new oxidase catalyzed the aerobic oxidation of mono-phenols, and the final product of tyrosine oxidation was melanin [1].

Tyrosinase is a tetramer which weights about 130,000 Daltons. Its active site contains two copper atoms which exist in three states; met, oxy, and deoxy. It occurs extensively in the pytogenetic scale. The enzyme is commonly found in yeast, mushrooms, grapes, bananas, apples, potatoes, frogs and mammals.

Tyrosinase catalyzes two reactions via separate active sites:

(1) the orthohydroxylation of monophenols, commonly referred to as the cresolase activity and
(2) the oxidoreduction of orthodiphenols to orthoquinones, commonly referred to as the catecholase activity.

Tyrosinase catalyzes the synthesis of melanin through the hydoxylation of tyrosine to dihydoxyphenylalanine (DOPA) and the subsequent oxidation of DOPA to dopaquinone. The unstable dopaquinone will polymerize and precipitate into melanin. However, in the presence of a reductor, the reaction will stop at the diphenol level [2]. The cresolase activity of tyrosinase is of particular importance because it synthesizes DOPA. DOPA is a precursor of dopamine, an important neural message transmitter.

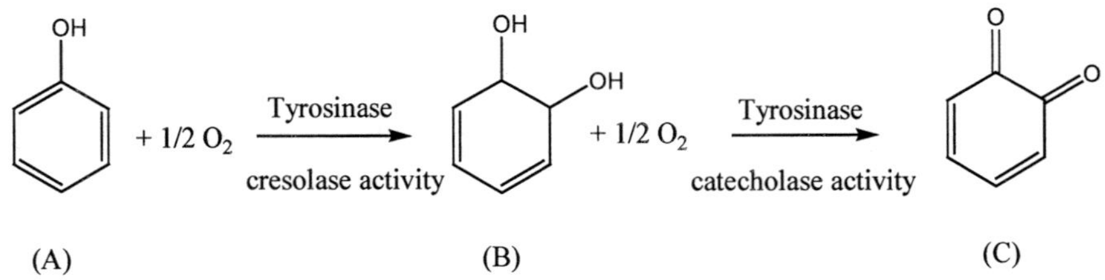


A:phenol B:catechol C:o-quinone

Scheme 1. Schematic representation of tyrosinase activity.

Patients who suffer from Parkinson's disease show a significant decrease in the concentration of dopamine found in the substantia nigra of the brain. [3].

The catecholase activity also has important applications, since this activity can be used in the analysis of phenols and its derivatives. Phenolic pollutants are frequently found in surface waters and in the effluent of industrial discharge sources. Some of the industrial sources of phenol discharge include oil refineries, coke and coal conversion plants, plastics and petrochemical companies, dyes, textiles, timber, mining, and the pulp and paper industries. Virtually all phenols are toxic. Moreover, they have a high oxygen demand and can deplete the oxygen in the body of water [4]. As a consequence, this may affect the ecosystem of water sources where phenols are discharged.

Atlow et al. [5] have reported a successful application of soluble tyrosinase in the "cleansing" of polluted waters. Tyrosinase causes the precipitation of phenols, which can then be filtered out from surface waters and industrial discharge sources. The enzyme has also been used as a sensor to detect the concentration of phenols in waste water [6,7].

The detection of phenols is not only of importance in industry but also in the medical field. Tyrosinase has been used as part of an enzyme-electrode system to detect catechols and assess catecholamines in the urine of patients with neural crest tumors [8].

Tyrosinase also has applications in the food industry, as it is responsible for the browning of fruits and vegetables. Interest in the enzyme has been demonstrated by tannin oil companies due to the role that it plays in melanogenesis. Also, it has been considered for use in melanin-related disorders, such as albinism, vitiligo, and melanoma [9].

Although tyrosinase has widespread applications, its use is limited by its inherent instability and rapid inactivation. By using enzyme immobilization technology, good operational stability and long-term stability can be achieved for tyrosinase.

Immobilization is the conversion of enzymes from a water-soluble, mobile state to a water-insoluble, immobile state. It prevents enzyme diffusion in the reaction mixtures and facilitates their recovery from the product stream by solid liquid separation techniques. The advantages of immobilization are (1) multiple and repetitive use of a single batch of enzymes; (2) creation of buffer by the support against changes in pH, temperature and ionic strength in the bulk solvent, as well as protection from shear forces; (3) no contamination of processed solution with enzyme; and (4) analytical considerations, especially with respect to long-life for activity and predictable decay rates. Generally, an enzyme is attached to a solid support material so that substrate can be continually converted to product. Thus, enzymes can be recycled and used many times. The goals are to increase the enzyme's stability, to increase the ability to recycle the enzyme, and to separate the enzyme easily from the product [10].

Previous studies [11-19] and this study includes the immobilization of enzyme via entrapment during the electrolysis performed for the synthesis of conducting polymers. Advantages of immobilization of enzymes in a conducting polymer by electropolymerization are easy one-step procedure, accurate control of the polymer thickness via the electrical charge passed during the film formation process, localization of the electrochemical reaction exclusively on the electrode surface allowing precise modification of micro-electrodes and surfaces of complex geometry and the possibility to build-up multi-layer structures [20].

By using the enzyme electrodes obtained at the end of the immobilization procedure, one can determine the amount of the phenolic compounds naturally found in wines.

Wine contains many phenolic substances. The phenolics have a number of important functions. In wine, especially in red wines, affect the tastes of bitterness and astringency. Second, the color of red wine is caused by phenolics. They are also bactericidal agents and impart antioxidant properties, being especially found in the skin and seeds of the grapes.

There are two types of phenols in wines. "flavonoids" and "non-flavonoids". The flavonoids are composed of three benzene rings and react readily, binding to other molecules and there are between 6,000 and 8,000 species of flavonoids. A group of flavonoids, called the flavon-3-ols, have been well characterized in wine. Flavon-3-ols are usually concentrated in grape seeds, stems and skin. When these parts of the grape are left in for as long as possible during the wine-making process, more flavon-3-ols end up in the resulting wine than the one where the seeds, stems and skin are removed earlier.

The non-flavonoids in wine comprise many classes of chemicals including hydroxycinnamates, benzoates, and stilbenes. Much ado has been made in the media about the health benefits of a particular kind of stilbene, called "resveratrol," which is unique to grapes and is not found in other fruits or vegetables.

The chemical composition of a wine is influenced by the climatic and atmospheric conditions, soil type, vine cultivation and the treatment to which it is subjected. Due to this reason amount of phenolics vary from one brand and type of wine to another. Process difference causes red wines to contain almost ten times higher amount of phenolics.

The typical methods for the determination of phenolic compounds are gas and liquid chromatography. These methods involve complex sample pre-treatment procedures and are unsuitable for on site or field based analyses.

A biosensing approach with advantages of high specificity, high sensitivity and rapid detection mechanism may provide a solution [21, 22].

In the previous studies immobilization of polyphenol oxidase in different conducting polymer matrices was studied as an alternative method for the determination of phenolic compounds [23-25].

In this study results obtained for the phenolic amounts in red wines tell us that we should consider the inhibitory effect of some compounds naturally found in red wines.

Tyrosinase is inhibited by various substances, among which are aromatic carboxylic acids. The inhibitory character of these compounds is linked to the presence of the benzene ring [26]. Inhibition by members of benzoic and cinammic acid series was studied in the literature [27-31]. It is known that benzoates and cinnamates are naturally found in red wines with high concentrations. So during the analysis performed for the determination of amount of phenolic compounds, we should consider the effect of these substances on the measurements.

This study includes the characterization of fabricated enzyme electrodes by using a copolymer of terepthalic acid bis-(2-thiophen-3-yl ethyl) ester (TATE) with pyrrole. Beside this, inhibition effect of benzoic acid was studied both for free and immobilized enzymes.

# Experimental

## Materials

Tyrosinase (PPO) [E.C 1.14.18.1] was purchased from Sigma. Pyrrole was purchased from Aldrich and sodium dodecyl sulfate (SDS) from Sigma. Pyrrole was distilled before use. MBTH, acetone and sulfuric acid used in spectrophotometric activity determination of PPO were also obtained from Sigma. For preparation of citrate buffer, tri-sodium citrate-2 hydrate and citric acid were used as received.

Catechol was purchased from Sigma. All catechol solutions were prepared in citrate buffer.

2-Thiophen-3-yl-ethanol (Aldrich), terepthaloyl chloride (Aldrich), triethylamine (Merck), dichloromethane (Merck) and tetrabutylammonium tetrafluoroborate (TBAFB) (Aldrich) were used without further purification.

## Preparation of Enzyme Electrodes

Prior to copolymerization, the precursor monomer, TATE should be synthesized with the procedure explained in a previous study [32]. 2-Thiophen-3-yl-ethanol was dissolved in dichloromethane containing triethylamine. To this solution terepthaloyl chloride in dichloromethane was added dropwise in half an hour, by cooling in ice bath and under nitrogen atmosphere. The esterification was carried out for overnight at 0°C. Then the solution was washed with HCl 1% solution and water. The organic layer was dried over $Na_2SO_4$ and the solvent was removed via rotaevaporatory. Twice recrystalization form ethanol provided white crystals of TATE (Scheme 2).

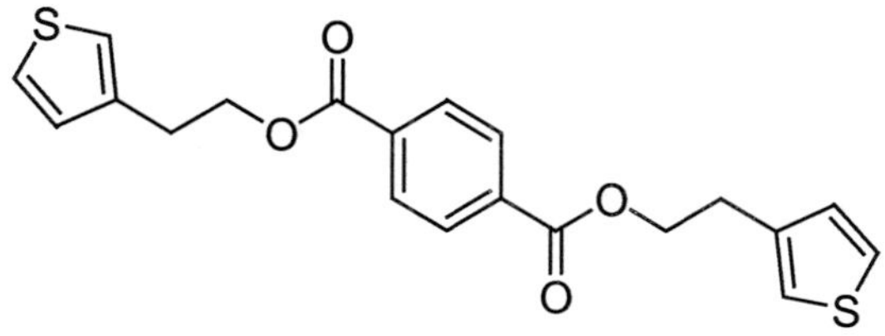

Scheme 2. Terepthalic acid bis-(2-thiophen-3-yl ethyl) ester (TATE)

Homopolymerization of TATE was achieved via constant current electrolysis in a one compartment cell. 0.01M TATE dissolved in dichloromethane/TBAFB solvent-supporting electrolyte couple. Electrolysis was performed while applying 20 mA constant current for 10 minutes.

Immobilization of polyphenol oxidase (PPO) was achieved by electropolymerization of pyrrole on a previously poly (TATE) coated platinum electrode. The solution consists 0.3 mg $ml^{-1}$ polyphenol oxidase, 0.6 mg $ml^{-1}$ supporting electrolyte (sodium dodecyl sulfate) and 0.01M pyrrole and 10 ml citrate buffer (pH=6.5). Immobilization was performed in a typical three electrode cell, consisting of Pt working and counter electrodes and a $Ag/Ag^+$ reference electrode. Immobilization was carried out at a constant potential of +1.0 V for 20 min at room temperature. Enzyme electrodes were kept at 4°C in citrate buffer solution when not in use.

## Determination of PPO Activity

Determination of phenolic groups was achieved by Besthorn's Hydrazone Method [33] which includes spectrophotometric measurements. In this method 3-methyl-2-benzothiozolinone (MBTH) interacts with the quinones produced by the enzyme to yield red products instead of brown colored pigments in the absence of the color reagent [34]. The pathway proposed by Rodriguez et al is shown in Scheme 3 [35].

For determination of free PPO activity, different concentrations of catechol were prepared. Solutions contain 1.0 ml citrate buffer, 0.5 ml MBTH (0.3% in ethanol) and 0.5 ml catechol. 1 minute of reaction time was given after the addition of 0.5 ml enzyme solution (0.1mg/ml). 0.5 ml sulfuric acid (5% v/v) was added to stop the enzymatic reaction. Quinone produced reacts with MBTH to form a red color complex and this complex was dissolved by adding 3.0 ml acetone in the test tube. After mixing, absorbance was measured at 495 nm by using a Shimadzu UV-Visible spectrophotometer.

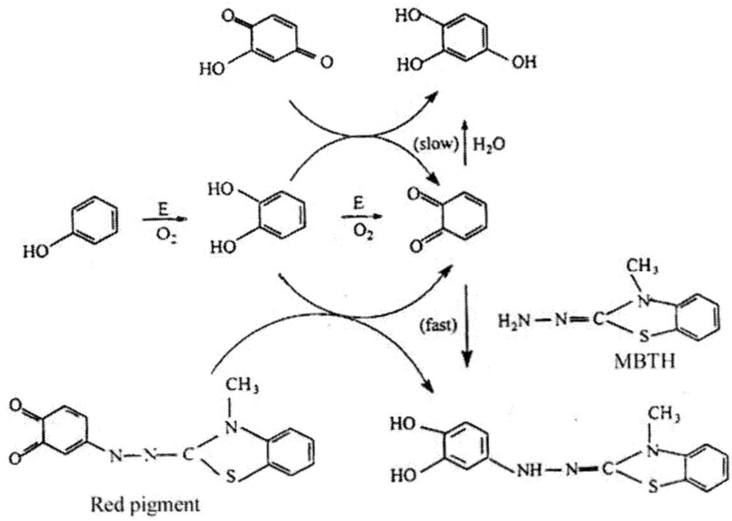


Scheme 3 The pathway of Besthorn's Hydrazone Method

For immobilized PPO, different concentrations of catechol were prepared (3.0 ml) and put in water bath at 25°C. 1 ml MBTH solution was added. Enzyme electrode was immersed in the solution and shaken for 5 minutes. 1 ml sulfuric acid and 1 ml acetone were added for a total volume of 6 ml. After mixing, absorbances were measured at 495 nm.

## Determination of Optimum Temperature and pH

Optimum temperature and pH determinations were carried out by changing incubation temperature and pH between 10°C-90°C and 2.5-12 respectively. The rest of the procedure was the same as the determination of PPO activity.

## Determination of Operational and Storage Stability of Immobilized PPO

Enzymes can easily lose their catalytic activity and denatured, so careful storage and handling are essential. To determine the stability against repetitive use and shelf-life of enzyme electrodes, the activity of electrode was checked. 20 measurements on the same day were done to perform the operational stability. For storage stability activity measurements were performed for 30 days.

## Protein Determination

Protein determination measurements were performed by Bradford's Method. Bradford reagent was prepared by mixing 25 ml phosphoric acid, 12.5 ml ethanol and 25 mg Coomassie Brilliant Blue (G-dye). The mixture was diluted to 50 ml with distilled water. During measurements, a solution of Bradford reagent was prepared by mixing 1 volume stock solution with 4 volumes of distilled water.

For the preparation of protein calibration curve, bovine serum albumin (BSA) was used. Different concentrations of BSA were prepared as 1 ml, and 2 ml of diluted Bradford reagent were added. The absorbance of these solutions was measured at 595 nm.

## Inhibitory Effect of Benzoic Acid on Free end Immobilized Enzyme

All steps explained in the section for determination of enzyme activity was performed also for determination of inhibitory effect of benzoic acid on free and immobilized enzyme activity. The only difference of this procedure was the presence of benzoic acid with various concentrations within the activity assay. The concentrations of benzoic acid were 0.001M, 0.005M and 0.01M. As it was described in the section for determination of PPO activity different concentrations of catechol were prepared, this time 2.5 ml, and 0.5 ml benzoic acid were added and put in water bath at 25°C. After adding 1 ml MBTH solution, enzyme electrode was immersed in the solution and shaken for 5 minutes. 1 ml sulfuric acid and 1 ml acetone were added for a total volume of 6 ml. After mixing, absorbances were measured at 495 nm.

# Results

## Kinetic Studies

Kinetic parameters include the maximum reaction rate ($V_{max}$) of the enzymatic reaction and the Michaelis-Menten constant ($K_m$) which defines the affinity of enzyme toward its substrate. Lower the Km value means higher its affinity against the substrate. These parameters were obtained from Lineweaver- Burk Plot [36] which is a plot of $1/v_o$ against $1/[S_0]$ for systems obeying the Michaelis-Menten equation. The graph being linear can be extrapolated at anywhere approximating to a saturating substrate concentration, even if no experiment has been performed and from the extrapolated graph, the values of $K_m$ and $V_{max}$ can be determined.

Free PPO has a maximum reaction rate of 17.6 μmol/min.mg protein and $K_m$ of 0.6 mM. $K_m$ is a parameter which is inversely proportional to the affinity of enzyme to substrate. Large $K_m$ values indicate that substrate and enzyme do not prefer to stay together for a long time. For the TATE matrice PPO has a $K_m$ value of 22 mM while free enzyme has a $K_m$ value of 0.6 mM. As expected, the affinity of the enzyme to substrate decreased when it is entrapped in a polymer matrix. However, TATE electrode provides a better environment to PPO compared to PPy electrode which has a $K_m$ value of 100 mM.

$V_{max}$ for PPy was found as 0.11μmol/min.electrode. For TATE enzyme electrode,it was obtained as 0.068μmol/min.electrode. When we compared this result with that of PPy matrix it is seen that $V_{max}$ of immobilized enzyme in TATE electrode was much lower than that of the PPy matrice. These results were also confirmed by the protein amount entrapped in the electrodes which were $9.6x10^{-3}$ mg protein for PPy and $6.0x10^{-3}$ mg protein for the PTATE matrice.

## Effect of Temperature on Enzyme Activity

Behavior of enzyme electrodes against temperature is another characteristic property that gives information about the stability of electrodes towards environmental conditions. The effect of temperature on the enzyme activity was investigated and illustrated in Figure 1. For free enzyme, maximum enzyme activity was observed at 40°C and it lost almost all of its activity after 50°C. Immobilized PPO in PPy electrode revealed a maximum activity at about same temperature and again it starts to lose its activity after 50°C, only this time gradually. On the other hand, TATE electrode revealed maximum activity at 70°C and after this temperature up to 80°C it lost only 40 % of its activity. These results imply that enzyme entrapped within this matrice showed high stability against temperature.

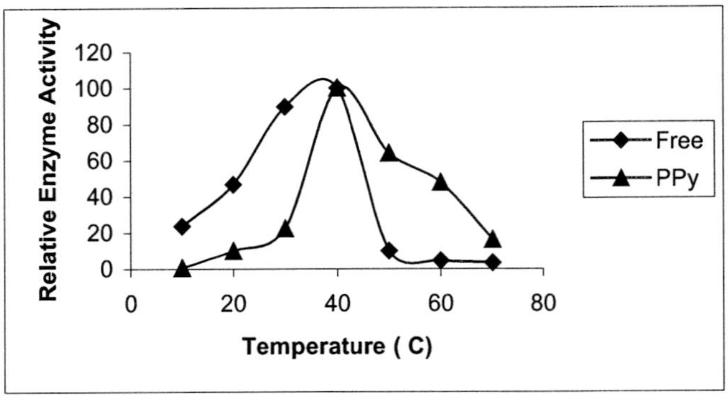


(a)

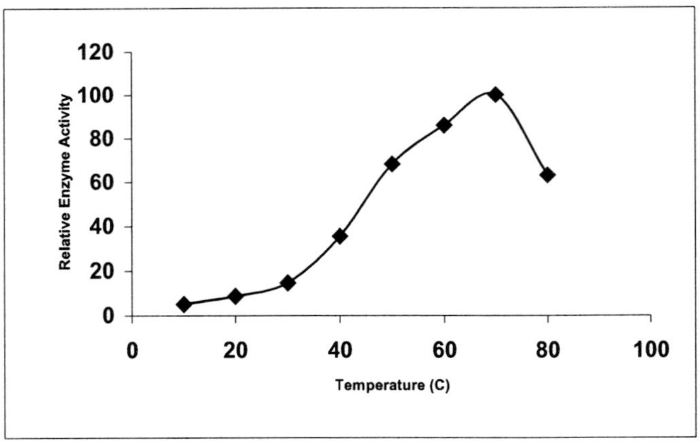


(b)

Figure1. Temperature effect on the activity of (a) free PPO and immobilized PPO in PPy/PPO electrode (b) immobilized PPO in PTATE/PPy/PPO electrode.

## Effect of pH on Enzyme Activity

Free PPO had an optimum pH of 5. Immobilized PPO revealed an optimal pH for PPy electrode as 7 which imply that pH stability is increased. For both PPy and TATE electrodes at levels of pH lower than 6, the activity of immobilized PPO was less sensitive to the pH change, indicating a better resistance of the immobilized protein molecules to the ionization, since the ionic state of the functional groups in or close to the active center has a great effect on enzyme activity.

The optimum pH value of both enzyme electrodes was shifted towards the alkaline side when compared with the free enzyme. This might be explained by the partitioning of protons. The negative groups of the matrix protons are concentrated around the enzymes and protect them against the high concentration of $OH^-$. This tendency makes the pH around enzyme lower than that of the bulk. Higher the tendency of matrix to concentrate protons in it, higher the pH stability of enzyme in this matrix.

At pH levels greater than the maximum points, the activity of entrapped PPO in PPy matrice was much affected by the pH change than that of the free enzyme. The positive charges on the matrix should decrease with increasing pH and so does the net charge of PPO where isoelectric point is 6.1. Therefore, the repulsive forces might have caused a split in the enzyme electrodes resulting in a less favorable conformation of the enzyme molecules.

The same behavior of shifting towards alkaline side in pH dependence was observed for TATE electrode, but these electrodes exhibit greater stability towards high pH (Figure 2). From pH 7 to 11 there is almost no change in the enzyme activity. This shows that this electrode can protect enzymes against high -OH concentration.

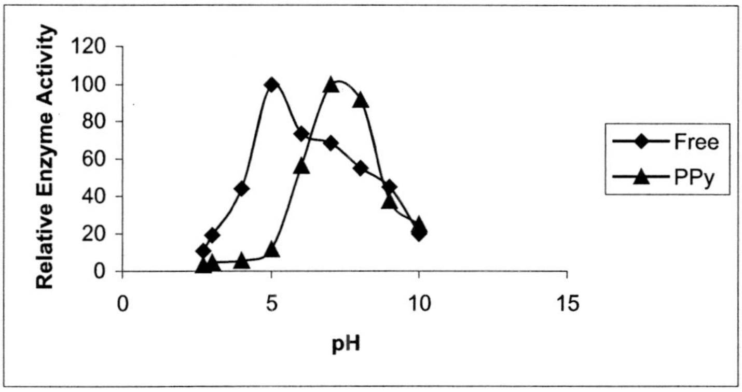


(a)

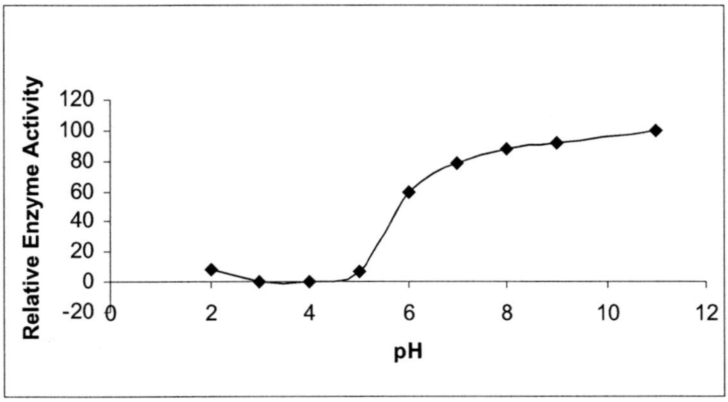


(b)

Figure2. pH effect on the activity of (a) free PPO and immobilized PPO in PPy/PPO electrode, (b) immobilized PPO in PTATE/PPy/PPO electrode.

## Operational Stability and Shelf-life of Enzyme Electrodes

Enzymes can easily lose their catalytic activity and denatured, thus, careful storage and handling are essential. Stability of electrodes in terms of repetitive uses was performed in 20 successive measurements. PPy electrode showed activity that gradually decreased up to 10$^{th}$ use and then stayed constant at 60% activity. In the case of TATE electrode, it exhibits a gradual decrease up to 10$^{th}$ use and stays constant at 20% enzyme activity until the twentieth use (Figure 3).

The activity of immobilized PPO was determined as a function of time over a period of 30 days. After 5$^{th}$ day an increase was observed which could be explained by the reorganization of enzyme molecules in the matrice. Up to 20$^{th}$ day, the activity was almost constant but this electrode lost 60% of its activity till the end of the experiment (Figure 4).

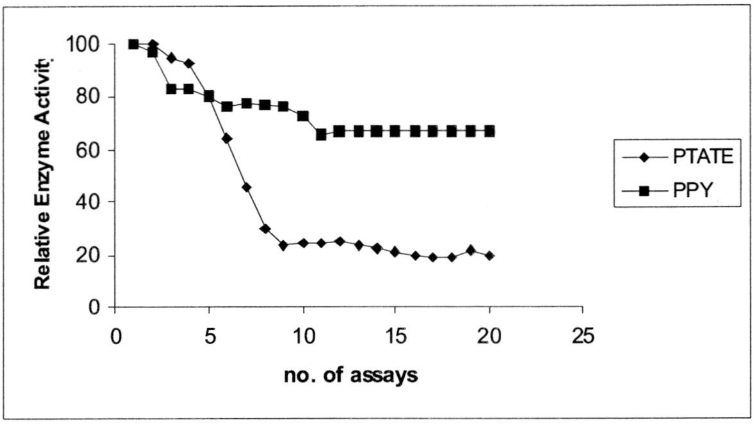


Figure 3. Operational stability for immobilized PPO in PTATE/PPy/PPO electrode.

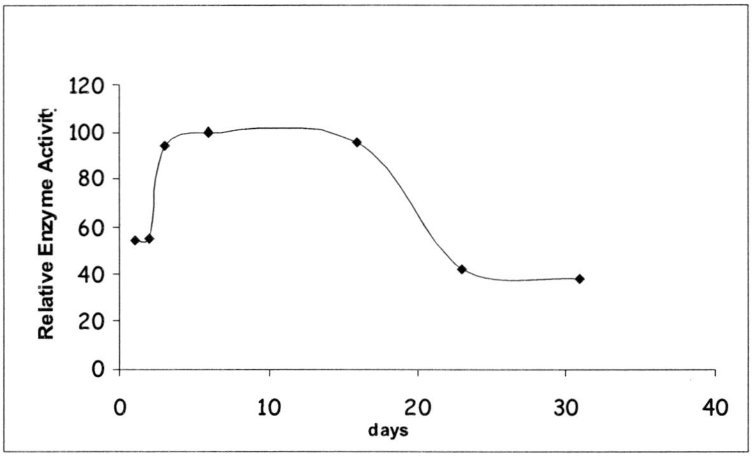


Figure 4. Shelf-life for immobilized PPO in PTATE/PPy/PPO electrode.

## Determination of Phenolic Compounds in Red Wines

This study was designated to determine total phenolic compounds by using the fabricated enzyme electrodes. Total phenolic compounds in Turkey wines were reported as 2000-3000 mg/L [37-39].

Polyphenol oxidase enzyme act on –OH groups on phenolic compounds. Via activity determination of enzyme electrodes in red wines we obtain the total –OH groups.

Two Turkish red wines (Brand K and Brand D) were analyzed for their concentration of phenolic compounds by using free and immobilized enzyme in pyrrole and PTATE matrices.

Results for phenolic determination by using free PPO enzyme give very small values when compared with both enzyme electrodes. As it is known from the literature, benzoates act as inhibitors for free PPO [40, 41]. So PPO is inhibited by benzoates, found naturally in wines, before it complete its enzymatic reactions. In previous studies, immobilized PPO was protected by the matrice that it entrapped and was not affected by the inhibitors found in the wine [23-25]. However, in this study, Table 1 shows that this microenvironment could not prevent the entry of the inhibitors into the matrice thus, enzyme was affected in terms of activity where the obtained results of phenolic amount in wines do not reflect the actual values.

Table1. Total phenolics in two different red wines, determined by the two enzyme electrodes.

| | *Free PPO** | *PPy/PPO** | *PTATE/PPy/PPO** |
|---|---|---|---|
| *Brand K* | 220mg/l | 4000mg/l | 830 mg/l |
| *Brand D* | 270mg/l | 2200mg/l | 920 mg/l |

*Total phenolics are expressed as Gallic acid equivalents [35].

Ki value is the inhibition constant which reflects the affinity of enzyme towards the inhibitors which was the benzoic acid in this case. It is known from the literature that benzoic acid is a competitive inhibitor toward the tyrosinase. Competitive inhibitors often closely resemble, in some respects, the substrates they inhibit and because of this structural similarity they may compete for the same binding-site on the enzyme. The enzyme bound inhibitor then either lacks the appropriate reactive group or is held in an unsuitable position with respect to the catalytic-site of enzyme or to other potential substrates for a reaction to take place. In either case, a dead-end-complex is formed and the inhibitor must dissociate from the enzyme [36].

The effect of a competitive inhibitor depends on the inhibitor concentration, the substrate concentration and the relative affinities of the substrate and the inhibitor for the enzyme. In general, at a particular inhibitor and enzyme concentration, if the substrate concentration is low, the inhibitor will compete favorably with the substrate and the degree of inhibition will be great. However, if the substrate concentration is high, inhibitor will be much less successful and the degree of inhibition will be less marked. At very high substrate concentrations, molecules of substrate will greatly outnumber molecules of inhibitor and the effect of inhibitor will be negligible. Hence $V_{max}$ for the reaction is unchanged. However, the apparent $K_m$ is clearly increased as a result of the inhibition and given the symbol $K_m'$.

Figure 5 exhibits the Lineweaver- Burk plots showing the effect of different inhibitor concentrations on free enzyme. A plot of $K_m'$ against inhibitor concentration $[I_0]$ will be linear with the intercept on the $[I_0]$ axis giving $-K_i$ (Figure 6).

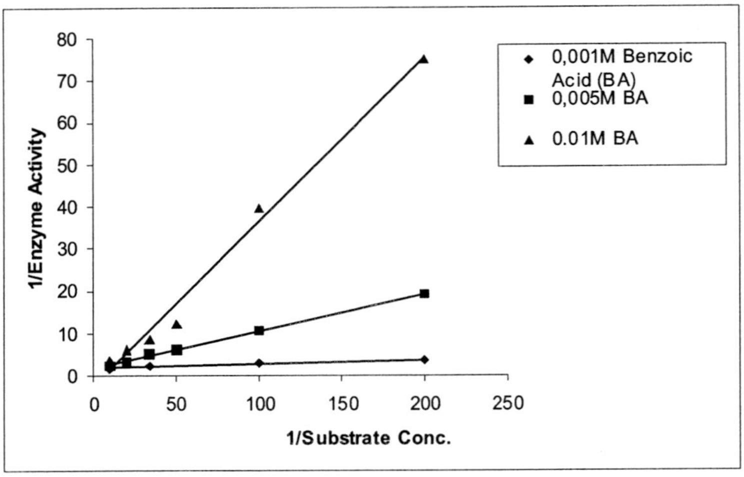


Figure 5. Lineweaver- Burk plots showing the effect of different inhibitor concentrations on free enzyme

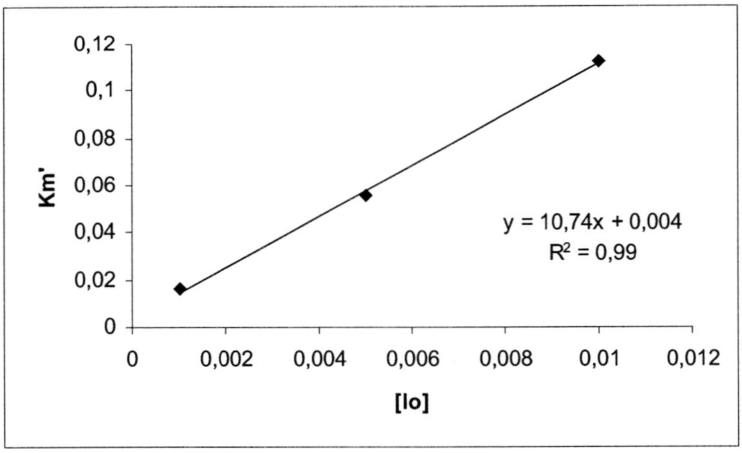


Figure 6. Secondary plot for inhibition of benzoic acid

Table 2 shows the comparison of $K_m$ and $K_i$ values for the free enzyme and immobilized enzyme in PPy and PTATE enzyme electrodes. As seen from the Table 2, as Km value increases which means the decrease of affinity of enzyme towards substrate, $K_i$ also increases. An increase in $K_i$ value means a decrease in the inhibitory effect of the inhibitor. According to these values, when compared with free enzyme, PTATE electrode protects the enzyme against inhibitors. However, polypyrrole electrode does this job much better than the PTATE electrode. Due to that reason, the values obtained in the phenolic analysis did not reflect the actual values in the case of PTATE electrode.

Table 2. Michaelis-Menten and Inhibition constants for free and immobilized tyrosinase.

| | $K_m$ *(mM)* | $K_i$ *(mM)* |
|---|---|---|
| *Free Tyrosinase* | 0.6 | 0.4 |
| *PPy/PPO* | 100 | 35 |
| *PTATE/PPy/PPO* | 22 | 4 |

## Conclusion

This study shows that immobilization of PPO can be successfully performed in PTATE matrice. As regards to the kinetic parameters, temperature and pH analysis, the results, imply that the fabricated enzyme electrode is successful. Especially strange pH behavior of enzyme electrode which protects enzymes against high pH, is a treatment which is a required property in enzyme immobilization studies. Usage stability, either operational stability or shelf life experiments, may exhibit a little inefficient results compared to polypyrrole matrice, but an overall evaluation of PTATE matrice reveal that it is suitable for the purpose.

Immobilization of polyphenol oxidase enzyme in a conducting polymer electrode was studied as an alternative method for determination of phenolic compounds in wines and results reveal that, by considering the inhibitory effects of substances found naturally in wines, this significant development can successfully replace the classical methods.

## References


[1] Nelson, JM; Dowson, CR. *Advances in Enzymology*, 1944 4, 99.

[2] Maddaluno, JF; Faull, KF. Fast Enzymatic Preparation of L-DOPA from Tyrosine and Molecular Oxygen: A Potential Method for Preparing [$^{15}$O] L-DOPA. *Applied Radiation and Isotopes*, 1990 41, 873-878

[3] Stocchi, F; Quinn, NP; Barbato, L; Patsalos, PN; O'Connel, MT; Ruggieri, S; Marsden, CD. Comparison between a Fast and a Slow Release Preparation of Levodopa and a Combination of the Two: A Clinical and Pharmacokinetic Study. *Clinical Neuropharmacology,* 1994 17, 38-44

[4] Lanouette, KH. Treatment of Phenolic Wastes. *Chemical Engineering*, 1977 84, 99-106.

[5] Atlow, SC; Banadonna-Aparo, L; Klibanov, AM. Dephenolization of Industrial Wastewaters Catalyzed by Polyphenol Oxidase. *Biotechnology and Bioengineering,* 1984 26, 599-603.

[6] Schiller, JG; Chen, AK; Liu, CC. Determination of Phenol Concentrations by an Electrochemical System with Immobilized Tyrosinase. *Analytical Biochemistry*, 1978 85, 25-33.

[7] Zachariah, K; Mottola, HA. Continuous-Flow Determination of Phenol with Chemically Immobilized Polyphenol Oxidase (Tyrosinase). *Analytical Letters*, 1989 22, 1145-1158.

[8] Tiliyer, CR; Gobin, PT. The Development of a Catechol Enzyme Electrode and its Possible use for Diagnosis and Monitoring of Neural Crest Tumors. *Biosensors and Bioelectronics*, 1991 6, 569-573
[9] Winder, AJ; Harris, H. New Assays for the Tyrosine Hydoxilase and Dopa Oxidase Activities of Tyrosinase. *European Journal of Biochemistry*, 1991 198, 317-326.
[10] Mittal, GS. Food Biotechnology. USA: Technomic Publishing Company; 1992.
[11] Selampinar, F; Akbulut, U; Toppare, L. Immobilization of Invertase in Conducting Polymer Matrices. *Biomaterials*, 1997 18, 1163-1169.
[12] Erginer, R; Toppare, L; Alkan, S; Immobilization of Invertase in Functionalized Copolymer Matrices. *Reactive and Functional Polymers*, 2000 45, 227-233.
[13] Tirkes, S; Toppare, L; Önen, A; Yagci, Y. Immobilization of Glucose Oxidase in Polypyrrole/Polytetrahydrofuran Graft Copolymers. *International Journal of Biological Macromolecules,* 2002 30, 81-87.
[14] Alkan, , S; Toppare, L; Bakir, U; Yagci, Y. Immobilization of Urease in Conducting Thiophene-Capped Poly(methyl methacrylate)/Pyrrole Matrices. *Synthetic Metals*, 2001 123, 95-99.
[15] Gursel, A; Alkan, S; Toppare, L; Yagci, Y. Immobilization of Invertase and Glucose Oxidase in Conducting H-type Polysiloxane/Polypyrrole Block Copolymers. *Reactive and Functional Polymers,* 2003 57, 57-65.
[16] Kizilyar, N; Akbulut, U; Toppare, L; Ozden, MY; Yagci, Y. Immobilization of Invertase in Conducting Polypyrrole/Polytetrahydrofuran Graft Polymer Matrices. *Synthetic Metals*, 1999 104, 45-50.
[17] Isik, S; Alkan, S; Toppare, L; Cianga, I; Yagci, Y. Immobilization of Invertase and Glucose Oxidase in Poly 2-methylbutyl-2(3-thienyl) acetate/Polypyrrole Matrices. *European Polymer Journal*, 2003 39, 2375-2381.
[18] Balci, Z; Akbulut, U; Toppare, L; Alkan, S; Bakir, U; Yagci, Y. Immobilization of Yeast Cells in Several Conducting Polymer Matrices. *Journal of Macromolecular Science – Pure and Applied Chemistry,* 2002 39, 183-197.
[19] Cirpan, A; Alkan, S; Toppare, L; Cianga, I; Yagci, Y. Immobilization of Cholesterol Oxidase in a Conducting Copolymer of Thiophene-3-yl Acetic Acid Cholesteryl Ester with Pyrrole. *Designed Monomers and Polymers,* 2003 6, 237-243.
[20] Cirpan, A; Alkan, S; Toppare, L; Hepuzer, Y; Yagci, Y. Immobilization of Invertase in Conducting Copolymers of 3-Methylthienyl Methacrylate. *Bioelectrochemistry,* 2003 59, 29-33.
[21] Vinas P; Lopez-Erroz C; Marin-Hernandez JJ; Hernandez-Cordoba M. Determination of phenols in wines by liquid chromatography with photodiode array and fluorescence detection. *Journal of Chromatography A*, 2000 871, 85-93.
[22] Zhang S; Zhao H; John R. A dual-phase biosensing system for the determination of phenols in both aqueous and organic media. *Analytical Chimica Acta*; 2001 441, 95-105.
[23] Kiralp, S; Toppare, L; Yağci Y. Immobilization of Polyphenol Oxidase in Conducting Copolymers and Determination of Phenolic Compounds in Wines with Enzyme Electrodes. *International. Journal of Biological Macromolecules,* 2003 33, 37-41.
[24] Kıralp, S, Toppare, L; Yağcı, Y. Determination of Phenolic Compounds in Wines with Enzyme Electrodes Fabricated by Immobilization of Polyphenol Oxidase in Conducting Copolymers. *Designed Monomers and Polymers,* 2004 7, 3-7.

[25] Kıralp, S; Cirpan, A; Toppare L. Enzyme Electrodes for Determination of Total Phenolic Capacity of Red Wines. *Journal of Applied Polymer Science. (submitted)*
[26] Robert, C; Rouch, C; Cadet F. Inhibition of palmito (Acanthophoenix rubra) polyphenol oxidase by carboxylic acids *Food Chemistry,* 1997 59, 355-361.
[27] Gunata, YZ, Sapis, JC; Moutonet, M. Substrates and Aromatic Carboxylic acid Inhibitors of Grape Polyphenol Oxidases. *Phytochemistry,* 1987 26, 1573-1575.
[28] Janovitz-Klapp, AH; Richard, FC; Goupy, PM; Nicolas, J. Inhibition Studies on Apple Polyphenol Oxidase. *Journal of Agricultural Food Chemistry,* 1990 38,926-931.
[29] Kermasha, S; Goetghebeur, M; Monfette, A. Studies on Inhibition of Mushroom Polyphenol Oxidase using Chlorogenic Acid as Substrate. *Journal of Agricultural Food Chemistry,* 1993 41, 526-531.
[30] McEvily, AJ; Iyengar, R; Gross, AT. Inhibition of Polyphenol Oxidase by Phenolic Compounds. In P*henolic compounds in Food and Their Effects on Health.* Vol.I, *Analysis, Occurance and Chemistry*, Washington, DC: eds Chi-Tang Ho,Chang Y. Lee and Mou-Tang Huang. American Chemical Society; 1992.
[31] Walker, JRL. The Enzymatic Control of Browning in Fruit Juices by Cinnamic Acids. *Journal of Food Technology*, 1976 11, 341-345.
[32] Coskun, Y; Cirpan, A; Toppare, L. Conducting polymers of Terepthalic Acid bis-(2-thiophen-3-yl ethyl) Ester and Their Electrochromic Properties. *Polymer,* 2004 45, 4989-4995.
[33] Mazzocco, F; Pifferi, PG. An Improvement of the Spectrophotometric Method for the Determination of Tyrosinase Catecholase Activity by Besthorn's Hydrazone. *Analytical Biochemistry* 1976 72, 643-647.
[34] Russell, IM; Burton, SG. Development and Demonstration of an Immobilized-Polyphenol Oxidase Bioprobe for the Detection of Phenolic Pollutants in Water. *Analytical Chimica Acta,* 1999 389, 161-170.
[35] Rodriguez-Lopez, JN; Escribano, J; Garcia-Canovas, FA. A Continuous Spectrophotometric Method for the Determination of Monophenolase Activity of Tyrosinase Using 3-methyl-2-benzothiazolinone hydrazone. *Analytical Biochemistry,* 1994 216, 205-212.
[36] Palmer, T. Understanding Enzymes. 4th ed. London: Prentice Hall; 1995.
[37] Lopez, M; Martinez, F; Del Valle, C; Orte, C; Miro, M. Analysis of phenolic constituents of biological interest in red wines by high-performance liquid chromatography. *Journal of Chromatography A,* 2001 922, 359-363.
[38] Sakkiadi, AV; Stavrakakis, MN; Haroutounian, SA. Direct HPLC assay of five biologically interesting phenolic antioxidants in varietal Greek red wines. *Lebensmittel-Wissenschaft und-Technologie,* 2001 34, 410-413.
[39] Karakaya, S; El, SN; Taş AA. Antioxidant activity of some foods containing phenolic compounds. *International Journal of Food Sciences and Nutrition,* 2001 52, 501-508.
[40] Tricand de la Gôutte, J; Khan, JA; Vulfson, EN. Identification of novel polyphenol oxidase inhibitors by enzymatic one-pot synthesis and deconvolution of combinatorial libraries. *Biotechnology and Bioengineering,* 2001 75, 93-99.
[41] Stanca, SE; Popescu, IC. Amperometric Study of the Inhibitory Effect of Carboxilic Acids on Tyrosinase. *Journal of Molecular Catalysis,* 2004 27, 221-225.

In: Frontiers in Polymer Research
Editor: Robert K. Bregg, pp. 171-209 ISBN: 1-59454-824-2


*Chapter 7*

# On Compatibility of Polymer Blends

***Fatemeh Sabzi*[*] *and Ali Boushehri***
Department of Chemistry, College of Sciences, Shiraz University,
Shiraz 71454, Iran

## Abstract

Due to their technological importance, polymer blends have attracted considerable attention during the past decade. For thermodynamic reasons, most polymer pairs are immiscible and their degree of compatibility is of underlying importance to the microphase structure and consequently, to the mechanical properties of the blend. The Flory-Huggins $\chi$ interaction parameter for the polymer pair plays a dominant role in explaining critical phase behavior of a compatible pair and in estimating interfacial tension and interfacial thickness for semicompatible or incompatible pairs. Direct measurement of this parameter is not always possible, thus the obtained information, in conjunction with suitable theoretical models of polymer solutions may lead to an assessment of the interaction parameters for the actual polymeric case.

In this work we present a theoretical discussion regarding this interaction parameter for 10 polymer-polymer-solvent systems, 4 copolymer-solvent systems along with their corresponding polymer pairs. Our polymer blends are real mixtures of 5 homopolymers consist of poly(N,N-dimethyl methacrylamide) (PDMAA), poly(2-dimethyl aminoethyl methacrylate) (PDMAEMA), poly(acrylic acid) (PAA), a typical membrane of commercial soft-contact lens i.e. poly(2-hydroxyethyl methacrylate) (PHEMA), and poly(N-vinyl-2-pyrrolidone) (PVP) all with water solvent. Copolymers studied are poly(acrylonitrile-co-butadiene) in acetonitrile, poly(styrene-co-acrylonitrile) in 1,2-dichloroethane, poly(acrylonitrile-co-butadiene) in hexane and poly(acrylonitrile-co-butadiene) in pentane.

For ternary systems, the results are expressed in terms of $\chi_{1,23}$ which reduces to the classical Flory-Huggins $\chi_{12}$ interaction parameter for the case of binary mixtures. The data on $\chi_{1,23}$ may be used for an approximate estimation of the $\chi'_{23}$ interaction parameter for the limiting case of zero solvent concentration. For this purpose, at the end of each subsection of the tables, the limiting value of $\chi^{\infty}_{1,23}$ is given. The limiting values of $\varphi^{\infty}_{2}$ , $\varphi^{\infty}_{3}$ and $\chi'_{23}$ are also appeared at the end of each table. It should be noted that these values are obtained by graphical extrapolation of the data to the zero concentration of solvent.

[*] E-mail address: fche1_sabzi@yahoo.com, Fax: ++98-711-2280926

# 1 Introduction

Continuing demand for polymer blends with desired properties has been propelling extensive research on liquid-liquid equilibria of polymer blends [1-5]. One motivation behind these studies is the search for miscible pairs of polymers, including copolymers, because such blends can exhibit a variety of mechanical, electrical, thermal, optical and other properties. In addition to their potential for new materials, random copolymers are also used as compatibilizers for blends of homopolymers that are immiscible without compatibilizers.

Miscibility of polymer blends has been often defined as the capability of a mixture to form a single phase over certain ranges of temperature, pressure and composition. Whether or not a single phase exists depends on the chemical structure, molar mass distribution and molecular architecture of the components present. For a two-component mixture, a necessary and sufficient condition for stable or metastable equilibrium of a homogeneous, single-phase is

$$\left(\frac{\partial^2 \Delta_{mix} G}{\partial \varphi^2}\right)_{T,P} > 0$$

where $\Delta_{mix} G$ is the Gibbs energy of mixing per unit volume, and $\varphi$ the composition, where $\varphi$ is usually takes as the volume fraction of one of the component substances. The system is unstable if the above second derivative is negative and will demix.

An immiscible polymer blend that exhibits macroscopically uniform physical properties is called compatible. Compatibility means the capability of individual component substances in either an immiscible polymer blend or a polymer composite to show interfacial adhesion that in which interfaces between phases or components are maintained by intermolecular forces, chain entanglements, or both, across the interfaces. In the case of block copolymers, if their parent 's homopolymers A and B are incompatible, the copolymer may show *segregation*, i.e., the A-segments and the B-segments are then located in separate *domain* [6]. The term domain may be used for an entity of a material system that is uniform in chemical composition and physical state.

Compatibility of polymer blends is often achieved through favorable specific interaction such as hydrogen bonding. Although a fundamental understanding of the pertinent thermodynamics plays a crucial role in the preparation of blends, there are few useful molecular thermodynamic models for polymer blends with specific interactions, a major exception is the classical incompressible model developed by Flory and Huggins [7-8]. The objective of this work is to develop an approximate but theoretically based molecular model for predicting compatibility of polymer blends within the framework of a lattice model.

In this work, Compatibility parameters $\chi'_{23}$ in 11 random copolymer-solvent systems and in 30 polymer-polymer-solvent systems are successfully predicted by the classical Flory-Huggins model. Table 1 gives a summary of all polymers and copolymers with their solvents that are considered in this chapter. Also the interaction parameter $\chi_{12}$ for the corresponding polymer pairs is evaluated. For ternary systems, the results are expressed in terms of $\chi_{1,23}$

parameter which reduces to the classical Flory-Huggins $\chi_{12}$ interaction parameter for the case of binary mixtures.

Table 1. Summary of polymers with their relevant solvents

| Polymer | $V_{sp}$ ($cm^3g^{-1}$) | Solvent | T (°c) | $P_s$ (KPa) | $V_{sp}$ ($cm^3g^{-1}$) |
|---|---|---|---|---|---|
| Poly(acrylonitrile) | 0.845[a] | Acetonitrile | 60 | 50.73[d] | 1.374[d] |
| Poly(cis-1,4-butadiene) | 1.099[a] | Acetonitrile | 60 | 50.73 | 1.374 |
| Poly(acrylonitrile) | 0.845 | 1,2-Dichloroethane | 70 | 66.65[e] | 0.849[f] |
| Polystyrene | 0.810[a] | 1,2-Dichloroethane | 70 | 66.65 | 0.849 |
| Poly(acrylonitrile) | 0.845 | 1,2-Dichloroethane | 80 | 93.31[e] | 0.861[f] |
| Polystyrene | 0.810 | 1,2-Dichloroethane | 80 | 93.31 | 0.861 |
| Poly(acrylonitrile) | 0.845 | Hexane | 60 | 76.36[g] | 1.608[g] |
| Poly(cis-1,4-butadiene) | 1.099 | Hexane | 60 | 76.36 | 1.608 |
| Poly(acrylonitrile) | 0.845 | Pentane | 60 | 214.5[g] | 1.709[g] |
| Poly(cis-1,4-butadiene) | 1.099 | Pentane | 60 | 214.5 | 1.709 |
| Poly(N,N-dimethyl methacrylamid) | 0.869[b] | water | 35 | 5.627[h] | 1.000 |
| Poly(2-dimethyl aminoethylmethacrylate) | 0.909[b] | water | 35 | 5.627 | 1.000 |
| Poly(acrylic acid) | 0.869[b] | water | 35 | 5.627 | 1.000 |
| Poly(2-hydroxyethyl methacrylate) | 0.869[b] | water | 35 | 5.627 | 1.000 |
| Poly(N-vinyl-2-pyrrolidone) | 0.862[c] | water | 35 | 5.627 | 1.000 |

[a] These values are taken from [17].
[b] This value is taken from [15].
[c] These values are taken from [16].
[d] This value is taken from [18].
[e] This value is taken from [19].
[f] This value is taken from [20].
[g] This value is taken from [21].
[h] This value is taken from [22].

## 2 The Lattice Model

Ideal solution behavior over extended ranges in both composition and temperature requires that the following conditions be fulfilled [9]:

(i) the entropy of mixing must be given by

$$\Delta S_m = -k[n_1 \ln X_1 + n_2 \ln X_2] \tag{1}$$

where $n_1$ and $n_2$ are the numbers of molecules of solvent and solute, respectively, and $X_1$ and $X_2$ their mole fractions; and (ii) the heat of mixing $\Delta H_M$ must equal zero. Deviations from ideality may arise from failure of either these conditions. Early work [10] on polymer solutions revealed that the deviations from ideality were not strongly temperature-dependent; hence it was concluded that condition (i), at least, is not fulfilled, a conclusion abundantly confirmed by more recent work. We shall therefore first derive an expression for the entropy of mixing polymer and solvent with which to replace Eq. (1).

The ideal entropy of mixing expression, Eq. (1), may be derived by considering a binary solution consisting of two types of molecules virtually identical in size, spatial configuration, and external force field. In such a mixture a molecule of one type may be replaced by one of the other solution. The greater entropy of the solution as compared with the pure components arises entirely from the greater number of arrangements in so simple a binary system is easily calculated.

The molecules in the pure liquids and in their solution are considered to be solid-like, in a quasicrystalline state, where the molecules do not translate fully in a chaotic manner as in a gas, but where each molecule tends to stay in a small region, a more or less fixed position in space about which it vibrated back and forth. The quasicrystalline picture of the liquid state supposes molecules to sit in a regular array in space, called a *lattice*, and therefore liquid and liquid mixture models based on this simplified picture are called *lattice models*.

We consider a mixture of two simple liquids 1 and 2 are small and spherically symmetric and the ratio of their sizes is close to unity. We suppose that the arrangement of the molecules in each pure liquid is that of a regular array; all the molecules are situated on lattice points that are equidistant from one another. Molecular motion is limited to vibration about the equilibrium positions and is not affected by the mixing process. We suppose further that for a fixed temperature, the lattice spacing for the two pure liquids and for the mixture is the same, independent of composition. These assumption having been accepted as required, we consider the total number of ways of arranging the $n_1$ identical molecules of the solvent and $n_2$ identical molecules of the solute on the lattice comprising $n_0 = n_1 + n_2$ cells. This just the number of combinations of $n_0$ things taken $n_1$ at a time, or $\Omega = n_0!/n_1!n_2!$. Whereas the pure components may be arranged in their respective lattices in only one way, the number of arrangements possible in the solution is given by the very large number Ω. It follows according to the Boltzmann relation that the entropy of mixing should be given by $\Delta S_M = k \ln \Omega$. With the introduction of Stirling's approximation, $\ln n! = n \ln n - n$, for the factorials, we have [11]

$$\Delta S_M = k\left[(n_1 + n_2)\ln(n_1 + n_2) - n_1 \ln n_1 + n_2 \ln n_2\right] \tag{2}$$

which reduces by suitable rearrangement to Eq. (1).

In spite of the fact that the ideal expression has proven to be a most useful generalization for solutions of simple molecules but there are deviations from ideal behavior in liquid solutions due primarily to the following effects: First, forces of attraction between unlike molecules are quantitatively different from those between like molecules, giving rise to a nonvanishing enthalpy of mixing; second, if the unlike molecules differ significantly in size or shape, the molecular arrangement in the mixture may be appreciably different from that for the pure liquids, giving rise to mixing; and finally, in a binary mixture, if the forces of attraction between one of three possible pair interactions are very much stronger (or very much weaker) than those of the other two, there are certain preferred orientations of the molecules in the mixture that, in extreme cases, may induce thermodynamic instability and demixing (incomplete miscibility).

Thus, this treatment, resting essentially on the assumed approximate interchangeability of molecules of solvent and solute in the solution, can not possibly hold for polymer solutions in which the solute molecule may be a thousand or more times the size of the solvent. The long chain polymer may be considered to consist of $r$ chain segments, each of which is equal in size to a solvent molecule. $r$ is, of course, the ratio of the molar volumes of the solute and solvent. A segment and a solvent molecule may replace one another in the liquid lattice. In other respects the assumptions required are equivalent to those used above. The polymer solution differs from that containing an equal proportion of monomeric solute in the one important respect that sets of $r$ contiguous cells in the lattice are required for accommodation of polymer molecules, whereas no such restriction applies to the solution of the monomeric solute.

## 3 The Flory-Huggins Theory

The Gibbs energy of mixing consists of an enthalpy term and an entropy term. The theory of regular solutions for molecules of similar size assumes that the entropy term corresponds to that for an ideal solution and attention is focused on the enthalpy of mixing; however, when considering solutions of molecules of very different size, it is advantageous to assume, at first, that the enthalpy of mixing is zero and to concentrate on the entropy of mixing. Solutions with zero enthalpy of mixing are called *athermal* solutions because, when mixed at constant temperature and pressure, there is no liberation or absorption of heat. Athermal behavior is never observed exactly but it is approximated by mixtures of components that are similar in their chemical characteristics even if their sizes are different.

It is convenient to write the thermodynamic mixing properties as the sum of two parts: (1) a *combinatorial* or *configurational* contribution that appears in the entropy (and therefore in the Gibbs energy and in the Helmholtz energy) but not in the enthalpy of mixing; and (2) a *residual* contribution, determined by differences in intermolecular forces and in free volumes between the components. For the entropy of mixing, we write

$$\Delta_{mix} S = \Delta S^C + S^R \tag{3}$$

where superscript $C$ stands for configurational and superscript $R$ stands for residual.

Consider a mixing process where the molecules of fluids 1 and 2 have no difference in molecular interactions and in free volume. For this case, isothermal, isobaric mixing occurs also at constant volume; the residual mixing properties are zero and we are concerned only with combinatorial mixing properties. Using concept of a quasicrystalline lattice as a model for a liquid, an expression for the configurational entropy of mixing was derived independently by Flory [7] and by Huggins [8] for flexible chain molecules that differ significantly in size. The derivation, based on statistical arguments and several well-defined assumptions has been presented in several references [9]; we give here only a brief discussion.

We consider a mixture of two liquids 1 and 2. Molecules of type 1 (solvent) are single spheres. Molecules of type 2 (polymer) are assumed to behave like flexible chains, i.e., as if they consist of a large number of mobile segments, each having the same size as that of a solvent molecule. Further, it is assumed that each site of the quasilattice is occupied by either a solvent molecule or a polymer segment and that adjacent segments occupy adjacent sites. Let there be $n_1$ molecules of solvent and $n_2$ molecules of polymer and let there be $r$ segments in a polymer molecules. The total number of lattice sites is $(n_1 + rn_2)$. Fractions $\varphi_1$ and $\varphi_2$ of sites occupied by the solvent and by the polymer are given by

$$\varphi_1 = \frac{n_1}{n_1 + rn_2} \qquad \varphi_2 = \frac{n_2}{n_1 + rn_2} \tag{4}$$

Flory and Huggins have shown that if the amorphous (i.e. noncrystalline) polymer and the solvent mix without any energetic effects (i.e. athermal behavior), the change in Gibbs energy and entropy of mixing are given by the remarkably simple expression:

$$\Delta S^C = -k(n_1 \ln \varphi_1 + n_2 \ln \varphi_2) \tag{5}$$

The entropy change in Eq. (5) is similar in form to that of Eq. (1) for a regular solution except that segment fractions are used rather than mole fractions. For the special case $r = 1$, the change in entropy given by Eq. (5) reduces to that of Eq. (1), as expected. However, when $r > 1$, Eq. (5) always gives a configurational entropy larger than that given by Eq. (1) for the same $n_1$ and $n_2$.

To apply the theoretical result of Flory and Huggins to real polymer solutions, i.e., to solutions that are not athermal, it has become common practice to add to the configurational part of the entropy, a semiemperical part for the residual contribution. In other words, we add a term that, if there is no difference in free volumes, is given by the enthalpy of mixing which we recast it in the form of

$$\Delta H_M = kT\chi_{12} n_1 \varphi_2 \tag{6}$$

where $\chi_{12}$ is a dimensionless quantity which characterizes the interaction energy per solvent molecule. The quantity $kT\chi_{12}$ represents merely the difference in energy of a solvent molecule immersed in the pure polymer $(\varphi_2 \cong 1)$ compared with one surrounded by molecules of its own kind, i.e., in the pure solvent. The heat of mixing expression, like the entropy of mixing expression, retains no parameters of the hypothetical lattice.

If the configurational entropy is assumed to represent the total entropy change $\Delta S_M$ on mixing, the free energy of mixing is simply obtained by combining Eqs. (5) and (6), that is,

$$\Delta F_M = \Delta \mathrm{H}_M - T\Delta S_M$$

$$= kT\left(n_1 \ln \varphi_1 + n_2 \ln \varphi_2 + \chi_{12} n_1 \varphi_2\right) \quad (7)$$

This equation expresses the total free energy change for the formation of the solution from pure, disoriented polymer (i.e. amorphous or liquid polymer) and pure solvent. The effect of accepting the configurational entropy of mixing as a proper expression for the total entropy of mixing is to neglect possible contribution which may arise from specific interactions between neighboring components (solvent molecules and polymer segments) of the solution.

The chemical potential $\mu_1$ of the solvent in the solution relative to its chemical potential $\mu_1^0$ in the pure liquid is obtained by differentiating the free energy of mixing, $\Delta F_M$, with respect to the number $n_1$ of solvent molecules. Differentiation of Eq. (7) with respect to $n_1$ (bearing in mind that $\varphi_1$ and $\varphi_2$ are functions of $n_1$) and multiplication of the result by Avogadro's number $N$ in order to obtain the chemical potential per mole gives

$$\mu_1 - \mu_1^0 = RT\left[\ln \varphi_1 + (1 - \frac{1}{r})\varphi_2 + \chi_{12}\varphi_2^2\right] \quad (8)$$

From the chemical potential we may at once set down expressions for the activity $a_1$ of the solvent, using standard relation of thermodynamics.

$$(\mu_1 - \mu_1{}^0) = RT \ln a_1 \quad (9)$$

$$\ln a_1 = \ln \varphi_1 + \left(1 - \frac{1}{r}\right)\varphi_2 + \chi_{12}\varphi_2{}^2 \quad (10)$$

Since the pure solvent has been chosen as the standard state, $a_1 = P_1/P_1^0$, to the approximation that the vapor may be regarded as an ideal gas neglecting the vapor pressure of plymer.

The same principles can also be applied to mixtures of two amorphous polymers1 and 2 or to a ternary polymer-polymer-solvent system. The presence of polymer 1 reduces the possible arrangements of monomeric units of polymer 2: the molar entropy of mixing can never become as positive as in polymer-solvent systems. The resulting entropy term is only slightly negative and can no longer compensate the positive enthalpy term

$\Delta H_M = kT\chi_{12} n_1 \varphi_2$ if the interaction parameter, $\chi_{12}$, is positive [12]. The molar Gibbs energy of mixing becomes positive; the polymer-polymer system can not exist as one phase and demixes. For a ternary system the solvent activity is given by [13]

$$\ln a_1 = \ln \varphi_1 + \left(1 - \frac{1}{r_2}\right)\varphi_2 + \left(1 - \frac{1}{r_3}\right)\varphi_3$$

$$+ \left(\chi_{12}\varphi_2 + \chi_{13}\varphi_3\right)\left(1 - \varphi_1\right) - \chi'_{23}\,\varphi_2\varphi_3 \qquad (11)$$

However, in the general case, $\chi_{12}$ and $\chi_{13}$ vary with composition and without a knowledge of this composition dependence, Eq. (11) can not be used for the evaluation of $\chi'_{23}$ parameter. It is important to observe that Eq. (11) may be used in the limiting case $\varphi_1 \to 0$. If $X_{2P}$ indicates mole fraction of component 2 in the polymer mixture (zero solvent concentration) and similarly, $X_{3P}$ for component 3, we may define $r_{23}$ and $\varphi_{23}$ as:

$$r_{23} = X_{2P} r_2 + X_{3P} r_3 \qquad (12)$$

and

$$\varphi_{23} = \varphi_2 + \varphi_3 = 1 - \varphi_1 \qquad (13)$$

In addition, $\chi_{1,23}$ is defined as

$$\chi_{1,23} = \left(\left(\chi_{12}\varphi_2 + \chi_{13}\varphi_3\right)\left(1 - \varphi_1\right) - \chi'_{23}\,\varphi_2\varphi_3\right) / \varphi_{23}^2$$

With these definitions, the solvent activity in the ternary system may be written as [13]

$$\ln a_1 = \ln \varphi_1 + \left(1 - \frac{1}{r_{23}}\right)\varphi_{23} + \chi_{1,23}\varphi_{23}^{\;2} \qquad (14)$$

in direct analogy with Eq. (10). Equation (14) reduces to Eq. (10) for the binary case. The use of Eq. (14) has the advantage of allowing to reporting data on ternary systems regardless of the dependence of $\chi_{12}$ and $\chi_{13}$ on composition.

The data on $\chi_{1,23}$ may now be used for an approximate estimation of the $\chi'_{23}$ interaction parameter for the limiting case of zero solvent concentration. For this purpose, at the end of each subsection of the tables, the limiting value of $\chi_{1,23}^{\infty}$ is shown. These values were obtained by graphical extrapolation of the data on lower solvent concentration. It should be stressed that there is an uncertainty in this extrapolation, especially when the data do not show any clear trend. This is why values thus obtained should be considered to be an approximate estimation of $\chi'_{23}$.

In terms of this limiting quantities, $\chi'_{23}$ is given by [13]

$$\chi'_{23} = \frac{\chi_{12}^{\infty}\varphi_2^{\infty} + \chi_{13}^{\infty}\varphi_3^{\infty} - \chi_{1,23}^{\infty}}{\varphi_2^{\infty}\varphi_3^{\infty}} \left(\varphi_1^{\infty} \to 0\right) \tag{15}$$

In Eq. (15), $\varphi_2^{\infty}$ and $\varphi_3^{\infty}$ are the segment fractions of components 2 and 3 respectively in the polymeric mixture. From the limiting values of $\chi_{1,23}$ at zero solvent concentration and Eq. (15), the $\chi'_{23}$ parameter is approximately estimated.

# 4 Application to Experimental Data

## 4.1 System: Acetonitrile($CH_3CN$)(1)-Polyacrylonitrile(PAN)(2)-Poly(cis-1,4-butadiene)(cis-Bu)(3) at 60°C

This system was studied in order to see the influence of the solvent on the $\chi'_{23}$ interaction parameter. Experimental VLE data for poly(acrylonitrile-co-butadiene) and its parent homopolymers that are reported in Table 2 have been taken from Gupta et al. work [14]. As it as shown, at a given activity, solvent absorption in polyacrylonitrile is higher than in polybutadiene because polar acetonitrile molecules prefer polar segments of polyacrylonitrile to hydrocarbon segments of polybutadiene. Intuitively, one might expect that the copolymer curve should lie between the two homopolymer curves, but Table 2 shows that the copolymer curve lie beyond the bound of the two homopolymer curves.

Table 2. Vapor-pressure lowering data for the system: Acetonitrile(1)-Polyacrylonitrile(2)-Poly(cis-1,4-butadiene)(3) at 60°c.

| | $W_{2p}$=0.0 | | |
|---|---|---|---|
| $m_s/m_p$ | $P/P^0$ | $\varphi_{23}$ ($\varphi_3$) | $\chi_{1,23}$ ($\chi_{13}$) |
| 0.0101 | 0.1321 | 0.9875 | 1.4070 |
| 0.0101 | 0.2622 | 0.9875 | 2.1101 |
| 0.0173 | 0.3607 | 0.9788 | 1.9379 |
| 0.0246 | 0.4573 | 0.9702 | 1.8694 |
| 0.0309 | 0.6406 | 0.9628 | 2.0308 |
| 0.0406 | 0.7471 | 0.9517 | 1.9730 |

$\chi_{1,23}^{\infty}$ ($\chi_{13}^{\infty}$) = 1.7119

Table 2. Continued

| | | $W_{2p}$=0.21 | | |
|---|---|---|---|---|
| $m_s/m_p$ | $P/P^o$ | $\varphi_{23}$ | $\varphi_2$ | $\chi_{1,23}$ |
| 0.0111 | 0.1340 | 0.9854 | 0.1695 | 1.2678 |
| 0.0267 | 0.2700 | 0.9656 | 0.1661 | 1.1754 |
| 0.0695 | 0.5145 | 0.9152 | 0.1575 | 1.0597 |
| 0.0929 | 0.5933 | 0.8899 | 0.1531 | 1.0025 |
| 0.1261 | 0.6722 | 0.8561 | 0.1473 | 0.9349 |
| 0.2300 | 0.7747 | 0.7653 | 0.1317 | 0.7324 |

$\chi_{1,23}^{\infty}$ =1.2720

| | | $W_{2p}$=0.33 | | |
|---|---|---|---|---|
| $m_s/m_p$ | $P/P^o$ | $\varphi_{23}$ | $\varphi_2$ | $\chi_{1,23}$ |
| 0.0235 | 0.1321 | 0.9693 | 0.2647 | 0.5203 |
| 0.0449 | 0.2622 | 0.9429 | 0.2576 | 0.6545 |
| 0.0695 | 0.3607 | 0.9144 | 0.2498 | 0.6266 |
| 0.1001 | 0.4573 | 0.8812 | 0.2407 | 0.6010 |
| 0.1738 | 0.6406 | 0.8104 | 0.2214 | 0.6110 |
| 0.2594 | 0.7471 | 0.7411 | 0.2024 | 0.5800 |

$\chi_{1,23}^{\infty}$ =0.5958

| | | $W_{2p}$=0.51 | | |
|---|---|---|---|---|
| $m_s/m_p$ | $P/P^o$ | $\varphi_{23}$ | $\varphi_2$ | $\chi_{1,23}$ |
| 0.0214 | 0.1340 | 0.9714 | 0.4184 | 0.6064 |
| 0.0493 | 0.2701 | 0.9365 | 0.4034 | 0.5831 |
| 0.1249 | 0.5145 | 0.8535 | 0.3677 | 0.5530 |
| 0.1628 | 0.5933 | 0.8172 | 0.3520 | 0.5392 |
| 0.2225 | 0.6722 | 0.7658 | 0.3299 | 0.4922 |
| 0.4144 | 0.7747 | 0.6371 | 0.2744 | 0.2988 |

$\chi_{1,23}^{\infty}$ = 0.6600

| | $W_{2p}$=1.0 | | |
|---|---|---|---|
| $m_s/m_p$ | $P/P^o$ | $\varphi_{23}$ ($\varphi_2$) | $\chi_{1,23}$ ($\chi_{12}$) |
| 0.0111 | 0.1321 | 0.9822 | 1.0606 |
| 0.0225 | 0.2622 | 0.9647 | 1.1176 |
| 0.0331 | 0.3610 | 0.9489 | 1.1177 |
| 0.0427 | 0.4573 | 0.9349 | 1.1617 |
| 0.0650 | 0.6406 | 0.9044 | 1.2200 |
| 0.0822 | 0.7471 | 0.8819 | 1.2384 |

$\chi_{1,23}^{\infty}$ ($\chi_{12}^{\infty}$) = 1.0397

| $W_{2p}$ | $\varphi_2^\infty$ | $\varphi_3^\infty$ | $\chi'_{23}$ |
|---|---|---|---|
| 0.21 | 0.1720 | 0.8280 | 2.2770 |
| 0.33 | 0.2731 | 0.7269 | 4.6975 |
| 0.51 | 0.4307 | 0.5693 | 3.1093 |

At a given pressure, acetonitrile solubility in a copolymer is much higher than that in the corresponding homopolymers. This non-intuitive behavior is attributed to intramolecular repulsion between unlike segments of the copolymer. This repulsive interaction is weakend when acetonitrile molecules are in the vicinity of unlike copolymer segments, favoring copolymer+solvent miscibility.

Also, we can see that in spite of insolubility of parent homopolymers in acetonitrile ($\chi$>1), copolymers are soluble in this solvent and their solubility increases with enhancement of acrylonitrile weight fraction because of interaction between polar groups. From, the limiting values of $\chi_{1,23}$ at zero solvent concentration and Eq. (15), the pair interaction parameter $\chi'_{23}$ is obtained. It is evident from the result of $\chi'_{23}$ quantity that the two polymers are incompatible in all concentration ranges.

## 4.2 System: 1,2-Dichloroethane($ClC_2H_4Cl$)(1)- Polystyrene(PS)(2)- Poly(acrylonitrile)(PAN)(3) at 70 and 80°C

Tables 3 and 4 display the experimental VLE data for poly(styrene-co-acrylonitrile) and its parent homopolymers that are taken from Gupta et al .work [14]. At a given activity, the solubility of 1,2-dichloroethane decreases dramatically with replacement of polystyrene(PS) with polyacrylonitrile(PAN). 1,2-dichloroethane has negligible solubility in PAN due to strong repulsion between nonpolar 1,2-dichloroethane molecules and highly polar acrylonitrile segments. 1,2-dichloroethane also has some repulsive interaction with styrene segments, but these are not as strong as those with acrylonitrile segments. The solubility of 1,2-dichloroethane rises in presence of PS or poly(styrene-co-acrylonitrile).

Table 3. Vapor-pressure lowering data for the system: 1,2-Dichloroethane(1)-Polystyrene(2)-Polyacrylonitrile(3) at 70°c.

| | | $W_{2p}$=0.0 | |
|---|---|---|---|
| $m_s/m_p$ | $P/P^0$ | $\varphi_{23}$ ($\varphi_3$) | $\chi_{1,23}$ ($\chi_{13}$) |
| 0.0256 | 0.2446 | 0.9749 | 1.3687 |
| 0.0320 | 0.3181 | 0.9688 | 1.4431 |
| 0.0417 | 0.4006 | 0.9598 | 1.4539 |
| 0.0449 | 0.4621 | 0.9568 | 1.5435 |
| 0.0482 | 0.5476 | 0.9538 | 1.6691 |
| 0.0526 | 0.6106 | 0.9498 | 1.7160 |
| 0.0571 | 0.6587 | 0.9457 | 1.7337 |
| 0.0616 | 0.6902 | 0.9417 | 1.7252 |
| 0.0650 | 0.7367 | 0.9388 | 1.7565 |

$\chi_{1,23}^\infty(\chi_{13}^\infty) = 1.0657$

Table 3. Continued

| | | $W_{2p}$=0.70 | | |
|---|---|---|---|---|
| $m_s/m_p$ | $P/P^o$ | $\varphi_{23}$ | $\varphi_2$ | $\chi_{1,23}$ |
| 0.0091 | 0.2446 | 0.9918 | 0.7094 | 2.4430 |
| 0.0267 | 0.3181 | 0.9763 | 0.6983 | 1.6988 |
| 0.0504 | 0.4006 | 0.9561 | 0.6838 | 1.3728 |
| 0.1614 | 0.4621 | 0.8718 | 0.6236 | 0.5401 |
| 0.2438 | 0.5476 | 0.8183 | 0.5853 | 0.4256 |
| 0.3210 | 0.6106 | 0.7738 | 0.5534 | 0.3662 |
| 0.3908 | 0.6587 | 0.7375 | 0.5275 | 0.3355 |
| 0.4205 | 0.6902 | 0.7231 | 0.5172 | 0.3637 |
| 0.5129 | 0.7367 | 0.6816 | 0.4875 | 0.3386 |

$\chi_{1,23}^{\infty} = 2.2490$

| | | $W_{2p}$=1.0 | |
|---|---|---|---|
| $m_s/m_p$ | $P/P^o$ | $\varphi_{23}$ ($\varphi_2$) | $\chi_{1,23}$ ( $\chi_{12}$ ) |
| 0.0684 | 0.2446 | 0.9426 | 0.5696 |
| 0.1013 | 0.3181 | 0.9172 | 0.5092 |
| 0.1455 | 0.4006 | 0.8852 | 0.4654 |
| 0.1792 | 0.4621 | 0.8622 | 0.4682 |
| 0.2642 | 0.5476 | 0.8094 | 0.3754 |
| 0.3227 | 0.6106 | 0.7766 | 0.3796 |
| 0.4065 | 0.6587 | 0.7340 | 0.3208 |
| 0.4409 | 0.6902 | 0.7179 | 0.3429 |
| 0.5432 | 0.7367 | 0.6738 | 0.3102 |

$\chi_{1,23}^{\infty} (\chi_{12}^{\infty}) = 0.6610$

| $W_{2p}$ | $\varphi_2^{\infty}$ | $\varphi_3^{\infty}$ | $\chi'_{23}$ |
|---|---|---|---|
| 0.70 | 0.7152 | 0.2848 | -7.23034 |

Flory interaction parameter, $\chi_{13}$, indicates that 1,2-dichloroethane is a non-solvent for PAN $(\chi_{13} > 1)$, when styrene segments add to PAN to build a copolymer, $\chi$ decreases such that 1,2-dichloroethane can slightly dissolve PS and poly(styrene(0.70)-acrylonitrile) especially at high activities. As it is shown by the values of the $\chi_{12}$ and $\chi_{13}$ parameters, the nature of interaction between solvent and two polymers is different, but due to attraction between polar styrene and acrylonitrile segments, $\chi'_{23}$ is negative and two polymers show compatibility in all ranges of concentration and in both temperature.

Table 4. Vapor-pressure lowering data for the system: 1,2-Dichloroethane(1)-Polystyrene(2)-Polyacrylonitrile(3) at 80°c.

| $W_{2p}=0.0$ | | | |
|---|---|---|---|
| $m_s/m_p$ | $P/P^o$ | $\varphi_{23}$ ($\varphi_3$) | $\chi_{1,23}$ ($\chi_{13}$) |
| 0.0173 | 0.2401 | 0.9827 | 1.7051 |
| 0.0204 | 0.2926 | 0.9796 | 1.7558 |
| 0.0246 | 0.3890 | 0.9756 | 1.8826 |
| 0.0395 | 0.4898 | 0.9613 | 1.7065 |
| 0.0460 | 0.5369 | 0.9552 | 1.6753 |
| 0.0560 | 0.6173 | 0.9460 | 1.6662 |
| 0.0604 | 0.6377 | 0.9420 | 1.6398 |
| 0.0638 | 0.6505 | 0.9389 | 1.6185 |

$\chi^{\infty}_{1,23}$ ($\chi^{\infty}_{13}$) = 1.8455

| $W_{2p}=0.70$ | | | | |
|---|---|---|---|---|
| $m_s/m_p$ | $P/P^o$ | $\varphi_{23}$ | $\varphi_2$ | $\chi_{1,23}$ |
| 0.0091 | 0.2400 | 0.9917 | 0.7093 | 2.4110 |
| 0.0406 | 0.2926 | 0.9639 | 0.6894 | 1.2142 |
| 0.0989 | 0.3890 | 0.9163 | 0.6554 | 0.7389 |
| 0.3351 | 0.4898 | 0.7637 | 0.5462 | -0.0598 |
| 0.3870 | 0.5369 | 0.7368 | 0.5270 | -0.0441 |
| 0.5337 | 0.6173 | 0.6699 | 0.4791 | -0.0980 |
| 0.6026 | 0.6377 | 0.6425 | 0.4596 | -0.1545 |
| 0.6367 | 0.6505 | 0.6298 | 0.4504 | -0.1666 |

$\chi^{\infty}_{1,23} = 2.1261$

| $W_{2p}=1.0$ | | | |
|---|---|---|---|
| $m_s/m_p$ | $P/P^o$ | $\varphi_{23}$ ($\varphi_2$) | $\chi_{1,23}$ ( $\chi_{12}$ ) |
| 0.0753 | 0.2400 | 0.9363 | 0.4456 |
| 0.0989 | 0.2926 | 0.9180 | 0.4196 |
| 0.1429 | 0.3890 | 0.8857 | 0.4320 |
| 0.2837 | 0.4898 | 0.7959 | 0.1257 |
| 0.3459 | 0.5369 | 0.7619 | 0.0882 |
| 0.4948 | 0.6173 | 0.6910 | 0.0023 |
| 0.5625 | 0.6377 | 0.6630 | -0.0575 |
| 0.6051 | 0.6505 | 0.6465 | -0.0876 |

$\chi^{\infty}_{1,23}$ ( $\chi^{\infty}_{12}$ ) = 0.5816

| $W_{2p}$ | $\varphi^{\infty}_2$ | $\varphi^{\infty}_3$ | $\chi'_{23}$ |
|---|---|---|---|
| 0.70 | 0.7152 | 0.2848 | -5.81544 |

## 4.3 System: Hexane($C_6H_{14}$)(1)-Poly(acrylonitrile)(PAN)(2)-Poly(cis-butadiene)(cis-Bu)(3) at 60°C

Experimental VLE data [14] for poly(acrylonitrile-co-butadiene) and its parent homopolymers are shown in Table 5. At a given activity, progressing form PAN to poly(acrylonitrile-co-butadiene) to poly(cis-Bu), the solubility of hexane increases with the butadiene content in the copolymer. Hexane has negligible solubility in PAN due to strong repulsion between nonpolar cyclohexane segments and highly polar acrylonitrile segments. Flory interaction parameter, $\chi_{13}$, indicates that hexane is a fairly good solvent for poly(cis-Bu) with $\chi_{13}$ close to zero, but when acrylonitrile segments add to poly(cis-Bu) to build a copolymer, $\chi$ increases with rising percentage of acrylonitrile in copolymer composition. Therefore, hexane is a moderate solvent for poly(acrylonitrile(0.21)-butadiene) with $\chi_{1,23} < 1$ and is a non-solvent for poly(acrylonitrile(0.33)-butadiene), poly(acrylonitrile(0.51)-butadiene) and PAN. Poly(acrylonitrile(0.33)-butadiene) shows the largest value of $\chi_{1,23}$ at zero solvent concentration between three copolymers that is an indication of the least interaction of this copolymer with solvent and therefore the largest compatibility. Results obtained using hexane as a solvent suggest the poly(acrylonitrile-co-butadiene) to be more compatible than results obtained when acetonitrile is used as a solvent.

Table 5. Vapor-pressure lowering data for the system: Hexane(1)-Polyacrylonitrile(2)-Poly(cis-1,4-butadiene)(3) at 60°c.

| | $W_{2p}$=0.0 | | |
|---|---|---|---|
| $m_s/m_p$ | $P/P^o$ | $\varphi_{23}$ ($\varphi_3$) | $\chi_{1,23}$ ($\chi_{13}$) |
| 0.0373 | 0.1624 | 0.9482 | 0.2158 |
| 0.0846 | 0.3287 | 0.8898 | 0.2568 |
| 0.1299 | 0.4714 | 0.8402 | 0.3425 |
| 0.1710 | 0.5670 | 0.7998 | 0.3776 |

$\chi^{\infty}_{1,23}$ ($\chi^{\infty}_{13}$) = 0.1481

| | | **$W_{2p}$= 0.21** | | |
|---|---|---|---|---|
| **$m_s/m_p$** | **$P/P^o$** | **$\varphi_{23}$** | **$\varphi_2$** | **$\chi_{1,23}$** |
| 0.0256 | 0.1689 | 0.9615 | 0.1654 | 0.5606 |
| 0.0460 | 0.3352 | 0.9330 | 0.1605 | 0.7781 |
| 0.0893 | 0.5081 | 0.8777 | 0.1510 | 0.7094 |
| 0.1161 | 0.5828 | 0.8467 | 0.1457 | 0.6816 |
| 0.1547 | 0.6653 | 0.8055 | 0.1386 | 0.6541 |
| 0.2048 | 0.7530 | 0.7578 | 0.1304 | 0.6559 |
| 0.3089 | 0.8342 | 0.6748 | 0.1161 | 0.5869 |
| 0.3351 | 0.8499 | 0.6567 | 0.1130 | 0.5793 |

$\chi^{\infty}_{1,23}$ = 0.7063

Table 5. Continued

| | | $W_{2p}$= 0.33 | | |
|---|---|---|---|---|
| $m_s/m_p$ | $P/P^o$ | $\varphi_{23}$ | $\varphi_2$ | $\chi_{1,23}$ |
| 0.0020 | 0.1624 | 0.9968 | 0.2723 | 2.9648 |
| 0.0173 | 0.3287 | 0.9735 | 0.2659 | 1.6286 |
| 0.0267 | 0.4714 | 0.9596 | 0.2621 | 1.6267 |
| 0.0384 | 0.5670 | 0.9429 | 0.2575 | 1.5216 |
| | | $\chi^{\infty}_{1,23} = 2.7696$ | | |

| | | $W_{2p}$= 0.51 | | |
|---|---|---|---|---|
| $m_s/m_p$ | $P/P^o$ | $\varphi_{23}$ | $\varphi_2$ | $\chi_{1,23}$ |
| 0.0183 | 0.1689 | 0.9714 | 0.4184 | 0.8516 |
| 0.0204 | 0.3352 | 0.9682 | 0.4170 | 1.4805 |
| 0.0299 | 0.5081 | 0.9542 | 0.4110 | 1.5944 |
| 0.0341 | 0.5828 | 0.9480 | 0.4083 | 1.6337 |
| 0.0427 | 0.6653 | 0.9357 | 0.4030 | 1.5996 |
| 0.0460 | 0.7530 | 0.9311 | 0.4011 | 1.6842 |
| 0.0661 | 0.8342 | 0.9039 | 0.3893 | 1.5388 |
| 0.0661 | 0.8499 | 0.9039 | 0.3893 | 1.5617 |
| | | $\chi^{\infty}_{1,23} = 1.1846$ | | |

| | $W_{2p}$= 1.0 | | |
|---|---|---|---|
| $m_s/m_p$ | $P/P^o$ | $\varphi_{23}$ ($\varphi_2$) | $\chi_{1,23}$ ($\chi_{12}$) |
| 0.0010 | 0.1624 | 0.9981 | 3.4619 |
| 0.0070 | 0.3287 | 0.9867 | 2.2849 |
| 0.0070 | 0.4714 | 0.9867 | 2.6553 |
| 0.0091 | 0.5670 | 0.9830 | 2.6123 |
| | $\chi^{\infty}_{1,23}$ ($\chi^{\infty}_{12}$) = 3.5108 | | |

| $W_{2p}$ | $\varphi^{\infty}_2$ | $\varphi^{\infty}_3$ | $\chi'_{23}$ |
|---|---|---|---|
| 0.21 | 0.1720 | 0.8280 | 0.1417 |
| 0.33 | 0.2731 | 0.7269 | -8.5794 |
| 0.51 | 0.4307 | 0.5693 | 1.6795 |

## 4.4 System: Pentane($C_5H_{12}$)(1)-Poly(acrylonitrile)(PAN)(2)-Poly(cis-butadiene)(cis-Bu)(3) at 60°C

Experimental VLE data [14] for poly(acrylonitrile-co-butadiene) and its parent homopolymers are shown in Table 6. Solvent absorption in the copolymer increases as its butadiene content rises. This rise is expected because the hydrocarbon segments of pentane are better liked by hydrocarbon segments of butadiene, whereas polar segments of acrylonitrile repulse nonpolar pentane molecules. Once again, Flory interaction parameter, $\chi$, implies that with rising acrylonitrile concentration in copolymer composition, polymer/solvent interaction weakens such that pentane can highly dissolve poly(cis-Bu) with $\chi_{13}$ close to zero and be a border solvent for poly(acrylonitrile(0.21)-butadiene) and dissolves poly(acrylonitrile(0.33)-butediene) at low activities.

Table 6. Vapor-pressure lowering data for the system: Pentane(1)-Polyacrylonitrile(2)-Poly(cis-1,4-butadiene)(3) at 60°c.

| | **$W_{2p}$=0.0** | | |
|---|---|---|---|
| **$m_s/m_p$** | **$P/P^o$** | **$\varphi_{23}$ ($\varphi_3$)** | **$\chi_{1,23}$ ($\chi_{13}$)** |
| 0.0121 | 0.0620 | 0.9815 | 0.2343 |
| 0.0246 | 0.1254 | 0.9632 | 0.2822 |
| 0.0482 | 0.2494 | 0.9302 | 0.3971 |
| 0.0650 | 0.3016 | 0.9082 | 0.3414 |
| 0.0893 | 0.3613 | 0.8780 | 0.2694 |
| 0.1050 | 0.4182 | 0.8596 | 0.3140 |
| 0.1325 | 0.4741 | 0.8291 | 0.2783 |
| 0.1547 | 0.5394 | 0.8060 | 0.3334 |
| 0.1834 | 0.6065 | 0.7780 | 0.3752 |

$\chi^{\infty}_{1,23}$ ( $\chi^{\infty}_{13}$ ) = 0.2762

| | | **$W_{2p}$= 0.21** | | |
|---|---|---|---|---|
| **$m_s/m_p$** | **$P/P^o$** | **$\varphi_{23}$** | **$\varphi_2$** | **$\chi_{1,23}$** |
| 0.0081 | 0.1063 | 0.9868 | 0.1698 | 1.1291 |
| 0.0163 | 0.1678 | 0.9737 | 0.1675 | 0.9296 |
| 0.0214 | 0.2182 | 0.9656 | 0.1661 | 0.9471 |
| 0.0341 | 0.3054 | 0.9464 | 0.1628 | 0.8868 |
| 0.0482 | 0.3916 | 0.9260 | 0.1593 | 0.8628 |
| 0.0582 | 0.4536 | 0.9120 | 0.1569 | 0.8750 |
| 0.0753 | 0.5282 | 0.8890 | 0.1529 | 0.8493 |
| 0.0846 | 0.5818 | 0.8770 | 0.1509 | 0.8800 |
| 0.0977 | 0.6415 | 0.8606 | 0.1481 | 0.8990 |

$\chi^{\infty}_{1,23}$ = 1.0092

| | | **$W_{2p}$= 0.33** | | |
|---|---|---|---|---|
| **$m_s/m_p$** | **$P/P^o$** | **$\varphi_{23}$** | **$\varphi_2$** | **$\chi_{1,23}$** |
| 0.0091 | 0.0620 | 0.9850 | 0.2690 | 0.4484 |
| 0.0142 | 0.1254 | 0.9768 | 0.2668 | 0.7434 |
| 0.0256 | 0.2494 | 0.9588 | 0.2619 | 0.9161 |
| 0.0277 | 0.3016 | 0.9556 | 0.2610 | 1.0512 |
| 0.0331 | 0.3613 | 0.9475 | 0.2588 | 1.0937 |
| 0.0373 | 0.4182 | 0.9411 | 0.2570 | 1.1509 |
| 0.0460 | 0.4741 | 0.9284 | 0.2536 | 1.1163 |
| 0.0526 | 0.5394 | 0.9190 | 0.2510 | 1.1565 |
| 0.0582 | 0.6066 | 0.9112 | 0.2489 | 1.2163 |

$\chi^{\infty}_{1,23}$ = 0.5184

Table 6. Continued

| | | $W_{2p}$= 0.51 | | |
|---|---|---|---|---|
| $m_s/m_p$ | $P/P^o$ | $\varphi_{23}$ | $\varphi_2$ | $\chi_{1,23}$ |
| 0.0101 | 0.1063 | 0.9830 | 0.4234 | 0.8812 |
| 0.0091 | 0.1678 | 0.9847 | 0.4242 | 1.4554 |
| 0.0132 | 0.2182 | 0.9780 | 0.4213 | 1.3753 |
| 0.0194 | 0.3054 | 0.9679 | 0.4169 | 1.3729 |
| 0.0225 | 0.3916 | 0.9630 | 0.4148 | 1.5050 |
| 0.0225 | 0.4536 | 0.9630 | 0.4148 | 1.6635 |
| 0.0267 | 0.5282 | 0.9564 | 0.4119 | 1.6806 |
| 0.0277 | 0.5818 | 0.9547 | 0.4112 | 1.7537 |
| 0.0352 | 0.6415 | 0.9432 | 0.4063 | 1.6656 |

$\chi_{1,23}^{\infty} = 0.9992$

| | | $W_{2p}$= 1.0 | |
|---|---|---|---|
| $m_s/m_p$ | $P/P^o$ | $\varphi_{23}$ ($\varphi_2$) | $\chi_{1,23}$ ($\chi_{12}$) |
| 0.0050 | 0.0620 | 0.9899 | 0.8448 |
| 0.0070 | 0.1254 | 0.9859 | 1.2364 |
| 0.0111 | 0.2494 | 0.9780 | 1.5154 |
| 0.0101 | 0.3016 | 0.9799 | 1.8033 |
| 0.0101 | 0.3613 | 0.9799 | 1.9913 |
| 0.0101 | 0.4182 | 0.9799 | 2.1435 |
| 0.0101 | 0.4741 | 0.9799 | 2.2743 |
| 0.0111 | 0.5394 | 0.9779 | 2.3218 |
| 0.0111 | 0.6065 | 0.9779 | 2.4444 |

$\chi_{1,23}^{\infty}(\chi_{12}^{\infty}) = -0.2578$

| $W_{2p}$ | $\varphi_2^{\infty}$ | $\varphi_3^{\infty}$ | $\chi'_{23}$ |
|---|---|---|---|
| 0.21 | 0.1720 | 0.8280 | -5.7918 |
| 0.33 | 0.2731 | 0.7269 | -1.9547 |
| 0.51 | 0.4307 | 0.5693 | -3.8866 |

From the limiting values of $\chi_{1,23}$ at zero solvent concentration the pair interaction parameter $\chi'_{23}$ is estimated. Incidentally, the system poly(acrylonitrile-co-butadiene) is predicted to form a compatible blend. Results obtained using pentane as a solvent suggest the poly(acrylonitrile-co-butadiene) to be more compatible than results obtained when hexane is used as solvent, but between three copolymers, poly(acrylonitrile(0.33)-butadiene) shows the most compatibility in hexane and the least compatibility in the presence of pentane.

## 4.5 System: Water(1)-Poly(N,N-Dimethylmethacrylate)(PDMAA)(2)-Poly(2-Dimethyl Aminoethyl Methacrylate)(PDMAEMA)(3) at 35°C

Experimental VLE data [15] for PDMAA and for PDMAEMA have been brought in Table 7. VLE data for the blends have been calculated from their homopolymer values. In PDMAA, $\chi_{12}$ shows a small variation with water concentration. This value is less than 0.5, the upper limit for complete miscibility when $\chi$ is independent of composition, indicating that water is a good solvent for this polymer. In less hydrophilic PDMAEMA, $\chi_{13}$ is higher than 1. It is evident that the limiting values of $\chi_{1,23}^{\infty}$ is a decreasing function of PDMAA concentration.

Table 7. Vapor-pressure lowering data for the system: Water (1)-PDMAA (2)-PDMAEMA (3) at 35°c.

**$W_{2p}$=0.0**

| $m_s/m_p$ | $P/P^o$ | $\varphi_{23}$ ($\varphi_3$) | $\chi_{1,23}$ ($\chi_{13}$) |
|---|---|---|---|
| 0.0040 | 0.14 | 0.9956 | 2.4867 |
| 0.0081 | 0.24 | 0.9912 | 2.3567 |
| 0.0194 | 0.38 | 0.9790 | 2.0060 |
| 0.0331 | 0.49 | 0.9649 | 1.7954 |
| 0.0504 | 0.58 | 0.9474 | 1.6196 |
| 0.0822 | 0.68 | 0.9170 | 1.4110 |
| 0.1261 | 0.76 | 0.8782 | 1.2351 |
| 0.2106 | 0.83 | 0.8119 | 1.0202 |

$\chi_{1,23}^{\infty}$ ($\chi_{13}^{\infty}$) = 2.4473

**$W_{2p}$= 0.25**

| $m_s/m_p$ | $P/P^o$ | $\varphi_{23}$ | $\varphi_2$ | $\chi_{1,23}$ |
|---|---|---|---|---|
| 0.0051 | 0.14 | 0.9944 | 0.2404 | 2.2448 |
| 0.0103 | 0.24 | 0.9887 | 0.2390 | 2.1163 |
| 0.0243 | 0.38 | 0.9737 | 0.2354 | 1.7883 |
| 0.0412 | 0.49 | 0.9562 | 0.2312 | 1.5952 |
| 0.0624 | 0.58 | 0.9351 | 0.2261 | 1.4357 |
| 0.1007 | 0.68 | 0.8993 | 0.2174 | 1.2497 |
| 0.1527 | 0.76 | 0.8548 | 0.2067 | 1.0957 |
| 0.2511 | 0.83 | 0.7817 | 0.1890 | 0.9064 |

$\chi_{1,23}^{\infty}$ = 2.2178

**$W_{2p}$= 0.50**

| $m_s/m_p$ | $P/P^o$ | $\varphi_{23}$ | $\varphi_2$ | $\chi_{1,23}$ |
|---|---|---|---|---|
| 0.0069 | 0.14 | 0.9922 | 0.4851 | 1.9318 |
| 0.0141 | 0.24 | 0.9844 | 0.4813 | 1.8045 |
| 0.0327 | 0.38 | 0.9645 | 0.4715 | 1.5124 |
| 0.0546 | 0.49 | 0.9421 | 0.4606 | 1.3451 |

| | | $W_{2p}$= 0.50 | | |
|---|---|---|---|---|
| **$m_s/m_p$** | **$P/P^o$** | **$\varphi_{23}$** | **$\varphi_2$** | **$\chi_{1,23}$** |
| 0.0818 | 0.58 | 0.9158 | 0.4477 | 1.2088 |
| 0.1298 | 0.68 | 0.8727 | 0.4266 | 1.0540 |
| 0.1934 | 0.76 | 0.8214 | 0.4016 | 0.9289 |
| 0.3108 | 0.83 | 0.7410 | 0.3623 | 0.7716 |

$\chi^{\infty}_{1,23} = 1.9207$

Table 7. Continued

| | | $W_{2p}$= 0.75 | | |
|---|---|---|---|---|
| **$m_s/m_p$** | **$P/P^o$** | **$\varphi_{23}$** | **$\varphi_2$** | **$\chi_{1,23}$** |
| 0.0109 | 0.14 | 0.9877 | 0.7325 | 1.4826 |
| 0.0225 | 0.24 | 0.9750 | 0.7231 | 1.3547 |
| 0.0499 | 0.38 | 0.9463 | 0.7018 | 1.1290 |
| 0.0811 | 0.49 | 0.9155 | 0.6789 | 1.0053 |
| 0.1187 | 0.58 | 0.8810 | 0.6534 | 0.9059 |
| 0.1825 | 0.68 | 0.8282 | 0.6142 | 0.7982 |
| 0.2637 | 0.76 | 0.7693 | 0.5705 | 0.7146 |
| 0.4077 | 0.83 | 0.6833 | 0.5067 | 0.5999 |

$\chi^{\infty}_{1,23} = 1.4896$

| | $W_{2p}$= 1.0 | | |
|---|---|---|---|
| **$m_s/m_p$** | **$P/P^o$** | **$\varphi_{23}$ ($\varphi_2$)** | **$\chi_{1,23}$ ($\chi_{12}$)** |
| 0.0256 | 0.14 | 0.9714 | 0.6522 |
| 0.0560 | 0.24 | 0.9395 | 0.4973 |
| 0.1050 | 0.38 | 0.8923 | 0.4627 |
| 0.1574 | 0.49 | 0.8467 | 0.4400 |
| 0.2165 | 0.58 | 0.8006 | 0.4169 |
| 0.3072 | 0.68 | 0.7389 | 0.4000 |
| 0.4144 | 0.76 | 0.6772 | 0.3906 |
| 0.5924 | 0.83 | 0.5948 | 0.3456 |

$\chi^{\infty}_{1,23}\,(\chi^{\infty}_{12}) = 0.6354$

| **$W_{2p}$** | $\varphi^{\infty}_2$ | $\varphi^{\infty}_3$ | **$\chi'_{23}$** |
|---|---|---|---|
| 0.25 | 0.2418 | 0.7582 | -1.1379 |
| 0.50 | 0.4889 | 0.5111 | -1.4377 |
| 0.75 | 0.7416 | 0.2584 | -2.0143 |

By using the limiting quantities, $\chi^{\infty}_{1,23}$, and Eq. (15) the values of $\chi'_{23}$ have been calculated and show that these two polymers are compatible in all ranges of concentration and their degree of compatibility enhances with increasing PDMAA content in the blend.

## 4.6 System:Water(1)-Poly(N,N-Dimethylmethacrylate)(PDMAA)(2)-Poly(Acrylic Acid)(PAA)(3) at 35°C

Experimental VLE data [15] for PDMAA and for PAA have been reported in Table 8. VLE data for the blends have been calculated from their homopolymer quantities. The values of $\chi_{13}$ less than 1 in PAA refer to good solvency power of water for this polymer like in the case of PDMAA. The limiting values of $\chi_{1,23}^{\infty}$ decreases with increasing PDMAA content in the blend. Results obtained in this system for $\chi'_{23}$ suggest the PDMAA-PAA blend to be less compatible than the PDMAA-PDMAEMA blend.

Table 8. Vapor-pressure lowering data for the system: Water (1)-PDMAA (2)-PAA (3) at 35°c.

**$W_{2p}$=0.0**

| $m_s/m_p$ | $P/P^o$ | $\varphi_{23}$ ($\varphi_3$) | $\chi_{1,23}$ ($\chi_{13}$) |
|---|---|---|---|
| 0.0712 | 0.14 | 0.9243 | -0.3623 |
| 0.0463 | 0.24 | 0.9494 | 0.6736 |
| 0.0513 | 0.38 | 0.9443 | 1.0946 |
| 0.0886 | 0.49 | 0.9075 | 0.9227 |
| 0.1456 | 0.58 | 0.8565 | 0.7367 |
| 0.2469 | 0.68 | 0.7788 | 0.5677 |
| 0.3698 | 0.76 | 0.7016 | 0.4739 |
| 0.5245 | 0.83 | 0.6238 | 0.4303 |

$\chi_{1,23}^{\infty}$ ($\chi_{13}^{\infty}$) = 1.3457

**$W_{2p}$= 0.25**

| $m_s/m_p$ | $P/P^o$ | $\varphi_{23}$ | $\varphi_2$ | $\chi_{1,23}$ |
|---|---|---|---|---|
| 0.0493 | 0.14 | 0.9463 | 0.2366 | 0.0140 |
| 0.0484 | 0.24 | 0.9472 | 0.2368 | 0.6327 |
| 0.0588 | 0.38 | 0.9367 | 0.2342 | 0.9746 |
| 0.0995 | 0.49 | 0.8973 | 0.2243 | 0.8267 |
| 0.1586 | 0.58 | 0.8457 | 0.2114 | 0.6691 |
| 0.2596 | 0.68 | 0.7701 | 0.1925 | 0.5299 |
| 0.3800 | 0.76 | 0.6959 | 0.1740 | 0.4543 |
| 0.5400 | 0.83 | 0.6169 | 0.1542 | 0.4106 |

$\chi_{1,23}^{\infty}$ = 1.2191

**$W_{2p}$= 0.50**

| $m_s/m_p$ | $P/P^o$ | $\varphi_{23}$ | $\varphi_2$ | $\chi_{1,23}$ |
|---|---|---|---|---|
| 0.0377 | 0.14 | 0.9584 | 0.4792 | 0.2787 |
| 0.0507 | 0.24 | 0.9449 | 0.4724 | 0.5898 |
| 0.0689 | 0.38 | 0.9266 | 0.4633 | 0.8356 |
| 0.1134 | 0.49 | 0.8846 | 0.4423 | 0.7178 |
| 0.1741 | 0.58 | 0.8331 | 0.4166 | 0.5947 |
| 0.2738 | 0.68 | 0.7606 | 0.3803 | 0.4896 |

| | | $W_{2p}$= 0.50 | | |
|---|---|---|---|---|
| $m_s/m_p$ | $P/P^o$ | $\varphi_{23}$ | $\varphi_2$ | $\chi_{1,23}$ |
| 0.3909 | 0.76 | 0.6899 | 0.3449 | 0.4339 |
| 0.5564 | 0.83 | 0.6098 | 0.3049 | 0.3899 |

$\chi_{1,23}^{\infty} = 1.0593$

Table 8. Continued

| | | $W_{2p}$= 0.75 | | |
|---|---|---|---|---|
| $m_s/m_p$ | $P/P^o$ | $\varphi_{23}$ | $\varphi_2$ | $\chi_{1,23}$ |
| 0.0305 | 0.14 | 0.9661 | 0.7246 | 0.4840 |
| 0.0532 | 0.24 | 0.9423 | 0.7067 | 0.5447 |
| 0.0832 | 0.38 | 0.9127 | 0.6845 | 0.6698 |
| 0.1318 | 0.49 | 0.8684 | 0.6513 | 0.5915 |
| 0.1930 | 0.58 | 0.8183 | 0.6137 | 0.5115 |
| 0.2895 | 0.68 | 0.7502 | 0.5627 | 0.4465 |
| 0.4023 | 0.76 | 0.6837 | 0.5128 | 0.4127 |
| 0.5738 | 0.83 | 0.6025 | 0.4518 | 0.3683 |

$\chi_{1,23}^{\infty} = 0.8413$

| | $W_{2p}$= 1.0 | | |
|---|---|---|---|
| $m_s/m_p$ | $P/P^o$ | $\varphi_{23}$ ($\varphi_2$) | $\chi_{1,23}$ ($\chi_{12}$) |
| 0.0256 | 0.14 | 0.9714 | 0.6522 |
| 0.0560 | 0.24 | 0.9395 | 0.4973 |
| 0.1050 | 0.38 | 0.8923 | 0.4627 |
| 0.1574 | 0.49 | 0.8467 | 0.4400 |
| 0.2165 | 0.58 | 0.8006 | 0.4169 |
| 0.3072 | 0.68 | 0.7389 | 0.4000 |
| 0.4144 | 0.76 | 0.6772 | 0.3906 |
| 0.5924 | 0.83 | 0.5948 | 0.3456 |

$\chi_{1,23}^{\infty}(\chi_{12}^{\infty}) = 0.6354$

| $W_{2p}$ | $\varphi_2^{\infty}$ | $\varphi_3^{\infty}$ | $\chi'_{23}$ |
|---|---|---|---|
| 0.25 | 0.25 | 0.75 | -0.2719 |
| 0.50 | 0.50 | 0.50 | -0.2750 |
| 0.75 | 0.75 | 0.25 | -0.1511 |

## 4.7 System:Water(1)-Poly(N,N-Dimethylmethacrylate)(PDMAA)(2)-Poly(2-Hydroxyethyl Methacrylate)(PHEMA)(3) at 35°C

Table 9 reports experimental VLE data for PDMAA and for PAA that are taken from Prausnitz et al. work [15]. VLE data for the blends have been calculated from their homopolymer quantities. Hydrophilicity of PHEMA-lens material that is essential in contact-lens technology is evident here with interaction parameter $\chi_{13}>1$. Once again, we see that the limiting values of $\chi_{1,23}^{\infty}$ enhances with increasing PHEMA content in the blend.

In terms of these limiting quantities, $\chi'_{23}$ is given by using Eq. (15), indicating that this blend is compatible in all concentration ranges. The degree of compatibility depends on both polymer concentration and decreases with PDMAA percentage in the blend.

Table 9. Vapor-pressure lowering data for the system: Water (1)-PDMAA (2)-PHEMA (3) at 35°c.

**$W_{2p}=0.0$**

| $m_s/m_p$ | $P/P^o$ | $\varphi_{23}$ ($\varphi_3$) | $\chi_{1,23}$ ($\chi_{13}$) |
|---|---|---|---|
| 0.0184 | 0.14 | 0.9792 | 0.9694 |
| 0.0187 | 0.24 | 0.9789 | 1.5175 |
| 0.0300 | 0.38 | 0.9667 | 1.5696 |
| 0.0483 | 0.49 | 0.9474 | 1.4307 |
| 0.0701 | 0.58 | 0.9254 | 1.3145 |
| 0.1025 | 0.68 | 0.8946 | 1.2116 |
| 0.1355 | 0.76 | 0.8652 | 1.1544 |
| 0.1706 | 0.83 | 0.8360 | 1.1241 |

$\chi_{1,23}^{\infty}(\chi_{13}^{\infty}) = 1.8226$

**$W_{2p}= 0.25$**

| $m_s/m_p$ | $P/P^o$ | $\varphi_{23}$ | $\varphi_2$ | $\chi_{1,23}$ |
|---|---|---|---|---|
| 0.0198 | 0.14 | 0.9777 | 0.2444 | 0.8995 |
| 0.0224 | 0.24 | 0.9748 | 0.2437 | 1.3475 |
| 0.0365 | 0.38 | 0.9597 | 0.2399 | 1.3943 |
| 0.0584 | 0.49 | 0.9370 | 0.2343 | 1.2699 |
| 0.0843 | 0.58 | 0.9116 | 0.2279 | 1.1666 |
| 0.1229 | 0.68 | 0.8761 | 0.2190 | 1.0770 |
| 0.1629 | 0.76 | 0.8422 | 0.2105 | 1.0288 |
| 0.2075 | 0.83 | 0.8074 | 0.2018 | 1.0022 |

$\chi_{1,23}^{\infty} = 1.6301$

**$W_{2p}= 0.50$**

| $m_s/m_p$ | $P/P^o$ | $\varphi_{23}$ | $\varphi_2$ | $\chi_{1,23}$ |
|---|---|---|---|---|
| 0.0214 | 0.14 | 0.9759 | 0.4880 | 0.8240 |
| 0.0280 | 0.24 | 0.9688 | 0.4844 | 1.1403 |
| 0.0466 | 0.38 | 0.9491 | 0.4745 | 1.1777 |
| 0.0739 | 0.49 | 0.9217 | 0.4608 | 1.0733 |
| 0.1059 | 0.58 | 0.8914 | 0.4457 | 0.9870 |
| 0.1537 | 0.68 | 0.8498 | 0.4249 | 0.9145 |
| 0.2042 | 0.76 | 0.8098 | 0.4049 | 0.8774 |
| 0.2648 | 0.83 | 0.7665 | 0.3833 | 0.8541 |

$\chi_{1,23}^{\infty} = 1.3876$

Table 9. Continued

| | | $W_{2p}$= 0.75 | | |
|---|---|---|---|---|
| $m_s/m_p$ | $P/P^o$ | $\varphi_{23}$ | $\varphi_2$ | $\chi_{1,23}$ |
| 0.0234 | 0.14 | 0.9738 | 0.7304 | 0.7420 |
| 0.0374 | 0.24 | 0.9588 | 0.7191 | 0.8740 |
| 0.0646 | 0.38 | 0.9308 | 0.6981 | 0.8921 |
| 0.1006 | 0.49 | 0.8963 | 0.6722 | 0.8175 |
| 0.1422 | 0.58 | 0.8594 | 0.6446 | 0.7553 |
| 0.2049 | 0.68 | 0.8093 | 0.6070 | 0.7057 |
| 0.2736 | 0.76 | 0.7606 | 0.5705 | 0.6822 |
| 0.3660 | 0.83 | 0.7038 | 0.5278 | 0.6592 |

$\chi_{1,23}^{\infty} = 1.0554$

| | | $W_{2p}$= 1.0 | |
|---|---|---|---|
| $m_s/m_p$ | $P/P^o$ | $\varphi_{23}$ ($\varphi_2$) | $\chi_{1,23}$ ( $\chi_{12}$ ) |
| 0.0256 | 0.14 | 0.9714 | 0.6522 |
| 0.0560 | 0.24 | 0.9395 | 0.4973 |
| 0.1050 | 0.38 | 0.8923 | 0.4627 |
| 0.1574 | 0.49 | 0.8467 | 0.4400 |
| 0.2165 | 0.58 | 0.8006 | 0.4169 |
| 0.3072 | 0.68 | 0.7389 | 0.4000 |
| 0.4144 | 0.76 | 0.6772 | 0.3906 |
| 0.5924 | 0.83 | 0.5948 | 0.3456 |

$\chi_{1,23}^{\infty} ( \chi_{12}^{\infty} ) = 0.6354$

| $W_{2p}$ | $\varphi_2^{\infty}$ | $\varphi_3^{\infty}$ | $\chi'_{23}$ |
|---|---|---|---|
| 0.25 | 0.25 | 0.75 | -0.5563 |
| 0.50 | 0.50 | 0.50 | -0.6344 |
| 0.75 | 0.75 | 0.25 | -0.6571 |

## 4.8 System: Water(1)-Poly(2-Dimethyl Aminoethyl Methacrylate) (PDMAEMA)(2)-Poly(Acrylic Acid)(PAA)(3) at 35°C

Table 10 consists of experimental VLE data [15] for PDMAEMA and for PAA. VLE data for the blends have been calculated from their homopolymer quantities. This table compares the interaction parameter $\chi_{12}$>1 for an hydrophilic material like PDMAEMA and the interaction parameter $\chi_{13}$<1 for PAA that dissolves in water well. The limiting values of $\chi_{1,23}^{\infty}$ fall in proportion to the PAA concentrations in the blend. From these limiting values at zero solvent concentration and Eq. (15), the pair interaction parameter $\chi'_{23}$ is estimated. This parameter implies that PDMAEMA and PAA will be compatible in all concentration ranges and their degree of compatibility enhances along with PDMAEMA content in the blend.

Table 10. Vapor-pressure lowering data for the system: Water (1)-PDMAEMA (2)-PAA (3) at 35°c.

| $W_{2p}$=0.0 | | | |
|---|---|---|---|
| $m_s/m_p$ | $P/P^o$ | $\varphi_{23}$ ($\varphi_3$) | $\chi_{1,23}$ ($\chi_{13}$) |
| 0.0712 | 0.14 | 0.9243 | -0.3623 |
| 0.0463 | 0.24 | 0.9494 | 0.6736 |
| 0.0513 | 0.38 | 0.9443 | 1.0946 |
| 0.0886 | 0.49 | 0.9075 | 0.9227 |
| 0.1456 | 0.58 | 0.8565 | 0.7367 |
| 0.2469 | 0.68 | 0.7788 | 0.5677 |
| 0.3698 | 0.76 | 0.7016 | 0.4739 |
| 0.5245 | 0.83 | 0.6238 | 0.4303 |

$\chi_{1,23}^{\infty}$ ( $\chi_{13}^{\infty}$ ) = 1.3457

| $W_{2p}$= 0.25 | | | | |
|---|---|---|---|---|
| $m_s/m_p$ | $P/P^o$ | $\varphi_{23}$ | $\varphi_2$ | $\chi_{1,23}$ |
| 0.0137 | 0.14 | 0.9846 | 0.2544 | 1.2623 |
| 0.0212 | 0.24 | 0.9765 | 0.2523 | 1.4114 |
| 0.0363 | 0.38 | 0.9603 | 0.2482 | 1.4090 |
| 0.0624 | 0.49 | 0.9337 | 0.2413 | 1.2240 |
| 0.0989 | 0.58 | 0.8989 | 0.2323 | 1.0494 |
| 0.1646 | 0.68 | 0.8424 | 0.2177 | 0.8730 |
| 0.2494 | 0.76 | 0.7791 | 0.2013 | 0.7520 |
| 0.3822 | 0.83 | 0.6971 | 0.1801 | 0.6398 |

$\chi_{1,23}^{\infty}$ = 1.6193

| $W_{2p}$= 0.50 | | | | |
|---|---|---|---|---|
| $m_s/m_p$ | $P/P^o$ | $\varphi_{23}$ | $\varphi_2$ | $\chi_{1,23}$ |
| 0.0076 | 0.14 | 0.9915 | 0.5068 | 1.8439 |
| 0.0137 | 0.24 | 0.9848 | 0.5033 | 1.8289 |
| 0.0281 | 0.38 | 0.9693 | 0.4954 | 1.6477 |
| 0.0481 | 0.49 | 0.9486 | 0.4849 | 1.4522 |
| 0.0749 | 0.58 | 0.9223 | 0.4714 | 1.2791 |
| 0.1234 | 0.68 | 0.8781 | 0.4488 | 1.0908 |
| 0.1881 | 0.76 | 0.8254 | 0.4219 | 0.9474 |
| 0.3006 | 0.83 | 0.7474 | 0.3820 | 0.7916 |

$\chi_{1,23}^{\infty}$ = 1.9149

| $W_{2p}$= 0.75 | | | | |
|---|---|---|---|---|
| $m_s/m_p$ | $P/P^o$ | $\varphi_{23}$ | $\varphi_2$ | $\chi_{1,23}$ |
| 0.0053 | 0.14 | 0.9942 | 0.7538 | 2.2133 |
| 0.0102 | 0.24 | 0.9888 | 0.7498 | 2.1253 |
| 0.0229 | 0.38 | 0.9751 | 0.7394 | 1.8417 |
| 0.0392 | 0.49 | 0.9582 | 0.7266 | 1.6378 |

| $W_{2p}$= 0.75 | | | | |
|---|---|---|---|---|
| $m_s/m_p$ | $P/P^o$ | $\varphi_{23}$ | $\varphi_2$ | $\chi_{1,23}$ |
| 0.0603 | 0.58 | 0.9372 | 0.7106 | 1.4637 |
| 0.0987 | 0.68 | 0.9011 | 0.6832 | 1.2645 |
| 0.1510 | 0.76 | 0.8562 | 0.6492 | 1.1032 |
| 0.2477 | 0.83 | 0.7840 | 0.5945 | 0.9146 |

$\chi_{1,23}^{\infty} = 2.2179$

Table 10. Continued

| $W_{2p}$= 1.0 | | | |
|---|---|---|---|
| $m_s/m_p$ | $P/P^o$ | $\varphi_{23}$ ($\varphi_2$) | $\chi_{1,23}$ ( $\chi_{12}$ ) |
| 0.0040 | 0.14 | 0.9956 | 2.4867 |
| 0.0081 | 0.24 | 0.9912 | 2.3567 |
| 0.0194 | 0.38 | 0.9790 | 2.0060 |
| 0.0331 | 0.49 | 0.9649 | 1.7954 |
| 0.0504 | 0.58 | 0.9474 | 1.6196 |
| 0.0822 | 0.68 | 0.9170 | 1.4110 |
| 0.1261 | 0.76 | 0.8782 | 1.2351 |
| 0.2106 | 0.83 | 0.8119 | 1.0202 |

$\chi_{1,23}^{\infty} ( \chi_{12}^{\infty} ) = 2.4473$

| $W_{2p}$ | $\varphi_2^{\infty}$ | $\varphi_3^{\infty}$ | $\chi'_{23}$ |
|---|---|---|---|
| 0.25 | 0.2584 | 0.7416 | 0.0577 |
| 0.50 | 0.5111 | 0.4889 | -0.0247 |
| 0.75 | 0.7582 | 0.2418 | -0.2016 |

### 4.9 System: Water(1)-Poly(2-Dimethyl Aminoethyl Methacrylate) (PDMAEMA)(2)-Poly(2-Hydroxyethyl Methacrylate)(PHEMA)(3) at 35°C

Table 11 displays the experimental VLE data [15] for PDMAEMA and for PHEMA. VLE data for the blends have been calculated from their homopolymer quantities. Flory interaction parameter $\chi_{12}$>1 and $\chi_{13}$>1 for PDMAEMA and for PHEMA, respectively, indicate that water as a solvent does not play its role well for these two polymers. Similar to the previous system, the limiting values of $\chi_{1,23}^{\infty}$ fall in proportion to the PHEMA content in the blend.

In terms of these limiting quantities, the pair interaction parameter, $\chi'_{23}$, is calculated by using Eq. (15), indicating that blend of PDMAEMA-PHEMA is less compatible than when PDMAEMA and PHEMA mix with other three polymers, i.e., with PAA, PDMAA and PVP. As we mentioned in section 4.8 degree of compatibility is an increasing function of PDMAEMA percentage in the blend.

Table 11. Vapor-pressure lowering data for the system: Water (1)-PDMAEMA (2)-PHEMA (3) at 35°c.

| $W_{2p}=0.0$ | | | |
|---|---|---|---|
| $m_s/m_p$ | $P/P^o$ | $\varphi_{23}$ ($\varphi_3$) | $\chi_{1,23}$ ($\chi_{13}$) |
| 0.0184 | 0.14 | 0.9792 | 0.9694 |
| 0.0187 | 0.24 | 0.9789 | 1.5175 |
| 0.0300 | 0.38 | 0.9667 | 1.5696 |
| 0.0483 | 0.49 | 0.9474 | 1.4307 |
| 0.0701 | 0.58 | 0.9254 | 1.3145 |
| 0.1025 | 0.68 | 0.8946 | 1.2116 |
| 0.1355 | 0.76 | 0.8652 | 1.1544 |
| 0.1705 | 0.83 | 0.8360 | 1.1241 |

$\chi_{1,23}^{\infty}$ ($\chi_{13}^{\infty}$) = 1.8226

| $W_{2p}=0.25$ | | | | |
|---|---|---|---|---|
| $m_s/m_p$ | $P/P^o$ | $\varphi_{23}$ | $\varphi_2$ | $\chi_{1,23}$ |
| 0.0097 | 0.14 | 0.9891 | 0.2556 | 1.5962 |
| 0.0141 | 0.24 | 0.9843 | 0.2544 | 1.7961 |
| 0.0264 | 0.38 | 0.9709 | 0.2509 | 1.6952 |
| 0.0433 | 0.49 | 0.9531 | 0.2463 | 1.5333 |
| 0.0638 | 0.58 | 0.9323 | 0.2409 | 1.3987 |
| 0.0965 | 0.68 | 0.9011 | 0.2329 | 1.2646 |
| 0.1330 | 0.76 | 0.8686 | 0.2245 | 1.1749 |
| 0.1791 | 0.83 | 0.8308 | 0.2147 | 1.1005 |

$\chi_{1,23}^{\infty} = 1.9384$

| $W_{2p}=0.50$ | | | | |
|---|---|---|---|---|
| $m_s/m_p$ | $P/P^o$ | $\varphi_{23}$ | $\varphi_2$ | $\chi_{1,23}$ |
| 0.0066 | 0.14 | 0.9926 | 0.5073 | 1.9819 |
| 0.0112 | 0.24 | 0.9875 | 0.5047 | 2.0164 |
| 0.0235 | 0.38 | 0.9742 | 0.4979 | 1.8083 |
| 0.0392 | 0.49 | 0.9577 | 0.4895 | 1.6273 |
| 0.0586 | 0.58 | 0.9381 | 0.4795 | 1.4770 |
| 0.0912 | 0.68 | 0.9069 | 0.4635 | 1.3153 |
| 0.1307 | 0.76 | 0.8719 | 0.4456 | 1.1952 |
| 0.1885 | 0.83 | 0.8251 | 0.4217 | 1.0755 |

$\chi_{1,23}^{\infty} = 2.1158$

| $W_{2p}=0.75$ | | | | |
|---|---|---|---|---|
| $m_s/m_p$ | $P/P^o$ | $\varphi_{23}$ | $\varphi_2$ | $\chi_{1,23}$ |
| 0.0050 | 0.14 | 0.9945 | 0.7541 | 2.2635 |
| 0.0094 | 0.24 | 0.9896 | 0.7504 | 2.1995 |
| 0.0212 | 0.38 | 0.9769 | 0.7407 | 1.9112 |
| 0.0359 | 0.49 | 0.9616 | 0.7291 | 1.7143 |

| | | $W_{2p}$= 0.75 | | |
|---|---|---|---|---|
| $m_s/m_p$ | $P/P^o$ | $\varphi_{23}$ | $\varphi_2$ | $\chi_{1,23}$ |
| 0.0542 | 0.58 | 0.9431 | 0.7151 | 1.5504 |
| 0.0865 | 0.68 | 0.9122 | 0.6917 | 1.3641 |
| 0.1283 | 0.76 | 0.8751 | 0.6635 | 1.2153 |
| 0.1990 | 0.83 | 0.8188 | 0.6209 | 1.0487 |

$\chi_{1,23}^{\infty} = 2.2935$

Table 11. Continued

| | | $W_{2p}$= 1.0 | |
|---|---|---|---|
| $m_s/m_p$ | $P/P^o$ | $\varphi_{23}$ ($\varphi_2$) | $\chi_{1,23}$ ( $\chi_{12}$ ) |
| 0.0040 | 0.14 | 0.9956 | 2.4867 |
| 0.0081 | 0.24 | 0.9912 | 2.3567 |
| 0.0194 | 0.38 | 0.9791 | 2.0060 |
| 0.0331 | 0.49 | 0.9649 | 1.7954 |
| 0.0504 | 0.58 | 0.9474 | 1.6196 |
| 0.0822 | 0.68 | 0.9170 | 1.4110 |
| 0.1261 | 0.76 | 0.8782 | 1.2351 |
| 0.2106 | 0.83 | 0.8119 | 1.0202 |

$\chi_{1,23}^{\infty} ( \chi_{12}^{\infty} ) = 2.4473$

| $W_{2p}$ | $\varphi_2^{\infty}$ | $\varphi_3^{\infty}$ | $\chi'_{23}$ |
|---|---|---|---|
| 0.25 | 0.2584 | 0.7416 | 0.2381 |
| 0.50 | 0.5111 | 0.4889 | 0.1044 |
| 0.75 | 0.7582 | 0.2418 | 0.0150 |

## 4.10 System: Water(1)-Poly(Acrylic Acid)(PAA)(2)-Poly(2-Hydroxyethyl Methacrylate)(PHEMA)(3) at 35°C

Experimental VLE data [15] for PAA and for PHEMA have been demonstrated in Table 12. VLE data for the blends have been calculated from their homopolymer quantities. Once again, hydrophobicity of PHEMA-lens material is evident from $\chi_{13}$>1. PAA with interaction parameter $\chi_{12}$<1 can be solved in water well. The limiting value of $\chi_{1,23}^{\infty}$ is a decreasing function of PAA concentration, similar to the results of section 4.8. Calculated pair interaction parameter, $\chi'_{23}$, from Eq. (15) indicates that these two polymers are compatible in all ranges of concentration and their degree of compatibility enhances with increasing percentage of PAA in the blend.

Table 12. Vapor-pressure lowering data for the system: Water (1)-PAA (2)-PHEMA (3) at 35°c.

| $W_{2p}=0.0$ | | | |
|---|---|---|---|
| $m_s/m_p$ | $P/P^o$ | $\varphi_{23}\,(\varphi_3)$ | $\chi_{1,23}\,(\chi_{13})$ |
| 0.0184 | 0.14 | 0.9792 | 0.9694 |
| 0.0187 | 0.24 | 0.9789 | 1.5175 |
| 0.0300 | 0.38 | 0.9667 | 1.5696 |
| 0.0483 | 0.49 | 0.9474 | 1.4307 |
| 0.0701 | 0.58 | 0.9254 | 1.3145 |
| 0.1025 | 0.68 | 0.8946 | 1.2116 |
| 0.1355 | 0.76 | 0.8652 | 1.1544 |
| 0.1705 | 0.83 | 0.8360 | 1.1241 |

$\chi_{1,23}^{\infty}\,(\chi_{13}^{\infty}) = 1.8226$

| $W_{2p}= 0.25$ | | | | |
|---|---|---|---|---|
| $m_s/m_p$ | $P/P^o$ | $\varphi_{23}$ | $\varphi_2$ | $\chi_{1,23}$ |
| 0.0226 | 0.14 | 0.9746 | 0.2437 | 0.7727 |
| 0.0220 | 0.24 | 0.9753 | 0.2438 | 1.3667 |
| 0.0335 | 0.38 | 0.9629 | 0.2407 | 1.4718 |
| 0.0545 | 0.49 | 0.9410 | 0.2353 | 1.3285 |
| 0.0805 | 0.58 | 0.9152 | 0.2288 | 1.2034 |
| 0.1200 | 0.68 | 0.8787 | 0.2197 | 1.0947 |
| 0.1610 | 0.76 | 0.8437 | 0.2109 | 1.0368 |
| 0.2052 | 0.83 | 0.8091 | 0.2023 | 1.0091 |

$\chi_{1,23}^{\infty} = 1.7256$

| $W_{2p}= 0.50$ | | | | |
|---|---|---|---|---|
| $m_s/m_p$ | $P/P^o$ | $\varphi_{23}$ | $\varphi_2$ | $\chi_{1,23}$ |
| 0.0293 | 0.14 | 0.9674 | 0.4837 | 0.5242 |
| 0.0267 | 0.24 | 0.9703 | 0.4851 | 1.1873 |
| 0.0379 | 0.38 | 0.9583 | 0.4791 | 1.3624 |
| 0.0625 | 0.49 | 0.9329 | 0.4665 | 1.2131 |
| 0.0946 | 0.58 | 0.9019 | 0.4509 | 1.0755 |
| 0.1448 | 0.68 | 0.8572 | 0.4286 | 0.9575 |
| 0.1984 | 0.76 | 0.8143 | 0.4071 | 0.8970 |
| 0.2574 | 0.83 | 0.7716 | 0.3858 | 0.8713 |

$\chi_{1,23}^{\infty} = 1.6169$

| $W_{2p}= 0.75$ | | | | |
|---|---|---|---|---|
| $m_s/m_p$ | $P/P^o$ | $\varphi_{23}$ | $\varphi_2$ | $\chi_{1,23}$ |
| 0.0415 | 0.14 | 0.9544 | 0.7158 | 0.1847 |
| 0.0339 | 0.24 | 0.9625 | 0.7219 | 0.9656 |
| 0.0435 | 0.38 | 0.9523 | 0.7142 | 1.2383 |
| 0.0733 | 0.49 | 0.9222 | 0.6917 | 1.0801 |

| $W_{2p}$= 0.75 | | | | |
|---|---|---|---|---|
| $m_s/m_p$ | $P/P^o$ | $\varphi_{23}$ | $\varphi_2$ | $\chi_{1,23}$ |
| 0.1147 | 0.58 | 0.8835 | 0.6626 | 0.9242 |
| 0.1826 | 0.68 | 0.8265 | 0.6199 | 0.7895 |
| 0.2582 | 0.76 | 0.7710 | 0.5783 | 0.7212 |
| 0.3453 | 0.83 | 0.7157 | 0.5368 | 0.6946 |

$\chi_{1,23}^{\infty} = 1.4928$

Table 12. Continued

| $W_{2p}$= 1.0 | | | |
|---|---|---|---|
| $m_s/m_p$ | $P/P^o$ | $\varphi_{23}$ ($\varphi_2$) | $\chi_{1,23}$ ($\chi_{12}$) |
| 0.0712 | 0.14 | 0.9243 | -0.3623 |
| 0.0463 | 0.24 | 0.9494 | 0.6736 |
| 0.0513 | 0.38 | 0.9443 | 1.0946 |
| 0.0886 | 0.49 | 0.9075 | 0.9227 |
| 0.1456 | 0.58 | 0.8565 | 0.7367 |
| 0.2469 | 0.68 | 0.7788 | 0.5677 |
| 0.3698 | 0.76 | 0.7016 | 0.4739 |
| 0.5245 | 0.83 | 0.6238 | 0.4303 |

$\chi_{1,23}^{\infty}(\chi_{12}^{\infty}) = 1.3457$

| $W_{2p}$ | $\varphi_2^{\infty}$ | $\varphi_3^{\infty}$ | $\chi'_{23}$ |
|---|---|---|---|
| 0.25 | 0.25 | 0.75 | -0.1185 |
| 0.50 | 0.50 | 0.50 | -0.1310 |
| 0.75 | 0.75 | 0.25 | -0.1487 |

## 4.11 System: Water(1)-Poly(N,N-Dimethylmethacrylate)(PDMAA)(2)-Poly(N-Vinyl-2-Pyrrolidone)(PVP)(3) at 35°C

Table 13 displays experimental VLE data points for PDMAA that are taken from Prausnitz et al. work [15]. Also we have used the same data points here for PVP from Prausnitz et al. work [16]. VLE data points for their blends have been calculated from their homopolymer values.

Table 13. Vapor-pressure lowering data for the system: Water (1)-PDMAA (2)-PVP (3) at 35°c.

| $W_{2p}$=0.0 | | | |
|---|---|---|---|
| $m_s/m_p$ | $P/P^o$ | $\varphi_{23}$ ($\varphi_3$) | $\chi_{1,23}$ ($\chi_{13}$) |
| 0.0572 | 0.14 | 0.9378 | -0.1438 |
| 0.0928 | 0.24 | 0.9028 | 0.0012 |
| 0.1583 | 0.38 | 0.8449 | 0.0717 |
| 0.2258 | 0.49 | 0.7924 | 0.1059 |

| $W_{2p}$=0.0 | | | |
|---|---|---|---|
| $m_s/m_p$ | $P/P^o$ | $\varphi_{23}$ ($\varphi_3$) | $\chi_{1,23}$ ($\chi_{13}$) |
| 0.2948 | 0.58 | 0.7451 | 0.1390 |
| 0.3905 | 0.68 | 0.6882 | 0.1934 |
| 0.4859 | 0.76 | 0.6395 | 0.2601 |
| 0.5874 | 0.83 | 0.5947 | 0.3454 |

$\chi_{1,23}^{\infty}$ ( $\chi_{13}^{\infty}$ ) = -0.1866

Table 13. Continued

| $W_{2p}$= 0.25 | | | | |
|---|---|---|---|---|
| $m_s/m_p$ | $P/P^o$ | $\varphi_{23}$ | $\varphi_2$ | $\chi_{1,23}$ |
| 0.0437 | 0.14 | 0.9519 | 0.2395 | 0.1268 |
| 0.0797 | 0.24 | 0.9155 | 0.2304 | 0.1537 |
| 0.1404 | 0.38 | 0.8602 | 0.2164 | 0.1887 |
| 0.2037 | 0.49 | 0.8092 | 0.2036 | 0.2047 |
| 0.2704 | 0.58 | 0.7617 | 0.1916 | 0.2199 |
| 0.3657 | 0.68 | 0.7026 | 0.1768 | 0.2520 |
| 0.4658 | 0.76 | 0.6497 | 0.1635 | 0.2957 |
| 0.5886 | 0.83 | 0.5948 | 0.1497 | 0.3454 |

$\chi_{1,23}^{\infty}$ = 0.1202

| $W_{2p}$= 0.50 | | | | |
|---|---|---|---|---|
| $m_s/m_p$ | $P/P^o$ | $\varphi_{23}$ | $\varphi_2$ | $\chi_{1,23}$ |
| 0.0354 | 0.14 | 0.9607 | 0.4824 | 0.3359 |
| 0.0698 | 0.24 | 0.9254 | 0.4647 | 0.2834 |
| 0.1262 | 0.38 | 0.8728 | 0.4382 | 0.2906 |
| 0.1855 | 0.49 | 0.8235 | 0.4136 | 0.2917 |
| 0.2497 | 0.58 | 0.7762 | 0.3898 | 0.2920 |
| 0.3439 | 0.68 | 0.7157 | 0.3594 | 0.3054 |
| 0.4473 | 0.76 | 0.6593 | 0.3311 | 0.3292 |
| 0.5899 | 0.83 | 0.5948 | 0.2987 | 0.3455 |

$\chi_{1,23}^{\infty}$ = 0.3346

| $W_{2p}$= 0.75 | | | | |
|---|---|---|---|---|
| $m_s/m_p$ | $P/P^o$ | $\varphi_{23}$ | $\varphi_2$ | $\chi_{1,23}$ |
| 0.0297 | 0.14 | 0.9669 | 0.7267 | 0.5071 |
| 0.0621 | 0.24 | 0.9331 | 0.7014 | 0.3966 |
| 0.1146 | 0.38 | 0.8833 | 0.6639 | 0.3811 |
| 0.1703 | 0.49 | 0.8359 | 0.6283 | 0.3695 |
| 0.2319 | 0.58 | 0.7891 | 0.5931 | 0.3572 |
| 0.3245 | 0.68 | 0.7278 | 0.5470 | 0.3545 |
| 0.4302 | 0.76 | 0.6685 | 0.5025 | 0.3607 |
| 0.5911 | 0.83 | 0.5948 | 0.4471 | 0.3455 |

$\chi_{1,23}^{\infty}$ = 0.50

Table 13. Continued

| $W_{2p}$= 1.0 | | | |
|---|---|---|---|
| $m_s/m_p$ | $P/P^o$ | $\varphi_{23}$ ($\varphi_2$) | $\chi_{1,23}$ ($\chi_{12}$) |
| 0.0256 | 0.14 | 0.9714 | 0.6522 |
| 0.0560 | 0.24 | 0.9395 | 0.4973 |
| 0.1050 | 0.38 | 0.8923 | 0.4627 |
| 0.1574 | 0.49 | 0.8467 | 0.4400 |
| 0.2165 | 0.58 | 0.8006 | 0.4169 |
| 0.3072 | 0.68 | 0.7389 | 0.4000 |
| 0.4144 | 0.76 | 0.6772 | 0.3906 |
| 0.5924 | 0.83 | 0.5948 | 0.3456 |

$\chi_{1,23}^{\infty}(\chi_{12}^{\infty}) = 0.6354$

| $W_{2p}$ | $\varphi_2^{\infty}$ | $\varphi_3^{\infty}$ | $\chi'_{23}$ |
|---|---|---|---|
| 0.25 | 0.2516 | 0.7484 | -0.5310 |
| 0.50 | 0.5022 | 0.4978 | -0.4336 |
| 0.75 | 0.7516 | 0.2484 | -0.3684 |

It is clear from the interaction parameter $\chi_{12}<1$ and $\chi_{13}<1$ that water is a good solvent for both polymers. Also, a decreasing pace of the limiting value, $\chi_{1,23}^{\infty}$, can be observed when PVP percentage increases. By using these limiting quantities, the pair interaction parameter, $\chi'_{23}$, can be estimated from Eq. (15). In accordance with the results observed for this quantity, PDMAA and PVP are compatible in all ranges of concentration and their degree of compatibility is a decreasing function of PDMAA content in the blend.

## 4.12 System: Water (1)-Poly(2-Dimethyl Aminoethyl Methacrylate) (PDMAEMA)(2)-Poly(N-Vinyl-2-Pyrrolidone) (PVP)(3) at 35°C

Experimental VLE data for PDMAEMA [15] and for PVP [16] have been reported in Table 14. We use these quantities to calculate VLE data for their blends. The interaction parameter $\chi_{12}$ and $\chi_{13}$ values provide a good (inverse) indication of the solvency power of water for the polymers. In this case, water is a good solvent for PVP with $\chi_{13}<1$ and is a non-solvent for PDMAEMA with $\chi_{12}>1$. Similar to the results of section 4.11, the limiting values of $\chi_{1,23}^{\infty}$ fall in proportion to the PVP concentration in the blend. From these limiting quantities, pair interaction parameter, $\chi'_{23}$, can be calculated by using Eq. (15). The results obtained for this quantity imply that compatibility can occur in PDMAEMA-PVP blend in all concentration ranges but it will decrease with increasing PDMAEMA content in the blend.

Table 14. Vapor-pressure lowering data for the system: Water (1)-PDMAEMA (2)-PVP (3) at 35°c.

| $W_{2p}$=0.0 | | | |
|---|---|---|---|
| **$m_s/m_p$** | **$P/P^o$** | **$\varphi_{23}$ ($\varphi_3$)** | **$\chi_{1,23}$ ($\chi_{13}$)** |
| 0.0572 | 0.14 | 0.9378 | -0.1438 |
| 0.0928 | 0.24 | 0.9028 | 0.0012 |
| 0.1583 | 0.38 | 0.8449 | 0.0717 |
| 0.2258 | 0.49 | 0.7924 | 0.1059 |
| 0.2948 | 0.58 | 0.7451 | 0.1390 |
| 0.3905 | 0.68 | 0.6882 | 0.1934 |
| 0.4859 | 0.76 | 0.6395 | 0.2601 |
| 0.5874 | 0.83 | 0.5947 | 0.3454 |

$\chi_{1,23}^{\infty}$ ( $\chi_{13}^{\infty}$ ) = -0.1866

| $W_{2p}$= 0.25 | | | | |
|---|---|---|---|---|
| **$m_s/m_p$** | **$P/P^o$** | **$\varphi_{23}$** | **$\varphi_2$** | **$\chi_{1,23}$** |
| 0.0133 | 0.14 | 0.9850 | 0.2562 | 1.2897 |
| 0.0256 | 0.24 | 0.9715 | 0.2527 | 1.2299 |
| 0.0567 | 0.38 | 0.9391 | 0.2442 | 1.0113 |
| 0.0919 | 0.49 | 0.9049 | 0.2353 | 0.8967 |
| 0.1333 | 0.58 | 0.8676 | 0.2257 | 0.8102 |
| 0.2016 | 0.68 | 0.8125 | 0.2113 | 0.7209 |
| 0.2836 | 0.76 | 0.7549 | 0.1964 | 0.6613 |
| 0.4059 | 0.83 | 0.6828 | 0.1776 | 0.5987 |

$\chi_{1,23}^{\infty}$ = 1.3418

| $W_{2p}$= 0.50 | | | | |
|---|---|---|---|---|
| **$m_s/m_p$** | **$P/P^o$** | **$\varphi_{23}$** | **$\varphi_2$** | **$\chi_{1,23}$** |
| 0.0075 | 0.14 | 0.9916 | 0.5090 | 1.8524 |
| 0.0148 | 0.24 | 0.9835 | 0.5048 | 1.7522 |
| 0.0345 | 0.38 | 0.9625 | 0.4940 | 1.4606 |
| 0.0577 | 0.49 | 0.9389 | 0.4819 | 1.2960 |
| 0.0861 | 0.58 | 0.9114 | 0.4678 | 1.1645 |
| 0.1359 | 0.68 | 0.8670 | 0.4450 | 1.0172 |
| 0.2003 | 0.76 | 0.8156 | 0.4186 | 0.9027 |
| 0.3101 | 0.83 | 0.7406 | 0.3802 | 0.7704 |

$\chi_{1,23}^{\infty}$ = 1.8627

| $W_{2p}$= 0.75 | | | | |
|---|---|---|---|---|
| **$m_s/m_p$** | **$P/P^o$** | **$\varphi_{23}$** | **$\varphi_2$** | **$\chi_{1,23}$** |
| 0.0052 | 0.14 | 0.9942 | 0.7554 | 2.2157 |
| 0.0104 | 0.24 | 0.9885 | 0.75119 | 2.0968 |
| 0.0248 | 0.38 | 0.9731 | 0.7394 | 1.7686 |
| 0.0420 | 0.49 | 0.9553 | 0.7258 | 1.5762 |

| | | $W_{2p}$= 0.75 | | |
|---|---|---|---|---|
| $m_s/m_p$ | $P/P^o$ | $\varphi_{23}$ | $\varphi_2$ | $\chi_{1,23}$ |
| 0.0636 | 0.58 | 0.9338 | 0.7095 | 1.4183 |
| 0.1025 | 0.68 | 0.8975 | 0.6819 | 1.2350 |
| 0.1548 | 0.76 | 0.8529 | 0.6480 | 1.0850 |
| 0.2508 | 0.83 | 0.7815 | 0.5938 | 0.9057 |

$\chi_{1,23}^{\infty} = 2.1968$

Table 14. Continued

| | | $W_{2p}$= 1.0 | |
|---|---|---|---|
| $m_s/m_p$ | $P/P^o$ | $\varphi_{23}$ ($\varphi_2$) | $\chi_{1,23}$ ($\chi_{12}$) |
| 0.0040 | 0.14 | 0.9956 | 2.4867 |
| 0.0081 | 0.24 | 0.9912 | 2.3567 |
| 0.0194 | 0.38 | 0.9791 | 2.0060 |
| 0.0331 | 0.49 | 0.9649 | 1.7954 |
| 0.0504 | 0.58 | 0.9474 | 1.6196 |
| 0.0822 | 0.68 | 0.9170 | 1.4110 |
| 0.1261 | 0.76 | 0.8782 | 1.2351 |
| 0.2106 | 0.83 | 0.8119 | 1.0202 |

$\chi_{1,23}^{\infty} (\chi_{12}^{\infty}) = 2.4473$

| $W_{2p}$ | $\varphi_2^{\infty}$ | $\varphi_3^{\infty}$ | $\chi'_{23}$ |
|---|---|---|---|
| 0.25 | 0.2601 | 0.7399 | -4.3821 |
| 0.50 | 0.5133 | 0.4867 | -2.7912 |
| 0.75 | 0.7598 | 0.2402 | -2.0940 |

## 4.13 System: Water (1)-Poly(Acrylic Acid)(PAA)(2)-Poly(N-Vinyl-2-Pyrrolidone)(PVP)(3) at 35°C

Table 15 displays experimental VLE data points for PAA [15] and for PVP [16]. The same quantities for their blend can be calculated by using VLE data points of pure Polymers. The results obtained for $\chi_{12}$ and $\chi_{13}$ demonstrate that both polymers can be solved in water well. The limiting value of $\chi_{1,23}^{\infty}$ at zero solvent concentration is a decreasing function of PVP content in the blend. Results obtained here for pair interaction parameter, $\chi'_{23}$, suggest the PAA-PVP pair to be less compatible than results obtained when PDMAEMA is selected to mix with PVP. On the other hand, in both cases the compatibility is shown enhanced with increasing PVP concentration.

Table 15. Vapor-pressure lowering data for the system: Water (1)-PAA (2)-PVP (3) at 35°c.

**$W_{2p}$=0.0**

| $m_s/m_p$ | $P/P^o$ | $\varphi_{23}$ ($\varphi_3$) | $\chi_{1,23}$ ($\chi_{13}$) |
|---|---|---|---|
| 0.0572 | 0.14 | 0.9378 | -0.1438 |
| 0.0928 | 0.24 | 0.9028 | 0.0012 |
| 0.1583 | 0.38 | 0.8449 | 0.0717 |
| 0.2258 | 0.49 | 0.7924 | 0.1059 |
| 0.2948 | 0.58 | 0.7451 | 0.1390 |
| 0.3905 | 0.68 | 0.6882 | 0.1934 |
| 0.4859 | 0.76 | 0.6395 | 0.2601 |
| 0.5874 | 0.83 | 0.5947 | 0.3454 |

$\chi_{1,23}^{\infty}$ ($\chi_{13}^{\infty}$) = -0.1866

**$W_{2p}$= 0.25**

| $m_s/m_p$ | $P/P^o$ | $\varphi_{23}$ | $\varphi_2$ | $\chi_{1,23}$ |
|---|---|---|---|---|
| 0.0601 | 0.14 | 0.9349 | 0.2352 | -0.1933 |
| 0.0742 | 0.24 | 0.9209 | 0.2317 | 0.2227 |
| 0.1040 | 0.38 | 0.8925 | 0.2246 | 0.4651 |
| 0.1628 | 0.49 | 0.8414 | 0.2117 | 0.4052 |
| 0.2347 | 0.58 | 0.7864 | 0.1979 | 0.3434 |
| 0.3409 | 0.68 | 0.7170 | 0.1804 | 0.3107 |
| 0.4505 | 0.76 | 0.6572 | 0.1654 | 0.3219 |
| 0.5703 | 0.83 | 0.6024 | 0.1516 | 0.3680 |

$\chi_{1,23}^{\infty}$ = 0.7055

**$W_{2p}$= 0.50**

| $m_s/m_p$ | $P/P^o$ | $\varphi_{23}$ | $\varphi_2$ | $\chi_{1,23}$ |
|---|---|---|---|---|
| 0.0634 | 0.14 | 0.9317 | 0.4679 | -0.2459 |
| 0.0618 | 0.24 | 0.9333 | 0.4687 | 0.3993 |
| 0.0775 | 0.38 | 0.9179 | 0.4609 | 0.7289 |
| 0.1273 | 0.49 | 0.8718 | 0.4378 | 0.6174 |
| 0.1950 | 0.58 | 0.8162 | 0.4099 | 0.4999 |
| 0.3025 | 0.68 | 0.7410 | 0.3721 | 0.4087 |
| 0.4200 | 0.76 | 0.6734 | 0.3381 | 0.3774 |
| 0.5542 | 0.83 | 0.6097 | 0.3062 | 0.3896 |

$\chi_{1,23}^{\infty}$ = 0.9956

**$W_{2p}$= 0.75**

| $m_s/m_p$ | $P/P^o$ | $\varphi_{23}$ | $\varphi_2$ | $\chi_{1,23}$ |
|---|---|---|---|---|
| 0.0671 | 0.14 | 0.9282 | 0.6977 | -0.3020 |
| 0.0530 | 0.24 | 0.9424 | 0.7084 | 0.5467 |
| 0.0617 | 0.38 | 0.9336 | 0.7017 | 0.9304 |
| 0.1045 | 0.49 | 0.8925 | 0.6708 | 0.7842 |
| 0.1667 | 0.58 | 0.8388 | 0.6305 | 0.6278 |

| $m_s/m_p$ | $P/P^o$ | $W_{2p}$= 0.75<br>$\varphi_{23}$ | $\varphi_2$ | $\chi_{1,23}$ |
|---|---|---|---|---|
| 0.2719 | 0.68 | 0.7614 | 0.5723 | 0.4932 |
| 0.3933 | 0.76 | 0.6881 | 0.5172 | 0.4277 |
| 0.5389 | 0.83 | 0.6169 | 0.4636 | 0.4104 |

$\chi_{1,23}^{\infty} = 1.1934$

Table 15. Continued

| $m_s/m_p$ | $P/P^o$ | $W_{2p}$= 1.0<br>$\varphi_{23}$ ($\varphi_2$) | $\chi_{1,23}$ ( $\chi_{12}$ ) |
|---|---|---|---|
| 0.0712 | 0.14 | 0.9243 | -0.3623 |
| 0.0463 | 0.24 | 0.9494 | 0.6736 |
| 0.0513 | 0.38 | 0.9443 | 1.0946 |
| 0.0886 | 0.49 | 0.9075 | 0.9227 |
| 0.1456 | 0.58 | 0.8565 | 0.7367 |
| 0.2469 | 0.68 | 0.7788 | 0.5677 |
| 0.3698 | 0.76 | 0.7016 | 0.4739 |
| 0.5245 | 0.83 | 0.6238 | 0.4303 |

$\chi_{1,23}^{\infty} ( \chi_{12}^{\infty} ) = 1.3457$

| $W_{2p}$ | $\varphi_2^{\infty}$ | $\varphi_3^{\infty}$ | $\chi'_{23}$ |
|---|---|---|---|
| 0.25 | 0.2516 | 0.7484 | -2.6903 |
| 0.50 | 0.5022 | 0.4978 | -1.6507 |
| 0.75 | 0.7516 | 0.2484 | -1.2229 |

## 4.14 System: Water(1)-Poly(2-Hydroxyethyl Methacrylate)(PHEMA)(2)-Poly(N-Vinyl-2-Pyrrolidone)(PVP)(3) at 35°C

Experimental VLE data points for PHEMA [15] and for PVP [16] have been reported in table 16. We have used these quantities to calculate VLE data for their blends. Once again, the hydrophobicity of PHEMA-lens material is obvious from $\chi_{12}$>1. PVP with interaction parameter $\chi_{13}$<1 can be dissolved in water well. Like the past three sections, a decreasing pace of the limiting values, $\chi_{1,23}^{\infty}$, at zero solvent concentration, can be seen when PVP content increases in the blend. Results obtained here for pair interaction parameter, $\chi'_{23}$, imply that PHEMA-PVP blend is compatible in all concentration ranges but is less compatible than when PDMAEMA or PAA are used to mix with PVP. Finally and similar to last three sections, the decreasing pace of $\chi'_{23}$ can be observed when PVP content decreases in the blend.

Table 16. Vapor-pressure lowering data for the system: Water (1)-PHEMAA (2)-PVP (3) at 35°c.

| $W_{2p}=0.0$ | | | |
|---|---|---|---|
| $m_s/m_p$ | $P/P^o$ | $\varphi_{23}$ ($\varphi_3$) | $\chi_{1,23}$ ($\chi_{13}$) |
| 0.0572 | 0.14 | 0.9378 | -0.1438 |
| 0.0928 | 0.24 | 0.9028 | 0.0012 |
| 0.1583 | 0.38 | 0.8449 | 0.0717 |
| 0.2258 | 0.49 | 0.7924 | 0.1059 |
| 0.2948 | 0.58 | 0.7451 | 0.1390 |
| 0.3905 | 0.68 | 0.6882 | 0.1934 |
| 0.4859 | 0.76 | 0.6395 | 0.2601 |
| 0.5874 | 0.83 | 0.5947 | 0.3454 |

$\chi_{1,23}^{\infty}$ ($\chi_{13}^{\infty}$) = -0.1866

| $W_{2p}=0.25$ | | | | |
|---|---|---|---|---|
| $m_s/m_p$ | $P/P^o$ | $\varphi_{23}$ | $\varphi_2$ | $\chi_{1,23}$ |
| 0.0375 | 0.14 | 0.9511 | 0.2836 | 0.1114 |
| 0.0466 | 0.24 | 0.9399 | 0.2803 | 0.5029 |
| 0.0765 | 0.38 | 0.9050 | 0.2699 | 0.5880 |
| 0.1177 | 0.49 | 0.8610 | 0.2568 | 0.5382 |
| 0.1636 | 0.58 | 0.8167 | 0.2435 | 0.5024 |
| 0.2293 | 0.68 | 0.7607 | 0.2269 | 0.4901 |
| 0.2951 | 0.76 | 0.7118 | 0.2123 | 0.5090 |
| 0.3646 | 0.83 | 0.6666 | 0.1988 | 0.5524 |

$\chi_{1,23}^{\infty} = 0.7802$

| $W_{2p}=0.50$ | | | | |
|---|---|---|---|---|
| $m_s/m_p$ | $P/P^o$ | $\varphi_{23}$ | $\varphi_2$ | $\chi_{1,23}$ |
| 0.0279 | 0.14 | 0.9653 | 0.5410 | 0.4615 |
| 0.031 | 0.24 | 0.9614 | 0.5388 | 0.9372 |
| 0.0504 | 0.38 | 0.9390 | 0.5262 | 1.0092 |
| 0.0796 | 0.49 | 0.9070 | 0.5083 | 0.9174 |
| 0.1132 | 0.58 | 0.8726 | 0.4890 | 0.8448 |
| 0.1623 | 0.68 | 0.8270 | 0.4634 | 0.7920 |
| 0.2119 | 0.76 | 0.7854 | 0.4402 | 0.7769 |
| 0.2643 | 0.83 | 0.7459 | 0.4180 | 0.7868 |

$\chi_{1,23}^{\infty} = 1.2302$

| $W_{2p}=0.75$ | | | | |
|---|---|---|---|---|
| $m_s/m_p$ | $P/P^o$ | $\varphi_{23}$ | $\varphi_2$ | $\chi_{1,23}$ |
| 0.0222 | 0.14 | 0.9737 | 0. 7719 | 0.7380 |
| 0.0234 | 0.24 | 0.9724 | 0.7708 | 1.2580 |
| 0.0376 | 0.38 | 0.9563 | 0.7581 | 1.3188 |
| 0.0601 | 0.49 | 0.9319 | 0.7388 | 1.1996 |

| | | $W_{2p}$= 0.75 | | |
|---|---|---|---|---|
| $m_s/m_p$ | $P/P^o$ | $\varphi_{23}$ | $\varphi_2$ | $\chi_{1,23}$ |
| 0.0866 | 0.58 | 0.9048 | 0.7172 | 1.1019 |
| 0.1256 | 0.68 | 0.8675 | 0.6877 | 1.0207 |
| 0.1653 | 0.76 | 0.8327 | 0.6601 | 0.9817 |
| 0.2073 | 0.83 | 0.7987 | 0.6332 | 0.9687 |

$\chi^{\infty}_{1,23} = 1.5584$

Table 16. Continued

| | | $W_{2p}$=1.0 | |
|---|---|---|---|
| $m_s/m_p$ | $P/P^o$ | $\varphi_{23}$ ($\varphi_3$) | $\chi_{1,23}$ ($\chi_{13}$) |
| 0.0184 | 0.14 | 0.9792 | 0.9694 |
| 0.0187 | 0.24 | 0.9789 | 1.5175 |
| 0.0300 | 0.38 | 0.9667 | 1.5696 |
| 0.0483 | 0.49 | 0.9474 | 1.4307 |
| 0.0701 | 0.58 | 0.9254 | 1.3145 |
| 0.1025 | 0.68 | 0.8946 | 1.2116 |
| 0.1355 | 0.76 | 0.8652 | 1.1544 |
| 0.1705 | 0.83 | 0.8360 | 1.1241 |

$\chi^{\infty}_{1,23}$ ($\chi^{\infty}_{12}$) = 1.8226

| $W_{2p}$ | $\varphi^{\infty}_2$ | $\varphi^{\infty}_3$ | $\chi'_{23}$ |
|---|---|---|---|
| 0.25 | 0.2982 | 0.7018 | -1.7568 |
| 0.50 | 0.5604 | 0.4396 | -1.1806 |
| 0.75 | 0.7927 | 0.2073 | -0.9269 |

# 5 Conclusion

Fifty-six isothermal data sets for vapor-liquid equilibria (VLE) have been used for 15 polymer+solvent binaries, 11 copolymer+solvent binaries and for 30 polymer-polymer-solvent ternaries to study compatibility of polymer blends. The equilibrium solubility of a penetrant in a polymer depends on their mutual compatibility. Equations based on theories of polymer solution tend to be more successful when there is some kind of similarity between the penetrant and the monomer repeat unit in the polymer, e.g., for nonpolar penetrants in polymers which do not contain appreciable polar groups. Expected nonideal behavior has been observed for systems containing hydrocarbons and poly(acrylonitrile-co-butadiene). The role of intramolecular interaction in vapor-liquid equilibria of copolymer+solvent systems is well documented for poly(acrylonitrile-co-butadiene) that have higher affinity for acetonitrile than do polyacrylonitrile or polybutadiene.

Also, the Flory interaction parameter has been obtained for 5 homopolymers of interest, i.e., PDMAEMA, PHEMA, PAA, PDMAA and PVP for packaging, insulation, and biomedical applications such as drug delivery systems or for contact lens technology. For the systems studied in this chapter, it was shown that the Flory-Huggins interaction parameter is dependent on the concentration of polymers considered. This parameter also shows that water has little solubility in PDMAEMA and PHEMA, as expected for less hydrophilic polymers. Water solubility is much higher in hydrophilic PDMAA, PAA and PVP. Data reduction using Flory-Huggins theory shows that polymers with fairly good interaction with water, can be compatible when mix with other less hydrophilic polymers. The only situation that incompatibility has been observed is for these less hydrophilic polymers with positive values of $\chi'_{23}$.

Although, $\chi'_{23}$ data reported in this work yield useful information concerning compatibility of polymer blends, but the value of $\chi'_{23}$ should be taken with caution. Extrapolation of the $\chi_{1,23}$ values for the limiting case of zero solvent concentration is probably the main cause of uncertainty. In addition, as discussed in the text, the solvent used for the study may also affect the value of $\chi'_{23}$.

## References


[1] Paul, D. R.; Newman, S. Eds *Polymer Blends*, Academic Press: New York, NY, 1978.

[2] Olabisi, O.; Robenson, L. M.; Shaw, M. T. *Polymer-Polymer Miscibility*, Academic Press: New York, NY, 1979.

[3] Schmitt, B. J. *Angew Chem Int* Ed Engl 1979, 18, 273.

[4] Solc, K.; Ed *Polymer Compatibility and Incompatibility: Principles and Practices*; MMI Press Symposium Series; Harwood Academic Publishers Gmbh: New York, NY, 1982; Vol. 2.

[5] Krause, S. *Pure Appl Chem* 1986, 58, 1553.

[6] van der Vegt, A. K. *From Polymers to Plastics,* Delft University Press: Netherlands, 1999.

[7] Flory, P. J. *J Chem Phys* 1942, 10, 51.

[8] Huggins, M. L. *J Phys Chem* 1942, 46, 151.

[9] Flory, P. *J. Principles of Polymer Chemistry*; Cornell University Press: Ithaca, NY, 1953; Chap. 12.

[10] Meyer K. H., Luhdemann R. *Helv Chim Acta* 1935, 18, 307. Meyer K. H. Z *physic Chem* **B44** 1939, 383*; Helv Chim Acta* 1940, 23, 1063.

[11] Prausnitz, J. M.; Lichtenthaler, R. N.; Azevedo, E. G. *Molecular Thermodynamic of Fluid-Phase Equilibria*. Prentice-Hall Inc.: New Jersey, 1999.

[12] Elias, H. G. *An Introduction to Polymer Science*. VCH Publishers, Inc.: New York, NY, 1997.

[13] Panayiotou, C.; Vera, H. *Polymer J* 1984, 16, 89.

[14] Gupta R. B.; Prausnitz J. M. *J Chem Eng Data* 1995, 40, 784.

[15] Prausnitz, J. M.; Rodriguez, O.; Fornasiero, F.; Arce A.; Radke, C. *J. Phys Chem Chem Phys* 2004, 6, 103.

[16] Prausnitz, J. M.; Rodriguez, O.; Fornasiero, F.; Arce A.; Radke, C. *J. Polymer* 2003, 44, 6323.

[17] Vanzo, B. (2003). SP2. *Scientific Polymer Products*, Ontario, New York, NY, Website: www.scientificpolymer.com.

[18] Dojcansky, J.; Heinrich, J. Chem Zvesti 1974, 28, 157.

[19] Gerhartz W. Ullmann's Encyclopedia of Industrial Chemistry, VCH Publishers: New York, NY, 1986; Vol. A6, PP 263-271.

[20] Yaws, C. L. Yaws' *Handbook of Thermodynamic and Physical Properties of Chemical Compounds*, 2003; Electronic ISBN: 1-59-124-444-7.

[21] Vargaftik, N. B. *Handbook of Physical Properties of Liquids and Gases*; Begell House, Inc.: New York, NY, 1996; Chap. 4.

[22] Lide D. R. CRC *Handbook of Chemistry and Physics*, CRC Press LLC: Florida, 1999.

In: Frontiers in Polymer Research
Editor: Robert K. Bregg, pp. 211-219
ISBN: 1-59454-824-2


***Chapter 8***

# STABILIZATION OF VINYLIDENE CHLORIDE POLYMERS BY COMONOMER INCORPORATION

***B.A. Howell***
Center for Applications in Polymer Science, Central Michigan University,
Mt. Pleasant, MI 48859

## Abstract

Vinylidene chloride polymers find important application in the barrier plastics packaging industry. These materials display low permeability rates for both oxygen (and other small molecules) and for food aroma and taste constituents. On the one hand, they function to prevent spoilage of packaged food items and, on the other, to prevent the loss of flavor agents that make these items palatable. While these materials have excellent barrier properties they may be processed only with difficulty owing to the propensity to undergo thermally-induced degradative dehydrochlorination. In fact, the homopolymer cannot be processed. Incorporating simple acrylate comonomers into the polymer structure lowers the melt temperature and improves processibility. However, it is insufficient to prevent significant degradation at process temperatures. Incorporation into vinylidene chloride polymers of a series of comonomers which result in the formation of polymer pendant groups with the potential 1.) to react with hydrogen chloride as it is formed (and thus prevent its interaction with the walls of process equipment to form Lewis acids, principally iron(III) chloride, which accelerate the dehydrochlorination reaction) and 2.) to expose phenolic units (which may scavenge chlorine atoms and other radical species) on reaction with hydrogen chloride has been examined as a means of stabilizing these materials.

The thermal degradation of poly(vinylidene chloride) and vinylidene chloride (VDC) copolymers usually occurs with the evolution of hydrogen chloride at elevated temperature[1]. For the homopolymer, the degradation occurs rapidly when heated to its melting point, making it difficult to formulate through extrusion processes. As a consequence, only the copolymers with vinyl chloride, alkyl acrylate and acrylonitrile, etc., are of commercial prominence[1]. These polymers have a lower melting point, and are more soluble in some solvent mixtures than the homopolymer so that processing is possible.

In the packaging industry, VDC copolymers have assumed a position of great importance because of the extremely low permeability for a wide variety of gases including oxygen and flavor and aroma constituents[2],[3],[4]. To retain these unique properties which are essential for high barrier to the mass transport of small molecules, the VDC content of these polymers is usually greater than 85%. There have been a number of studies focused on thermal decomposition processes of these polymers[5-10]. It has been suggested that a defect site, most probably internal double bonds introduced during the preparation or processing of the polymer is responsible for the initiation of the degradation[5-9]. The initiation, propagation and termination phases are typical for degradative dehydrochlorination along the polymer mainchain and the "unzipping" results in the formation of polyene sequences comprised of chloroacetylene units[5]. The "unzipping" can be stopped by comonomer units present such as acrylate, i.e., the length of the polyene sequences may be limited by the level of comonomer present, however, as noted above, the level of comonomer present cannot exceed 10-15% if the copolymer is to retain its barrier properties.

Therefore the copolymer must be stabilized in some way to control degradation during thermal processing[11]. As previously noted, internal double bonds are the defect sites responsible for the initiation of degradation. This is because the indroduction of a double bond (via dehydrochlorination) into a VDC sequence generates an allylic dichloromethylene which is very prone to carbon-chlorine bond homolysis to propagate the unzipping dehydroclorination reaction.

One obvious approach to controlling the degradation sequence would be to treat the polymer with an agent which would replace allylic halogen with more thermally stable groups. This approach is successful for the stabilization of poly(vinyl chloride) (PVC) in which the polymer is treated with lead, tin, cadmium, or zinc carboxylates which convert allylic chloride groups to allylic esters wich degrade only above the temperature required for processing[12]. Treatment of vinylidene chloride polymers with these same compounds leads to rapid degradation of the polymers. This is due to the sensitivity of the dichloromethylene group to the presence of Lewis acids[13],[14].

For this approach to be effective the metal cation must be capable of coordinating allylic halogen to promote displacement by carboxylate but not sufficiently acidic so as to promote carbon-chlorine bond heterolysis (as depicted in Scheme 2). Copper(II) carboxylate seems to posses the proper balance between Lewis acidity of the cation and nucleophilicity of the anion to behave as an effective stabilizer in these systems[15],[16].

Similar stabilization can be effected with strongly nucleophilic but weakly basic additives. For example, amines, even highly hindered amines, are too basic to permit displacement of allylic halogen in the absence of E2-type dehydrohalogenation[17-19]. Trialkyl/aryl phosphates, on the other hand, have shown some utility as stabilizers in these polymers.

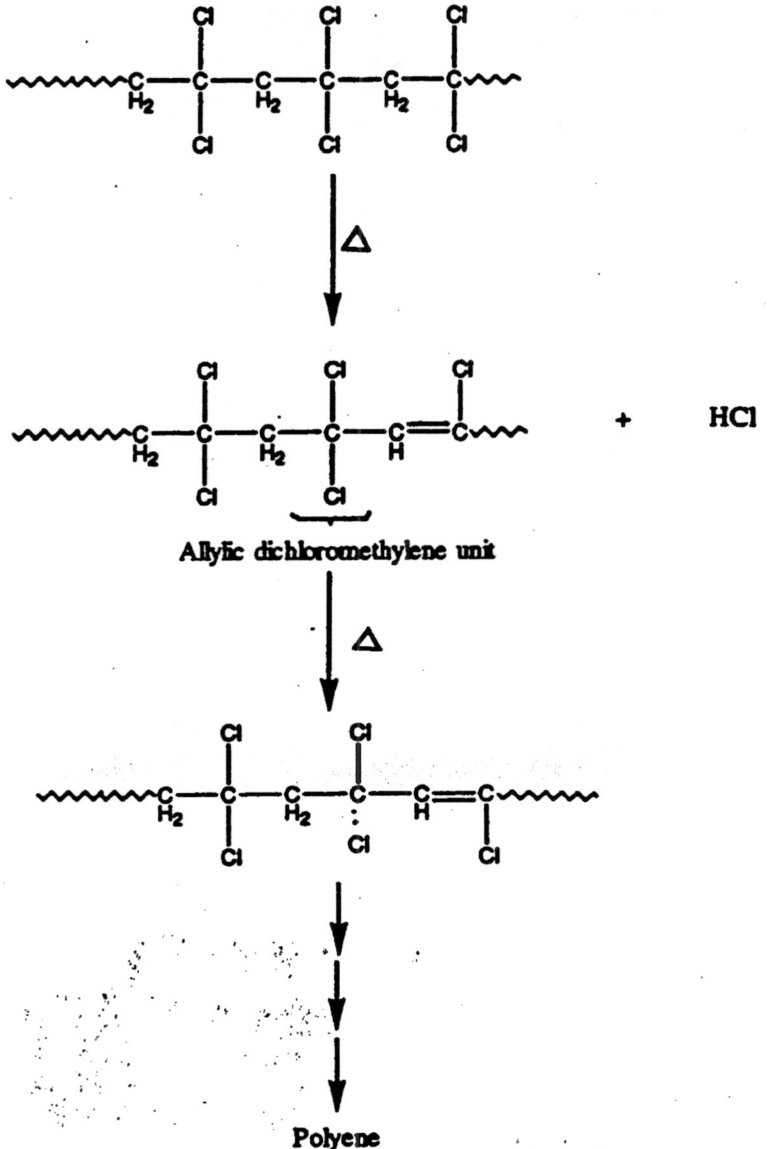


Scheme 1. Thermal Degradation of a Vinylidene Chloride Polymer

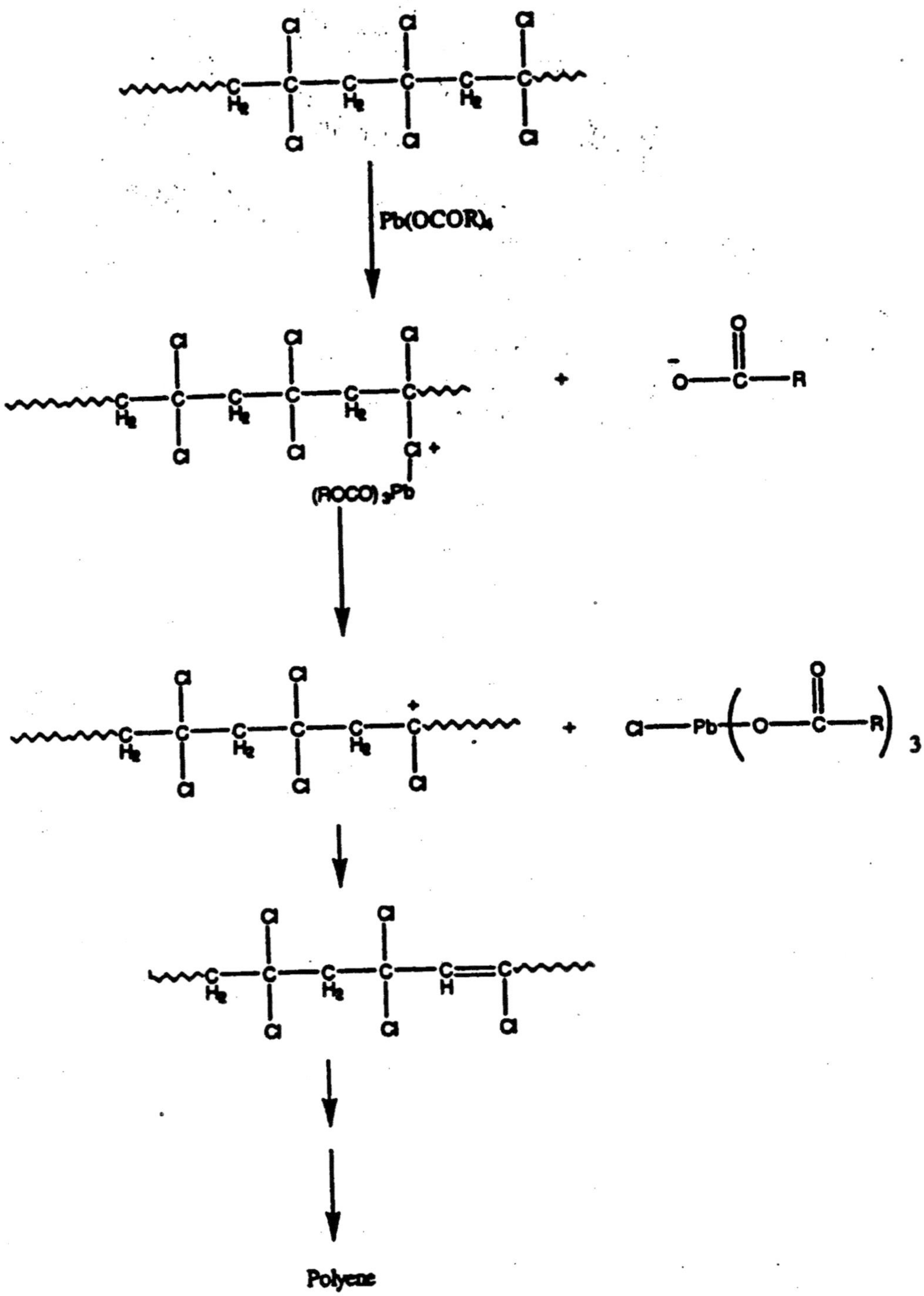


Scheme 2. Interaction of Metal Carboxylates with Vinylidene Chloride Polymers

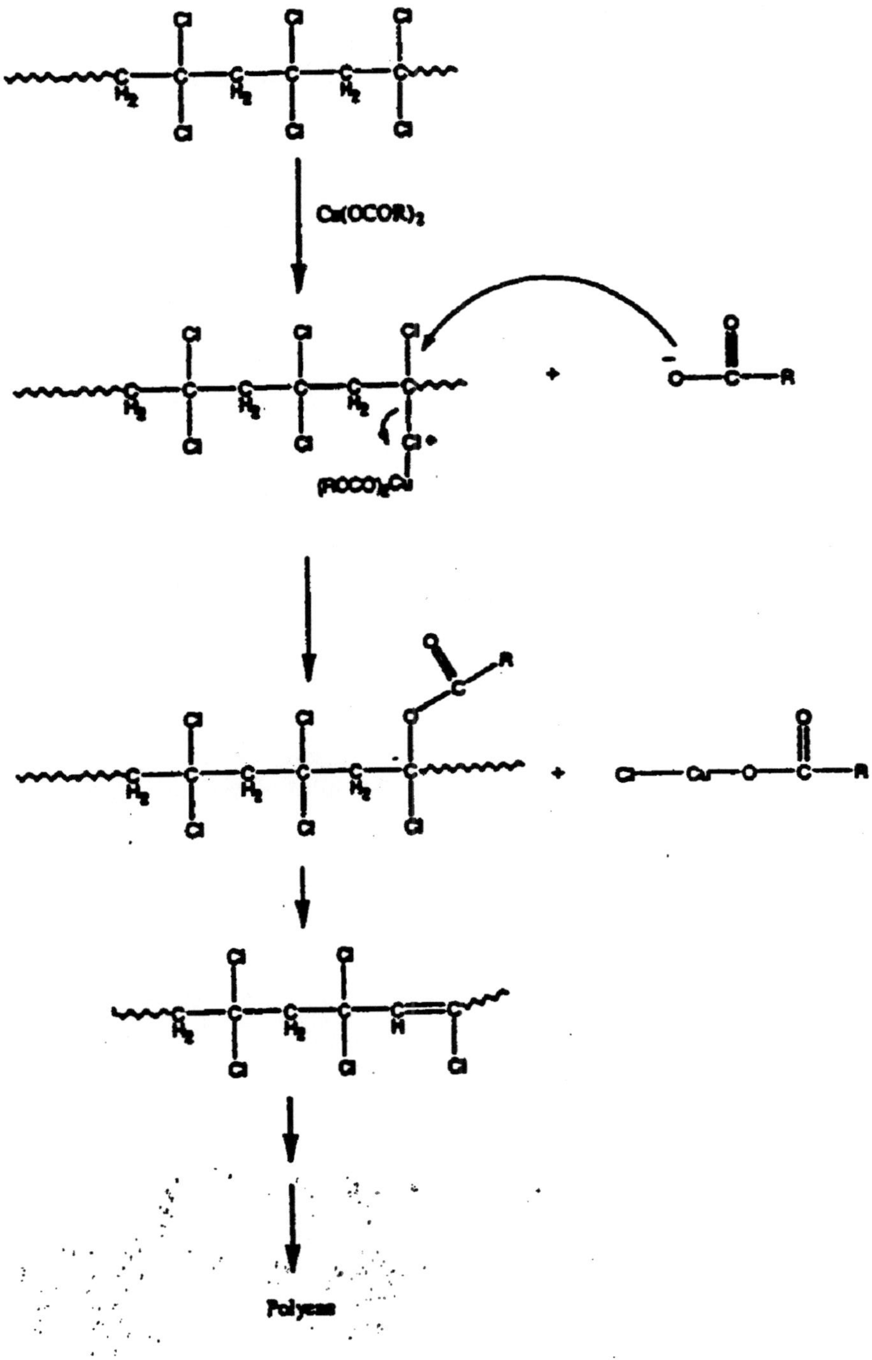


Scheme 3. Interaction of Copper(II) Carboxylate with a Vinylidene Chloride Polymer

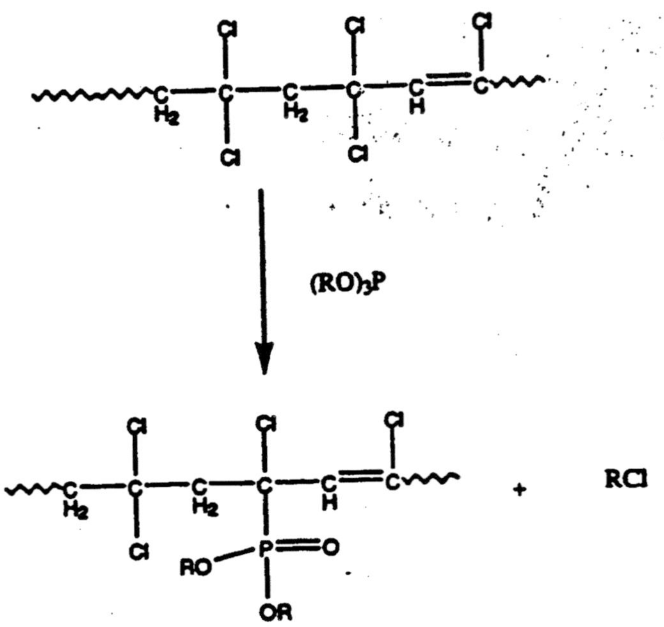


Scheme 4. Stabilization of Vinylidene Chloride Polymers by Trialkyl/aryl Phosphites

Other approaches to stabilization in these systems have involved the use of N-substitutedmaleimides to act as dienophiles to consume dienes as they are formed during propagation of the dehydrochlorination reaction or radical scavengers (hindered phenols) to trap chlorine atoms formed from the homolysis of allylic carbon-chlorine bonds[20],[21],[22]. A passive base such as magnesium oxide or tetrasodium pyrophosphate is often a component of a suitable stabilization package as well[23]. The purpose of these agents is to scavenge hydrogen chloride as it is formed and prevent its interaction with the walls of processing equipment to generate Lewis acids, particularly iron(III) chloride, which promote the dehydrochlorination reaction.

A potentially more useful approach which has received much recent attention has been the incorporation into the polymer of a comonomer which contains a substitutuent group suitable for reaction with hydrogen chloride to expose a phenolic unit potentially capable of scavenging chlorine atoms[24-27].

In the first instance, 4-acetoxystryrene was used as a comonomer to produce a copolymer containing acetoxy groups[24]. Reaction of the acetoxy group with evolved hydrogen chloride could consume a mole of hydrogen chloride to liberate acetic acid (a weak acid) and a phenol which might scavenge chlorine atoms. Unfortunately, under the conditions of thermal degradation of the polymer the interaction of anhydrous hydrogen chloride with the ester was not sufficient to effect cleavage. Therefore, the presence of acetoxystyrene units in the polymer did not impart stability. Indeed, the introduction of a benzylic group into the polymer mainchain provided a site for thermally-induced chain scission and led to a copolymer that was considerably less stable than the corresponding acrylate copolymer. Similarly, the generation of a copolymer containing pendant ether functionality (catechol cyclopentanone ketal) was not a useful approach for the synthesis of a stable vinylidene chloride copolymer[25]. Again the presence of anhydrous hydrogen chloride was not sufficient to promote cleavage of the ketal.

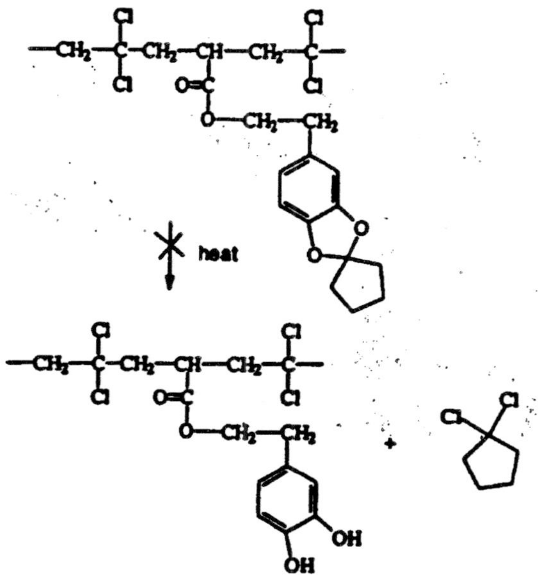


Scheme 5. Degradation of a Vinlyidene Chloride Copolymer Containing Pendant Catechol Cyclopentanone Ketal Functionality.

In contrast, the carbonate moiety is susceptible to cleavage under conditions of copolymer degradation. A copolymer containing a benzyl acrylate bearing a *t*-butoxycarbonyloxy group in 4-position of the phenyl nucleus undergoes thermal degradation to evolve several volatile products in addition to hydrogen chloride[26],[27]. Infrared analysis of the decomposing polymer suggested that the acrylate units were being converted to acrylic acid units, i.e., that the polymer was being converted to a copolymer of vinylidene chloride and acrylic acid. The decomposition may be depicted as shown below.

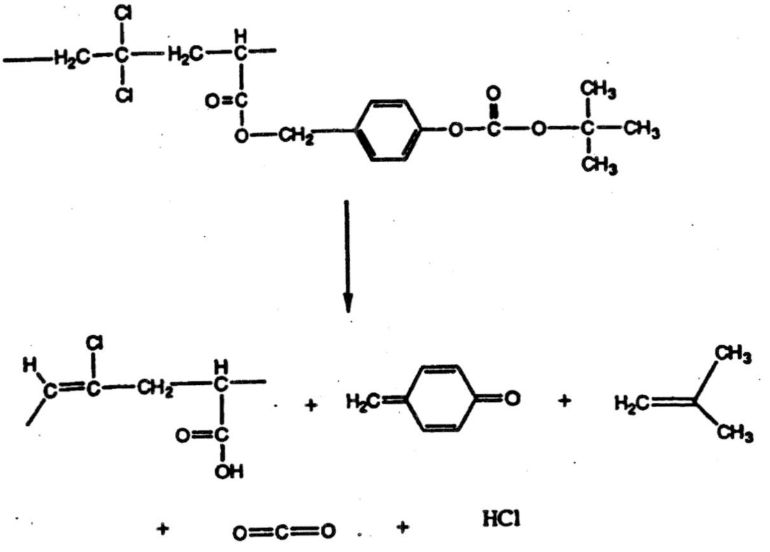


Scheme 6. Thermal Degradation of a Vinylidene Chloride/[4-(*t*-Butoxycarbonyloxy)phenyl]methyl Acrylate.

## Conclusions

Vinylidene chloride copolymers are important materials for use in barrier plastic packaging applications Because of the propensity to undergo thermally-induced degradative dehydrochlorination during processing these polymers must be stabilized. The homopolymer, poly(vinylidene chloride), undergoes catastrophic dehydrochlorination at its melt temperature (200° C) and is not a commercial material. Incorporation of an alkyl acrylate comonomer (4-10%) into the polymer lowers the melt temperature and limits the vinylidene chloride sequence length. The resulting polymers can be processed and are extremely good barrier resins for food packaging applications. The incorporation of acrylate units containing pendent groups capable of absorbing hydrogen chloride to expose a moiety that can scavenge chlorine atoms and other radical species has the potential to provide resins of unusual stability.

## References

[1] R.A. Wessling, D.S. Gibbs, P.T. Delassus, D.E. Obi, and B.A. Howell, *Kirk-Othmer Encyclopedia of Chemical Technology*, Vol. **24**, 2nd Ed., John Wiley and Sons, Inc., New York, NY, 1997, pp. 883-923.

[2] P.T. Delassus, W.E. Brown, and B.A. Howell in A.L. Brody ad K.S. Marsh, Ed., *Enclyclopedia of Packaging Technology*, **2nd Ed.**, John Wiley and Sons, Inc., New York, NY, 1997, pp. 958-961.

[3] G. Strandburg, P.T. Delassus, and B.A. Howell, in S.J. Risch and J.H. Hotckiss, Ed., *Food and Packaging Interaction II*, American Chemical Society (Symposium Series No. 473), Washington D.C., 1991, Ch.12.

[4] P.T. DeLassus, G.Strandburg and B.A. Howell, *Tappi J.*, **71**, 177 (1998).

[5] B.A. Howell, *J. Polym. Sci., Polym. Chem. Ed.*, **25**, 1681 (1987).

[6] B.A. Howell, P.T. Delassus and C. Gerig, *Polym. Prepr.,* **28**(1), 278 (1987).

[7] B.A. Howell and P.T. Delassus, *J. Polym. Sci., Polym. Chem. Ed.,* **25**, 1697 (1987).

[8] B.A. Howell and P.B. Smith, *J. Polym. Sci., Polym. Chem. Ed.,***26**, 1287 (1988).

[9] B.A. Howell, *Thermochim. Acta,***134**, 207 (1988).

[10] 10. S. Collins, K. Yoda, N. Anazawa, and C. Birkinshaw, *Polym. Degrad. Stability*, **66**(1), 87, 93 (1999).

[11] 11. B.A. Howell, B.S. Warner, C.V. Rajaram, S.I. Ahmed and Z Ahmed, *Polym. Adv. Tech.,* **5**, 485 (1994).

[12] T. Hjertberg, E. Martinddon and E. Sorvik, *Macromolecules,* **21**, 603 (1998).

[13] B.A. Howell and J. R. Keeley, *Thermochim. Acta,* **272**, 131 (1996).

[14] B.A. Howell and A.Q. Campbell, *Thermochim. Acta*, **340**, 231 (1996).

[15] B.A. Howell and C.V. Rajaram, *J. Thermal Anal.*, **40**, 575 (1993).

[16] B.A. Howell and C.V. Rajaram, *J. Vinyl Tech.,* **15**, 202 (1993).

[17] B.A. Howell and H. Liu, *J. Vinyl Tech.*, **13**, 187 (1991).

[18] B.A. Howell and H.Liu, *Thermochim. Acta*, **212**, 1 (1992).

[19] B.A. Howell and F.M. Uhl, *Thermochim. Acta,* **357**, 127 (2000).

[20] B.A. Howell and J. Zhang, *Polym. Prepr.*, **42**(2), 624 (2001).

[21] B.A. Howell, M.F. Debney, and C.V. Rajaram, *Thermochim. Acta*, **212**, 215 (1992).

[22] B.A. Howell, Z. Ahmed and S.I. Ahmed, *Thermochim. Acta*, **357**, 103 (2000).

[23] B.A. Howell, F.M. Uhl and D. Townsend, *Thermochim. Acta*, **357**, 127 (2000).
[24] B.A. Howell and R.C. Mason, unpublished results.
[25] B.A. Howell, B.B.S Sastry, S.I. Ahmed and P.B. Smith, *Thermochim. Acta*,**272**, 139 (1996).
[26] B.A. Howell, D.A. Spears and P.B. Smith, *Thermal Degradation of Vinylidene Chloride/[(4-t-Butoxycarbonyloxy)phenyl]methyl Acrylate Copolymers,* 24th North American Thermal Analysis Society Meeting, San Franscisco, CA, September, 1995, No. 60.
[27] B.A. Howell and B. Pan, *Polym. Mater. Sci. Eng.,* **76**, 401 (1997).

In: Frontiers in Polymer Research
Editor: Robert K. Bregg, pp. 221-256
ISBN: 1-59454-824-2


*Chapter 9*

# BIODEGRADABLE HYDROCARBON POLYMERS AN ENVIRONMENTALLY ACCEPTABLE SOLUTION TO PLASTICS WASTE AND LITTER

***Gerald Scott***
Aston University, Birmingham, UK

## Abstract

The polyolefins have an established position in packaging and in agriculture as a result of their technological properties, which include water and microbe resistance. However, it is generally accepted that commercial polyolefins for durable goods do not biodegrade rapidly enough in the environment where they accumulate as litter. Their behaviour in the environment can be compared with that of natural rubber (*cis*-poly(*iso*prene)), which, when fabricated to an automobile tyre, is resistant to biodegradation for decades, although it is oxo-biodegradable as produced by nature. This results in both cases from the addition of antioxidants during manufacture. There is convincing evidence to show that the rate-determining step in the biodegradation of hydrocarbon polymers is the rate of peroxidation. This process is accelerated by transition metal ions both thermally and by light so that abiotic peroxidation and oxo-biodegradation lead synergistically to the bioassimilation of polymers in the outdoor environment. Special antioxidants inhibit the formation of low molar mass highly biodegradable oxidation products and hence inhibit bio-degradation during use.

Contaminated mixed plastics wastes from domestic sources present a difficult challenge to traditional recycling techniques. On the other hand, the hydrocarbon portion of mixed domestic wastes can be made oxo-biodegradable by the incorporation of transition metal ions that accelerate both perooxidation and biodegradation. Oxo-biodegradable plastics thus make a realistic contribution to the recovery of value from waste packaging as fertilisers and soil-improvers for agriculture and horticulture.

Standards are essential to ensure the environmental safety of compost. Hydrocarbon plastics do not biodegrade rapidly in compost or in soil and it must be demonstrated that, like of nature's wastes, they do not accumulate in the soil. Standards for biodegradability and compostability of plastics must therefore address, not only the question of non-accumulation of any long-lasting plastics residues in the soil substances but also the safety of any non-degradable residues. These aspects will be discussed in the light of recent scientific studies.

**Key words:** polyolefins, peroxidation, biodegradation, oxo-biodegradation, recycling, bioplastics

# 1 Introduction

The polyolefins, unlike the polysaccharides are hydrophobic and are not permeated by water and microorganisms. Consequently, they are not biodegradable during use. However, it is this very advantage during service that causes problems when polyolefin packaging becomes litter since, unlike paper, commercial hydrocarbon polymers do not biodegrade rapidly in the environment [1]. Furthermore, the stability of commercial polyolefin packaging in industrial composting systems leads to clogging of machinery and to visible residues of non-degraded plastics in the final compost. Similarly when used in agriculture or horticulture as mulching films, they remain on the soil when the crop is harvested, causing clogging of automated machinery and accumulation in the soil, thus interfering with subsequent crop growth [2]. A major cause of visual pollution in the countryside is discarded silage-wrap film which may be carried by the wind for long distances before being caught on hedgerows and trees, often in areas of natural beauty [3].

Contrary to popular belief, the physical fragmentation of hydrocarbon polymers can be accelerated in a controlled manner by heat and light by the use of appropriate combinations of additives [2]. This results from peroxidation and chemical modification of the component polymers and is associated with the formation of low molecular weight carboxylic acids and alcohols. This encourages microbial colonisation at the surface of the plastic and leads to bioerosion of the film with mass loss.

# 2 Environmental Degradation of Hydrocarbon Polymers

## 2.1 Peroxidation and Biodegradation

The term "biodegradable" is frequently used by environmentalists to convey the idea that environmental degradation occurs exclusively by the attack of microorganisms. This is not so since most natural polymers, like their synthetic analogues, are susceptible to environmental influences other than microorganisms. A well studied example is (*cis*-poly(isoprene), (*cis*-PI), which is produced by the rubber tree to seal the wound caused by the "tapping" of latex. This polymer, like its synthetic equvalent, isoprene rubber, IR, undergoes rapid abiotic peroxidation and subsequent biodegradation in the environment. Consequently, when NR was identified as a valuable technological material in the $19^{th}$ century, it was realised that an urgent solution was required to protect the natural elasticity of rubber from the effects of the environment for industrial applications. This resulted in the empirical discovery of chemicals that inhibited environmental degradation and loss of useful mechanical properties. Subsequent research by the natural rubber industry showed that the chemistry of this process was a free radical chain reaction involving atmospheric dioxygen (reactions (1) and (2)) and that this could be readily inhibited by removing the chain-carrying free radicals (reaction (3)) [4-6].

$$P^{\cdot} + O_2 \rightarrow POO^{\cdot} \tag{1}$$

$$POO^{\cdot} + PH \rightarrow POOH + P^{\cdot} \tag{2}$$

$$POO^{\cdot} + AH \rightarrow POOH + A^{\cdot} \tag{3}$$

where PH is the polymer and $P^{\cdot}$ is the derived carbon-centred radical. AH is a hydrogen donor antioxidant, typically a hindered phenol or an aromatic amine. Unlike the carbon-centred radical $P^{\cdot}$, $A^{\cdot}$ is a stable free radical that cannot continue the kinetic chain [4-6]. The detailed chemistry of the radical chain reaction leading to low molar mass products has been extensively discussed in the literature [4,6-8] is shown in Scheme 1 for the peroxidation of *cis*-PI.

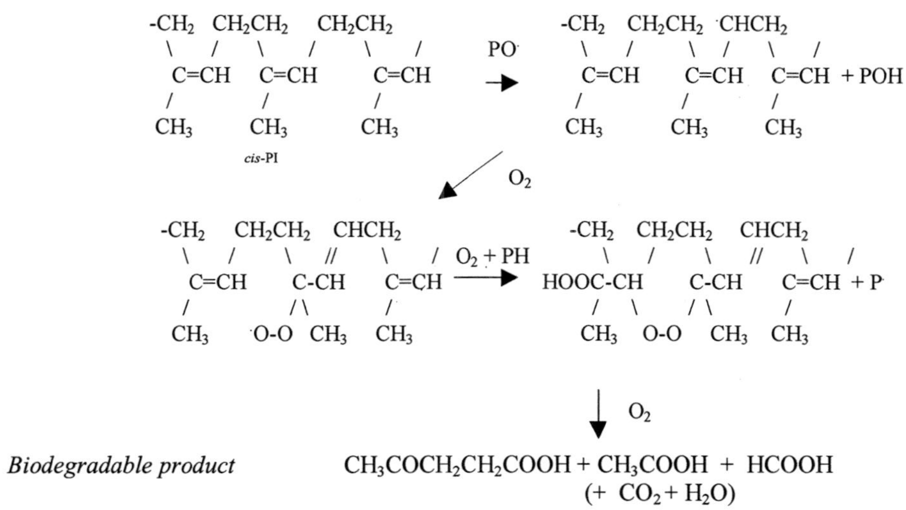


Scheme 1. Peroxidation of *cis-PI (cis*-poly(isoprene)) to low molar mass biodegradable products

The low molar mass oxidation products formed from *cis*-PI biodegrade rapidly in the natural environment as rapidly as they are formed with the formation of cell biomass. The synthetic analogue of *cis*-PI (IR), manufactured from petrochemical intermediates, peroxidises in the same way as NR and recently, IR has also been shown to undergo bioassimilation at a similar rate to NR (see Section 3).

The antioxidants added to rubber during the manufacture of industrial products are much more important in determining the rate at which hydrocarbon polymers degradae than the chemical structure of the polymer. Antioxidants not only inhibit abiotic peroxidation but they also retard oxo-biodegradation by scavenging the free radicals formed in both processes [7]. Consequently discarded tyres, unlike the polymer from which they have been fabricated can survive intact for decades in the environment without any sign of biodegradation by the incorporation of antidegradants at the fabrication stage. This is, of course fully justified in the case of tyres, which have to survive extreme conditions of stress and temperature during use, placing the motorist at great risk if premature blow-out occurs under modern motorway conditions. However, because of their resistance to biodegradation, discarded tyres that are still largely landfilled after use are increasingly being "recovered" from the waste stream by incinerated with heat recovery.

The poly(olefins) are less peroxidisable than their unsaturated homologues, the poly(dienes). Scheme 2 indicates the order of relative intrinsic oxidisability of the macromolecules [8]. However, all hydrocarbon polymers, as manufactured, are too oxidatively unstable to be useful in the manufacture of technological products. They begin to degrade with loss of mechanical properties during processing and thus require the addition of antioxidants to provide stability both during fabrication and in service. Consequently polymers intended for short-term applications, such as disposable packaging, are over-stabilised for their purpose, since like the poly(dienes), the antioxidants added at manufacture to provide processing stability interfere with their ultimate biodegradation in the environment [7]

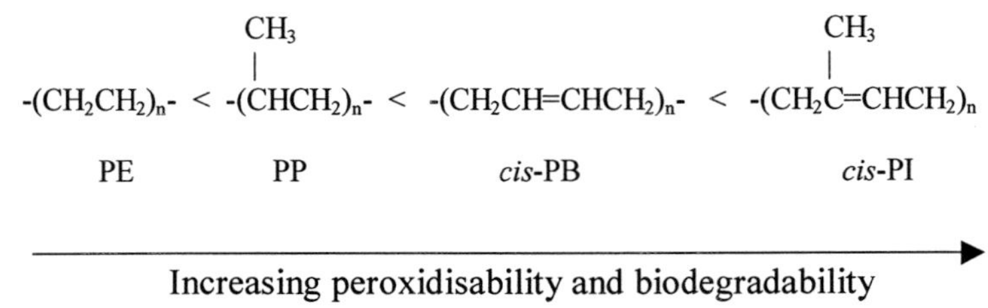


Scheme 2 Relative stability of hydrocarbon polymers in the environment[8]

As with the poly(dienes), the polyolefins peroxidise to biodegradable products through the intermediate unstable hydroperoxides. Scheme 3 illustrates this process for poly(ethene), (PE). Poly(propene) (PP) peroxidises much more rapidly than PE due to the tertiary carbon atom (Scheme 2) and the higher proportion of vicinal hydroperoxides formed [6,9]. In-chain hydroperoxide formation continues steadily in all the polyolefins provided oxygen is available and the rate of biodegradation correlates with the rate of abiotic peroxidation.

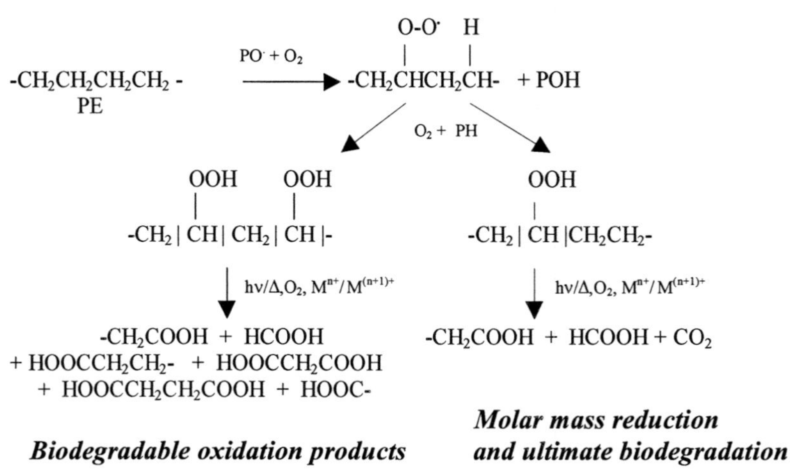


PH = polyethylene. In the above formulae, | indicates weak C-C bonds during hydroperoxide decomposition

Scheme 3. Formation and breakdown of hydroperoxides in polyethylene [6].

The processes described in Schemes 1 and 3 are catalysed by transition metal ions ($M^{n+}$), notably by Fe, Co and Mn. Both the decompostion (reactions (4) and (5)) and the consequent formation of hydroperoxides (reactions (6) and (7)) is accelerated, leading to the rapid loss of molar mass in a fraction of the time required for the same change in the absence of a prooxidant transition metal ion.

$$M^{n+} + POOH \rightarrow M^{(n+1)+} + PO^{\cdot} + OH^{-} \qquad (4)$$

$$M^{(n+1)+} + POOH \rightarrow M^{n+} + POO^{\cdot} + H^{+} \qquad (5)$$

$$POO^{\cdot} + PH \rightarrow POOH + P^{\cdot} \xrightarrow{nO_2/PH} nPOOH + P^{\cdot} \qquad (6)$$

$$PO^{\cdot} + PH \rightarrow POH + P^{\cdot} \xrightarrow{nO_2/PH} n\,POOH + P^{\cdot} \qquad (7)$$

The above reaction sequence leads to low molar mass biodegradable products similar to those produced in the abiotic acid catalysed hydrolysis of typical biodegradable aliphatic polyesters such as poly(lactic acid) (PLA) or poly(caprolactone) (PCL). Reaction (8) illustrates this for PLA [10].

$$\underset{\text{PLA}}{[\text{-O-}\overset{\overset{CH_3}{|}}{C}\text{H-CO-}]_n} \xrightarrow{nH_2O/H^+} \underset{\textit{Biodegradable lactic acid}}{n\ \text{HO-}\overset{\overset{CH_3}{|}}{C}\text{H-COOH}} \qquad (8)$$

## 2.2 Antioxidants

The term antioxidant is used to describe the inhibition of polymer peroxidation in both abiotic and biotic systems [6,8,11]. It embraces more specific technological terms such as antidegradants (thermoantioxidants), antifatigue agents (mechanooxidation inhibitors) and light stabilisers (photoantioxidants). It was seen above that peroxidation of carbon-chain polymers occurs by a free radical chain mechanism (reactions 1 and 2) and that the hydroperoxide products that are the intrinsic initiators for peroxidation dissociate to "oxyl" radicals under the influence of heat and light. Both hydroperoxide formation and hydroperoxide decompostion are thus catalysed by transition metal ions.

The chemical mechanisms involved in the action of antioxidants have been discussed in a number of reviews [8,11-18] and the reader is directed to these and the references they contain for more detailed information. Two complementary antioxidant mechanisms are frequently used synergistically in polyolefins. The first is the kinetic chain-breaking hydrogen donor process, (CB-D) summarised in reaction (3). The relatively stable radicals ($A^{\cdot}$) produced (e.g. phenoxyl from phenols and aminoxyl from aromatic amines) cannot continue the kinetic chain and disappear from the system by coupling with other or the same free radicals. However, it should be noted that this process is stoichiometric and hydroperoxides are produced in each inhibiting step (reaction (3), thus limiting their effectiveness, particularly at high temperatures.

The second mechanism, is the peroxidolytic or peroxide decomposing process (PD) removes the initiating hydroperoxides (POOH) in reactions that do not produce free radicals (reaction (9).

$$\text{POOH} \quad \rightarrow \quad \text{Non-radical products} \qquad (9)$$

CB-D and PD antioxidants thus have complementary functions and when used together they act synergistically, producing an effect greater than that of the sum of their individual effects. Many sulphur compounds (e.g. thiols, their metal complexes and dialkyl sulphides) are known to destroy hydroperoxides catalytically [17,18], whereas simple phosphite esters are stoichiometric hydroperoxide decomposers [11].

The metal dithiocarbamates represented by the general structure MDRC, below comprises an important group of catalytic PD antioxidants that have been used in rubbers and in polyolefins [11,14,17,18] for many years. Some, notably the transtition metal dithiocarbamates are effective light stabilisers (photoantioxidants) since they are also light stable. In this context, particular attention has been paid to the Ni and Co and Cu complexes, which have the ability to release sulphur acids slowly over a long period [16,18]. FeDRC and MnDRC by contrast are effective thermal antioxidants, that are rapidly decomposed in compost by heat or photolysed in sunlight to give the corresponding prooxidant transition metal ions (reactions 4-7). The effectiveness of the MRDCs as light stabilisers depends crucially on the photo- stability of the metal complexes [11,14,16-18].

```
       S     S
     //  \  /  \                    M = Zn, Ca, etc.,  thermal antioxidant
R2NC     M     CNR2     MDRC        M = Ni, Co, Cu, photoantioxidant
     \  / \  //                     M = Fe, Mn, thermal antioxidants;
       S     S                          photo-prooxidants
```

## 2.3 The Role Antioxidants in Degradable Polyolefins

Many applications of degradable polymers require stability against the environment from as little as a few weeks, as in the case of much retail packaging, to many months or even years for specialised applications such as protective films in agriculture. The oxo-biodegradable polyolefins, because of their ability to be stabilised against the oxygen of the environment have inherent advantages over biopolymers that are randomly attacked by microorganisms. The behaviour of the ideal degradable polymer is shown in Fig. 1 [19].

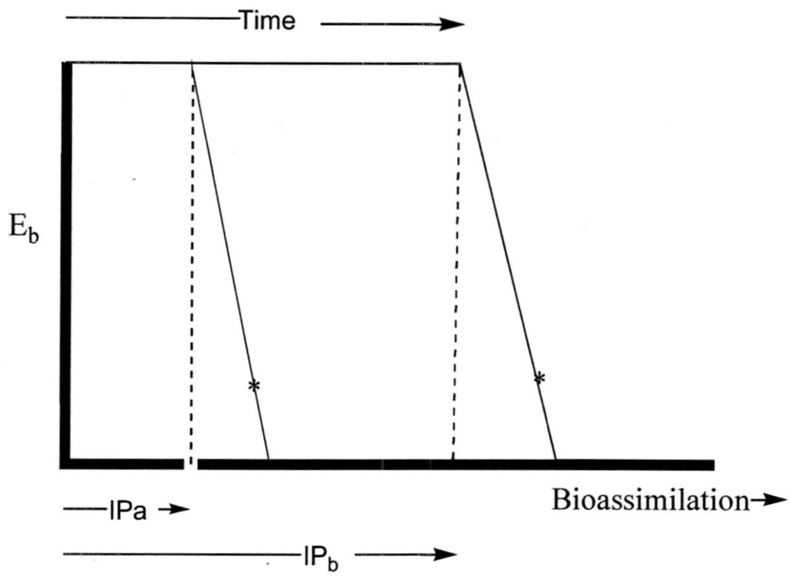


Fig. 1. Ideal degradable polymer for industrial applications [19]. $IP_{a,b}$ are the induction periods of two different formulations of the same polymer, during which no peroxidation or change of properties occurs.

Typical prooxidant transition metal compounds (e.g. iron, cobalt or manganese stearates) are used commercially to induce peroxidation in degradable plastics. However, transition metal prooxidants alone have no practical utility in commercial products unless they are deactivated during polymer fabrication, since oxidative degradation begins during processing [13], resulting in products with inferior mechanical performance and almost no resistance to the environmental degradation. Figure 4 shows the behaviour of a commercial degradable polyethylene film used in landfill covers and packaging applications (EPI's TDPA$^{TM}$) at composting temperatures and in a weatherometer. It is clear that at ambient temperatures, the shelf life of the polymer is adequate for the intended purpose, whereas at composting temperatures 60-70°C or in landfill, the polymer peroxidises rapidly as measured by the formation of carboxylic acids. Commercial hindered phenol antioxidants are inhibitors of peroxidation under conditions where peroxyl radicals are rapidly generated (e.g. in the presence of prooxidant transition metal ions or in UV light) but the induction time is relatively short under these conditions and, if the objective outlined in Fig. 1 is to be achieved, the induction time must be controllable [19]. This is especially true for agricultural applications of degradable plastics such a mulching films where the economics of the use of degradable plastics depend on retaining the film intact until just before harvest [2] followed by a rapid loss of properties so that the film disintegrates and can then be ploughed into the soil where it continues to degrade in the presence of oxygen, catalysed by transition metal ions and peroxidase enzymes associated oxidation products (C=O FTIR absorbance at 1715 $cm^{-1}$).

The transition metal dithiolates, of which the dithiocarbamates (MRDC) above are typical, combine antioxidant and prooxidant functions in the same molecule [16,18,21-23]. The MRDCs act as processing stabilisers during manufacture and as antioxidants during storage but they rapidly 'invert' from UV stabilisers to prooxidants in the outdoor environment. By varying the concentration of the additive, both the induction time and the rate of loss of mechanical properties at the end of the induction time (figure 1) can be varied by the combining different MRDC. This is the basis of the Scott-Gilead (S-G) photo-biodegradable polyolefins that are particularly valuable in agricultural applications where relatively long programmed lives are required [20,22-25].

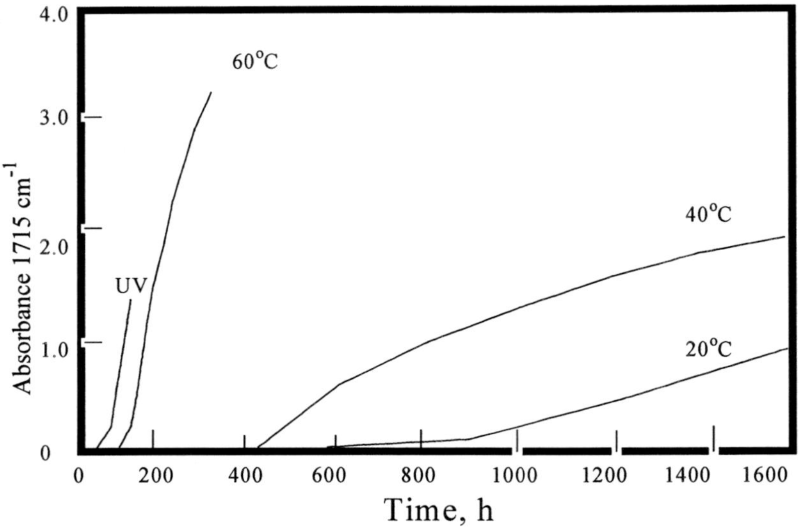


Fig. 2. Formation of oxidation products in degradable PE (TDPA™) during exposure to heat (20-60°C) and SEPAP accelerated weathering (UV) procedures [26]

Most commercial light stabilisers are not designed to give a sharp end to the peroxidation induction period, since the latter may result in catastrophic failure of polymer components. Thus for example, the UV absorbers, hydroxybenzophenones and hydroxybenzotriazoles) are destroyed relatively slowly by light. The hindered amine light stabilisers (HALS) are catalytic chain-breaking (CB) antioxidants [27,28] and are depleted very gradually in hydrocarbon polymers. Neither class shows the required rapid 'inversion' behaviour characterisitic of the transition metal thiolates. Applications of transition metal technology in packaging and in agriculture are discussed in later Section 4.

# 3 Biodegradation of Hydrocarbon Polymers – Experimental Studies

Detailed experiments have been carried out in order to understand the effect of environmental exposure on the chemical, physical and biological changes occurring in commercial degradable polyethylenes during service and on exposure to biotic environments [26,29]. PE films, after peroxidation at composting temperatures or after being subjected to photooxidation, were incubated with bacteria and fungi that had been isolated from soils that were adapted to the presence of degraded polyethylene. The peroxidised samples were used as the sole source of carbon for a period of six months. Three commercially available degradable polyolefins were compared.

(a) Photodegradable (photolytic) polymers made by copolymerisation of ethylene with carbon monoxide (E-CO). Union carbide technology [30].
(b) Conventional polyethylene containing a transition metal prooxidant blended with starch (E-St). Griffin technology [31].
(c) Oxo-biodegradable) polymers based on conventional polyethylene containing a transition metal compound in combination with an antioxidant. Scott-Gilead technology (Section 2).

Although E/CO initially photodegraded to fragments more rapidly than S-G and E-St, photo-degradation of the transition metal ion catalysed systems continued to a much lower molar mass. After fragmentation, the peroxidised polymers were incubated in the absence of any other source of carbon with three microorganisms isolated from soil in the vicinity of discarded polyethylene. Two were bacteria (*Nocardia asteroides* and *Rhodococcus rhodochrrous*) and one was a fungus (*Cladosporium cladosporioides*). Biofilm formation was very rapid on the surface of peroxidised polyethylene [26,29] and this is followed by attack into the surface of the polymer by exo-enzymes from the colonised bacteria.

Three parameters were used to measure the extent of biodegradation. The first was mass loss as measured by the decrease in the thickness of the films during incubation. *Nocardia asteroides* was particularly effective in bioassimilating all peroxidised polymers (Table 1), whereas *Rhodococcus rhodochrous* bioassimilated photooxidised PBD but had little effect on photooxidised E-CO. *Nocardia* is also very effective in bioassimilating *cis*-PI under similar conditions. The fungus, *Cladosporium* was least effective in reducing the mass of polyethylene samples but it did degrade the photooxidised starch-filled polymer.

Table 1 Effect of *Nocardia asteroides* on peroxidised degradable polyethylenes [29]

| | | **Control** | | **Nutrients only** | | ***Nocardia* no. 13** | | | |
|---|---|---|---|---|---|---|---|---|---|
| **Polymer** | **Abiotic Treatment** | **OD[a]** | **$M_w$ ($x10^{-3}$)** | **OD[a]** | **$M_w$ ($x10^{-3}$)** | **OD[a]** | **$M_w$ ($x10^{-3}$)** | **SEM[b]** | **Δth[c]** |
| **E-CO** | **Untreated** | **0.05** | **290** | **0.05** | **299** | **0.04** | **328** | **0** | **0** |
| | **Photoox.** | **0.12** | **16** | **0.10** | **14** | **0.11** | **12** | **2+** | **0** |
| | **Thermoox.** | **0.45** | **21** | **0.40** | **19** | **0.20** | **17** | **2+** | **-20** |
| **PBD** | **Untreated** | **0** | **248** | **0** | **280** | **0** | **282** | **0** | **0** |
| | **Photoox.** | **0.35** | **40** | **0.30** | **32** | **0.25** | **19** | **2+** | **0** |
| | **Thermoox.** | **1.05** | **16** | **0.90** | **16** | **0.75** | **15** | **3+** | **-20** |
| **E-St** | **Untreated** | **0.10** | **206** | **0.11** | **285** | **0.11** | **289** | **0** | **0** |
| | **Photoox.** | **1.15** | **16** | **1.08** | **15** | **0.70** | **12** | **4+** | **-27** |
| | **Thermoox.** | **.90** | **nd** | **1.95** | **nd** | **1.55** | **nd** | **2** | **-6** |

**[a] Optical density at 1715 $cm^{-1}$. [b] Visual rating of surface erosion; 0 = no erosion… .4 = severely eroded. [c] Change in thickness by change in absorbance at 1375 $cm^{-1}$**

The second parameter was surface colonisation of the plastic films followed by bioerosion. Epiflurescence spectroscopy, which evaluates changes in the surface of the plastic films [26], has shown that polymer films that have been subjected to thermal treatment or light exposure, are rapidly colonised by microorganisms in a biotic environment [26,29]. This technique can be used to monitor the complete colonisation of the polymer surface [26].

Surface colonisation by bacteria is a very rapid process in the case of pre-aged polyethylene but it also occurs even on polymer films containing transition metal ions that have not been deliberately pre-oxidised. Colonisation is followed rapidly by the bioerosion of he surface of the polymer (Fig. 3).

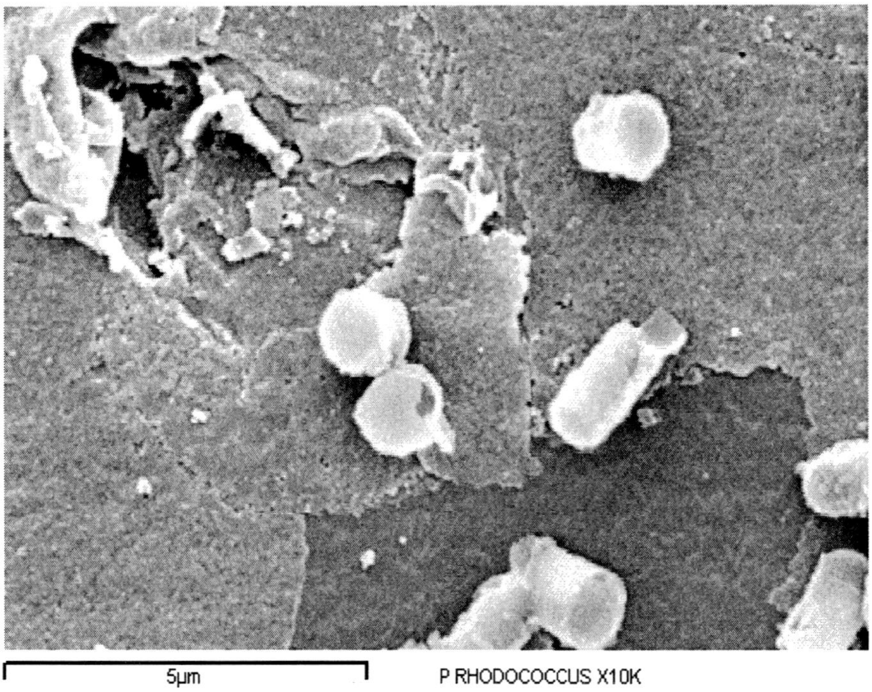


Fig. 3 Bioerosion of biodegradable PE (TDPA$^{TM}$) and the growth of *Rhodococcus Rhodochrous* in 1 month in the absence of any other source of carbon, measured by SEM.. Reproduced with prmission from S. Bonhomme, A. Cuer, A-M. Delort, J. Lemaire, M. Sancelme and G. Scott (2003) *Polym. Deg. Stab.,* **81**, 441-452 [26]

Bioerosion was demonstrated by scanning electron microscope examination (Table 1, SEM), after removing the biomass by sterilisation. The results broadly correlated with the mass-loss measurements, although in some cases, bioassimilation was greater by this measure than by mass loss.

Table 2 shows the time scale for these events and it can then be predicted by means of the Arrhenius equation [32] that ultimately the homogeneous polymer would disappear by bio-erosion.

Table 2 Time scale for the bioassimilation of aged oxo-biodegradable polyethylene [6]

| | |
|---|---|
| **Photooxidation (SEPAP, 60°C)** | **100 h** |
| **or** | |
| **Thermooxidation (60°C)** | **300 h** |
| **Biofilm formation** | **0.25 h** |
| **Surface disintegration (*R.rhodocrous)*** | **1 month** |
| **Mass loss (6 months)** | **15-20%** |

**Note: in compost and in soil, thermooxidation and bio-oxidation occur synergistically**

The most biodegradable of the commodity polyolefins is polypropylene (PP). Pandey and Singh have shown that polypropylene, after removal of antioxidants by solvent extraction, biodegrades much more rapidly than polyethylene by mass loss in compost [33]. PP lost over 60% mass in 6 months whereas LDPE lost about 10% in the same time. Ethylene-propylene (EP) co-polymers biodegraded at rates intermediate between PP and PE. As expected, prior UV irradiation (photooxidation) increased both the rate and extent of the bioassimilation. This is fully in accord with the rates of environmental peroxidation of these molecules [8] and it has been shown that PP acts as a sensitiser for the peroxidation of LDPE [34].

Mass loss has so far not been accepted as a test method for biodegradability, since it is argued that microscopically small particles of recalcitrant polymer may not be detected by mass loss alone. This may be true for heterogeneous polymers, where one component polymer is much more resistant to biodegradation than another. However, this argument does not apply to homogeneous polymers such as the polyolefins, which degrade by bioerosion from the surface and the rate of bioassimilation can be estimated by reduction of thickness of the item [26]. At present, however, measurement of evolved carbon dioxide is considered by the international Standards Organisations to be the only reliable criterion of biodegradability (see Section 6).

FTIR studies of the surface of degrading PE shows the development of species identified as polysaccharides and protein by photo-acoustic FTIR [26]. These are both characteristic of the growth of microorganisms [26,29]. Ikram and co-workers [35] have shown that in normal soils at 25°C, NR gloves showed 54% loss of thickness after 4 weeks and 94% mass loss after 48 weeks. On the other hand nitrile and neoprene rubbers showed insignificant loss in this time and plasticised PVC showed a smaller mass loss (11.6%) due entirely to biodegradation of the plasticiser. Ikram went on to show (see Table 3) that the rate of mass loss is strongly dependant upon the nutritional quality of the soil [36].

Table 3. Effect of added soil nutrients on the mass loss of rubber and plastic Films (%) after 40 weeks in soil. (Adapted from Ikram et al. [36] with permission.)

| Polymer | Nutrient treatment | | |
|---|---|---|---|
| | **High*** | **Low*** | **Control*** |
| NR | -82.4 | -38.5 | -29.7 |
| **Neoprene** | **+0.3** | **-13.0** | **-1.1** |
| **Nitrile** | **-4.3** | **-3.2** | **-3.5** |
| **Plasticised PVC** | **-26.1** | **-13.4** | **-11.1** |

*** Nutrients added: High 100 mg/l N and 150 mg/l P; Low 10 mg/l N. 15 mg/l P; Control nil**

After 24 weeks NR in the high N (100mg/l), P (150mg/l) system had lost 61.5% of its mass whereas in the low N (10mg/l), P (15mg/l) system, only 23.6% mass was lost. Control (unfertilised) soil produced least mass loss (17.3%). Microbial growth measured on the rubber pieces were in decreasing order as expected. Bacterial populations on the NR gloves (12317/mg) were higher than for fungi (441.47/mg), which were in turn significantly higher than actinomycetes (297.02/mg). Nevertheless, Heisey and Papadatos [37] isolated 10 actinomycetes (seven strains of *Streptomycetes*, two strains of *Amycolatopsis* and one strain of *Nocardia*) from soil that reduced the mass of NR gloves from 10-18% in 6 weeks.

Lee and co-workers [38], in a study designed to promote the use of starch- polyethylene blends (6% starch + mixed transition metal ion pro-degradants selected from Fe, Zn, Ni and Mn) for the disposal of garden waste in compost, examined the effect of a number of lignin-degrading microorganisms. The polymer films were first peroxidised either thermally at 70°C in an air oven for up to 20 days or by long wave UV irradiation for up to 8 weeks before being exposed to three cellulose-degrading bacteria (*Streptomyces viridosporus, Streptomyces badius* and *Streptomyces* serini) and one lignocellulose-degrading fungus (*Panerochaete chrysosporium*). Using a starch-agar assay, it was found that *S. setonii and P. chrysosporium* were unable to utilise corn starch but the former did biodegrade polyethylene. Mass-loss measurements were inconclusive due to the difficulty on removing microflora but GPC showed a reduction of polydispersity in the case of the samples incubated with *Streptomyces* spp., indicating the selective removal of lower molar mass species. The authors confirmed previous findings that prior peroxidation is an essential prerequisite to the biodegradation of polyethylene.

# 4 Commercial Applications of Biodegradable Polyolfins in Agriculture

The term plasticulture has been coined to describe the dominant position of plastics in agriculture during the past 25 years. The earliest use of polyolefins was in greenhouse films and since the relatively heavy gauge films used have to replace glass, they must remain tough and strong for several years in sunlight in order to compete with glass. At the opposite end of the stability spectrum are the degradable polymers which are used as programmed-life films to avoid the cost of removing partially degraded plastics from the soil after cropping. The following sections describe the main areas of application or potential application of polyolefins [2,3,19,20,23,25,39,40,41,43].

## 4.1 Mulching Films and Tunnels

The use of degradable plastics in tunnels and mulching films for the growing of soft fruits and vegetables has become an important economic tool in commercial horticulture. Scott-Gilead (SG) time-controlled degradable films have been utilised in soft fruit growing in Israel, Southern Europe and the USA since the early 1980s and the technology has been discussed by Gilead and Scott [2],Gilead [25], Fabbri [39] and Scott and Wiles [20]. Major advantages in early cropping, coupled with water and fertiliser conservation have been achieved [39]. In additional advantage of the techniques is the desalination of "sour" or saline soils by the refluxing of water from the interior surface of the films, thus taking the salts out of reach of the plant roots [25]. Protective films make possible the growth of crops such as chilli, sweetcorn and sweet potatoes in many parts of the world that until recently could only be grown in warmer climates. Degradable mulch is also used in cereal growing, notably maize and sweet corn and in forestry and environmental improvement schemes (e.g. growing of shrubs on road embankments). Mulching films are used in a variety of different ways and when used to their maximum potential, they greatly increase the value of commercial crops as a result of earlier and heavier cropping (Table 4).

Table 4 Ratio of increased income to cost of mulching film [25].

| Crop | Increased income: cost |
|---|---|
| Melons | 13.0 |
| **Vegetables** | **5.0** |
| **Peanuts** | **3.9** |
| **Sugar cane** | **3.6** |
| **Cotton** | **3.0** |
| **Maize** | **2.5** |

Mulching films and tunnels made from conventional plastics have to be removed from the fields before the next planting season since otherwise they interfere with root growth and reduce crop yields [25,39]. This is a labour intensive process since it is particularly difficult to remove all plastics from the soil due to the poor mechanical properties of the residual plastic. With degradable films disintegration is programmed to commence at the time of cropping. It must not occur too early or the micro-environment at the roots of the plants is destroyed and much reduced yields result. The use of mulching films reduces water and fertiliser usage to less than half that on open ground [39]. The same is of course true for degradable plastics mulch, but to optimise the benefits of this technology, they must remain intact until just before harvesting. This is particularly important in the case of soft fruits such as melons, bell peppers and sensitive vegetables that are normally irrigated with aqueous nutrients. Since mulching films create a microenvironment at the roots of the plant and the plants take up only the water and nutrients that they need, excess water from heavy rainfall can be just as damaging to sensitive crops as too little water. If then the films degrade prematurely, much of the benefit of the protective mulch may be lost with consequent loss of income to the farmer. In automated film laying, which is normally accomplished by turning under the 'tuck' to eradicate wind-lift, the young plants are rooted through pre-punched holes in the film in a single operation. Consequently, the film then has to be tough enough to resist this mechanical process.

With normal plastics, after cropping, the tough plastic residues clog the cropping machinery and manual removal is essential [25,39]. Modern mulching film technology uses low-micron films (8-10 μm) that are so fragile after environmental expose that it is virtually impossible to manually remove them from the soil [39]. They are, nevertheless, an impediment to plant growth if left on the soil and are also a potential hazard to animals when the land is subsequently put down to grass. Films made from regular polymers such as polyethylene and polypropylene tend to accumulate from year to year due to the durability of commercial products. Programmed-life polyolefins on the other hand degrade sharply at the end of their service life and do not accumulate in the soil. SG additives have been used for 15 years on the same fields in Israel and the USA and apart from "the tuck", which is not subjected to direct sunlight exposure during the first season, the photo-oxidised polyethylene can be ploughed into the soil, becoming part of the soil structure. This is followed by slower bioassimilation of the material that has not been exposed to sunlight but which has been heated in the soil surface, leaving no visible residues at the beginning of the next planting season. Ploughing also causes mechano-degradation, which in turn accelerates boassimilation during the following season.

Reduced water and fertiliser usage are particularly important in arid areas and degradable mulch has considerable potential in the recovery of desert land to agricultural use. The above requirements demand accurate time-control in the service-life of the plastic films and it is crucial that they remain tough just as long as is required but then rapidly fragment (lose elongation, $E_b$) just before cropping. This is achieved in the SG system(see Section 4) by the use of antioxidants that are effective during polymer processing but photo-transient for controlled periods in the environment.

SG technology for degradable polyethylene in agricultural applications provides the film manufacturer with a range of time-controlled concentrates. These are added to regular commercial PE generally at a standard addition rate. However, the additives give very different outdoor lifetimes before physical disintegration (Table 5) . In spite of the differences in user lifetime, once fragmentation of the films has started, bioassimilation commences and the bioassimilation time is not very different for all the grades.

Table 5 Standard SG grades [20].

| SG additive grade | | Time to embrittlement* |
|---|---|---|
| **1.** | **#221** | **6 weeks** |
| **2.** | **#131** | **3 months** |
| **3.** | **#19** | **4 months** |
| **4.** | **#12** | **6 months** |
| **5.** | **#112** | **12 months** |

*** Average times for mid-Europe or mid-USA for spring planting**

It has been found by experience that the SG grades listed above cover requirements in every situation encountered where mulching films and tunnels are used. Although there is a shift in fragmentation (embrittlement) time between the hot sunny southern climates to the more northerly cooler climates, it can normally be predicted fairly accurately which additive grade will give the desired result in each climate, based on UV incidence and temperature. However, the preferred protocol with time-controlled materials is that an initial small-scale trial is carried out in the field so that the results of field tests can be compared with the behaviour of the same films in laboratory accelerated weathering tests. SG programmed life degradable plastics are widely used in many counties throughout the world under the brand name of the film manufacturer.

The concentrate additive (masterbatch) approach to degradable agricultural films provides the most cost-effective technology for protective films in agriculture. In northern climes where silage film waste is a major problem, the longest SG formulation provides a user lifetime of 18 months before disintegration of film integrity occurs (see Section 7.2). The polyolefin additive concentrate technology - unlike bioplastics technology - does not require a separate and relatively small-scale manufacture for each grade of degradable polymer since the additive concentrate replaces the additive package that is normally added to conventional polyolefins by film manufacturers to provide the environmental durability.

Degradable plastics technology (EPI TDPA$^{TM}$) has also proved to be of value in the cultivation of forage maize in northern climes [20]. The value of the crop is restricted by the amount of heat it receives and it is necessary to keep the soil covered for approximately 6 – 8 weeks. Dry matter (DM), starch and in vitro digestibility (DOMD) and metabolisable energy (ME) may be calculated from DM and DOMD values. The degradable film products evaluated in this trial were: a PE film based on additive technology developed by EPI, and comparison among mulch film candidates (TDPA$^{TM}$ (transition metal ion PE), Degradyl (PVC) and IP plastic (LLDPE)) showed that TDPA$^{TM}$ had the lowest costs per unit weight for DM, for ME and for starch, and the lowest total costs for the trial. At harvest, visible residues of plastic were least for TDPA$^{TM}$, which began to break down after approximately 6 weeks of cover, as intended. The underground residue from TDPA$^{TM}$ became embrittled and was partly broken up by discing. In contrast, that part of the Degradyl and IP plastic were still intact on the surface at harvest and this resulted in numerous small pieces being blown about the fields, causing concern for environmental acceptability. In addition, of course, perceived environmental problems from the use of a chlorine-containing plastic would be a concern with the use of the Degradyl product.

The conventional way of using polyethylene mulching films is in a single cropping regime in which the film disintegrates at the time of cropping (see Fig. 4, SG #221). Time-controlled biodegradable SG films have been evaluated in Taiwan in 'dual crop' cultivation of fast-growing vegetables. The objective is to produce two crops in quick succession on the same degradable film. In this regime the film is applied late in the year and the second crop is planted in the holes left after the harvest of the first crop [40]. The film is timed to degrade as the second crop is being harvested (see Fig. 3, SG #131). This procedure offers considerable advantages in irrigated systems since quite apart from the lower costs, the irrigation tubes are not disturbed, ensuring a fast changeover. Thus, a late crop planted in September can harvested in December and this may be followed by a second very early cash crop in the next season by planting on the same film in December or January. For longer cropping times #19, #12# and #112 may be used. [25].

October 27 - 28 days after laying

December 30 – 92 days after laying

Fig. 4 Single (#221) and double (#131) cropping of fast-growing vegetables with Scott-Gilead formulations over the winter period in Tainan, Taiwan [40].

Another procedure that is also gaining acceptance is to sow seed directly into the soil under a complete plastic cover. This has the advantage of avoiding the 'shock' of transplanting with consequent earlier maturity of the crop. 'Mid-bed trenching' as this process is called [25,39] involves sowing the seed in a trench and laying the plastic over the growing plants (Fig. 5). The plastic film must be timed to break under the pressure of wind and weather (i.e. $E_b < 10\%$) when the leaves of the plant contact the cover. If the film breaks too early, the greenhouse effect will be lost and if it breaks too late the plant will be misshapen.

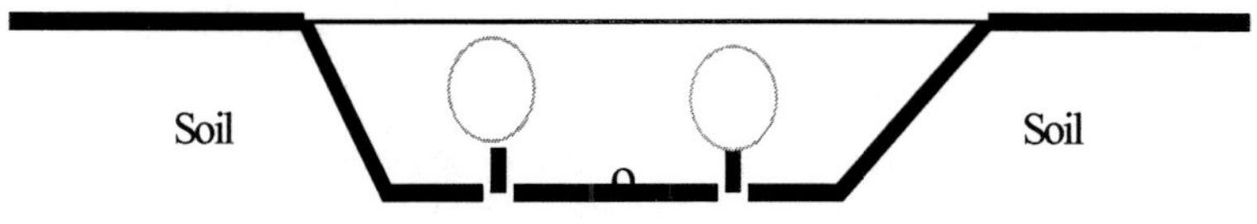


Fig. 5. Mid-bed trenching using photo-biodegradable polyethylene

Programmed-life films have also been evaluated as solar sterilisation films in tropical climates [25]. The principle involved is that photodegradable films are laid before the crops are planted. This results in the destruction of pathogenic bacteria, which tend to accumulate in intensively cultivated land. The high temperatures achieved under transparent films destroy pathogenic bacteria but leave untouched beneficial microorganisms. Normally, the films are allowed to fragment before the crop is grown on a new photodegradable film but if the initial film is designed to remain strong, as described above, the plants could be grown through holes punched in the original film.

## 4.2 Auxiliary Products

Biodegradable plastics are being increasingly used in auxiliary products for agriculture and horticulture that frequently end up in the environment as litter. These include irrigation tubing, plant pots, soil sterilisation films, polyolefin baler twines, fruit protection bags and crop-protection netting [25]. In some of these applications, for example plant pots that are intended to be used only once, there is a useful application for mechanically recyclable polymers, particularly in polypropylene.

Some auxiliary plastics are not at present degradable. Typically silage and hay wrap films, fertiliser and animal feed sacks and tree shelters that protect vulnerable young trees from attack by animals. Agricultural packaging persists in the environment for many years and the wind-blown plastics detritus is a particularly serious nuisance in areas of environmental sensitivity. There is little economic incentive to the plastics manufacturer to make use of biodegradability when there is no cost-benefit other than a cleaner environment. Some auxiliary products such as packaging materials can be recovered in a relatively clean form for recycling but the subsequent performance of the secondary products produced from them [36,48] is generally inferior to that of virgin materials. Most plastics detritus recovered from farms is not worth recycling at all because it is seriously contaminated by biological matter and by transition metal ions picked up from the soil. Mechanical recycling of farmyard plastics is in most cases less ecologically viable than making the same products from virgin polymers [34] and biocycling offers a better environmental solution to this problem.

### Twines and Protective Netting

Polyolefin baler twines, which have largely replaced sisal due to lower production costs, become heavily contaminated during service. These hold the bales together during storage and are split open when the fodder is required. Consequently they may become litter and are trampled underfoot by cattle. In the USA, biodegradable polyolefin twines are now being manufactured [25] as "Cleanfields" using SG technology because of the environmental benefits they bring. In the case of polypropylene twines, a lifetime of one year has been found to be adequate, followed by rapid disintegration and subsequent bioassimilation. The same technology can be applied to protective netting and fastenings for fruit bushes and vines, which are difficult to remove manually after use [25].

### Stretch-wrap Films for Silage and Hay-storage

It is now common practice to store hay for use as silage in an airtight bag so that the nitrogen produced by fermentation is contained within the forage. This involves completely sealing the hay after harvesting and the seal is not normally broken until the contents are fed to animals the following winter. The films have to remain tough and strong during the fermentation period but after the silage is fed to animals, the residual plastic is carried by the wind, sometimes for many miles, and becomes an environmental nuisance after being deposited in trees and hedges along river banks. This waste material is again highly contaminated through contact with the soil and is expensive to collect for disposal.

The cost of a clean environment is not easy to calculate but in a recent survey in UK [41] it was estimated that about 500 tonnes of film are used in just one area of outstanding natural beauty annually (about 65,000 tonnes nationally) and this accumulates from year to year since almost none is routinely collected for disposal. The cost of landfill disposal is at present £30/tonne and is increasing year-by-year. Controlled biodegradability is an obvious solution to this problem and the technology is now available to allow even black pigmented polyolefins to be made photo-biodegradable with a time delay of one year (or more if required). So far, the manufacturers of agricultural silage film have shown little enthusiasm for this environmentally acceptable technology.

### Controlled Release Systems for Fertilisers

An important development is the use of biodegradable polyolefins in controlled release of fertilisers by encapsulation [42]. This results in controlled release in leaching environments over an extended period of time compared with direct application. This in turn effectively reduces the pollution of streams and the eutrophication of lakes and watercourses. Controlled release of pesticides by encapsulation has also considerable potential by matching the application time to the life cycle of the pest.

Commercial polyolefins show no significant reduction in molar mass in typical ambient biometric tests over many months or even years. The resistance of hydrocarbon polymers to biodegradation during use is one of their major advantages in agricultural application over bio-based plastics. The useful life of photo-biodegradable polyolefins is controlled commercially by appropriate transition metal complexes in which the ligands are antioxidants. The induction (IP) to biodegradation is controlled by concentration to meet the needs of the user. The antioxidants are light and/or heat sensitive. Consequently, after the decay of antioxidant activity, the rate of molar mass reduction proceeds at a rate similar to that of the same polymer without antioxidants or light stabilisers. The rate controlling step in the biodegradation of oxo-biodegradable polymers is thus not attack by micro-organisms, but abiotic oxidation of the polymer, which as we have seen is powerfully influenced by the presence of transition metal ion prodegradants and antioxidants. The balance between durability during use in the outdoor environment and the rate of biodegradation is achieved by combinations of antioxidants and by adjusting the antioxidant-prooxidant ratio (Section 2.3).

**"Green" Waste Bags and Landfill Covers**

Biodegradable polyolefins are also now used commercially in compostable garden waste bags. Thus, EPI's TDPA® compost bags oxidise and fragment during normal composting procedures - normally at ~60-70°C [20]. They are also used as temporary landfill covers, which subsequently disintegrate releasing in the surface layers of the landfill thus releasing the contents to the landfill environment with much more rapid reduction in the volume of the landfill site [20].

# 5 The Disposal and Reclamation of Plastics Wastes

It was recognised over 30 years ago that the wastes arising from the use of plastics packaging posed a threat to the environment due to their persistence in the seas and in the countryside [45]. Unlike paper, they were not obviously biodegradable as manufactured and even at that time it was clear that that this was not a consequence of the intrinsic non-biodegradability of the synthetic polymer but of over-stabilisation during manufacture. Consequently, by changing the normal additive formulation to include prooxidants as well as antioxidants in the additive package plastic, time- controlled oxidation and biodegradation could be used to give the polyolefins the required durability during use but rapid disintegradation and biodegradation after discard in the environment [21,44].

Unfortunately the principle of programmed biodegradation as outlined in the earlier Sections was not understood by environmentalists and in a comprehensive dismissal of synthetic polymers based on fossil-derived intermediates, Commoner et al. on behalf of Greenpeace [45] formulated the simplistic view that "because petrochemical products are not the outcome of biological evolution, living things lack enzymes that can break them down into components that can be assimilated into the biological cycles" and "every pound of plastic that has been produced, if it has not been burned, is still with us". This concept ignores the fact that abiotic as well as biotic processes are involved in the bioasimilation of both natural and synthetic polymers. Even the naturally occurring hydrocarbon polymers of which *cis*-polyisoprene (*cis*-PI) is the best known, begin to biodegrade only after exposure to the environment, when peroxidation leads to low molar mass products that support microbial. The chemically synthesised *cis*-PI and its close relative cis-polybutadiene (*cis*-PB) have been since shown to behave in the same way [6]. The significance of this misunderstanding of polymer biodegradation will be discussed further in the following Sections. In spite of the prejudice against fossil-based plastics, oxo-biodegradable polyolefins rapidly gained acceptance in time-controlled mulching films and are now widely used in agriculture in most agricultural economies.

## 5.1 The Post-use Treatment of Plastcs for the Recovery of Value

In modern urban waste management, the emphasis has changed since the recognition of man's contribution to global warming from the prevention of litter in the environment to the recovery of value or the recovery of energy from wastes by recycling. The European Waste Framework Directive (March 1991) defines recovery as follows [1].

"Recycling/reclamation of organic substances.....use as fuel to generate energy and spreading on land resulting in benefit to agriculture and ecological improvement, including composting and other biological processes"

In principle some items of packaging may be re-used. However, it is not always clear that the energy involved in collecting and cleansing used plastics is ecologically acceptable [1,34]. High in popular esteem is mechanical recycling, since this word carries an aura of ecological sustainability. In practice reprocessing of plastics is only sustainable if it leads to an overall reduction in the use of fossil resources and hence to carbon dioxide reduction. This is possible in the case of clean, homogeneous wastes - generally from industrial sources but it is not a viable procedure for contaminated mixed waste plastics, particularly when it is collected from widely dispersed rural sites [1,34]. Furthermore, the reprocessing operation is itself energy intensive and there are generally no clear ecological or technical benefits in reprocessing mixed wastes [34]. On the other hand, the hydrocarbon portion of mixed plastics wastes has a high fuel value. For example polyethylene has the same calorific value as the oil from which it was manufactured [1,20] and thus the fossil resource used in the molecular composition of polyolefins is given a "second life" as a fuel in combined heat and power (CHP) incinerators [1].

## 5.2 Termination of Plastivs in Biologically Active Environments

Plastics packaging may end up in a variety of quite different environments [19] of which the following are the most important.

(1) Inland water courses, sewage systems and the oceans
(2) Compost
(3) Litter on or in the soil
(4) Landfill

Each of these processes requires a different time-scale for biodegradation (Fig. 6) [19].

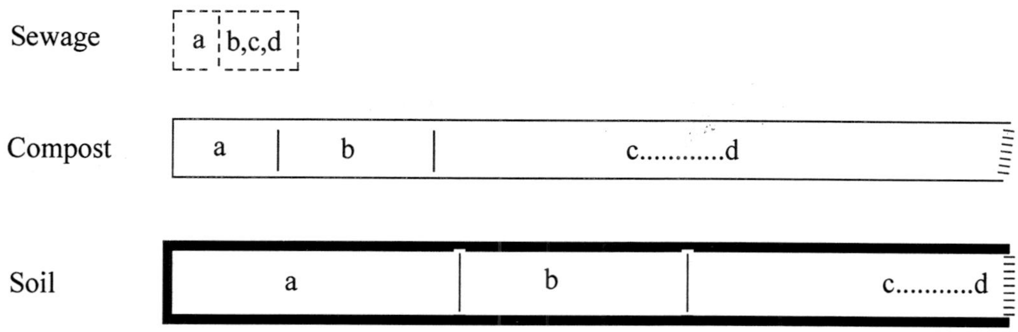


a No change in chemical or mechanical properties; a requirement of all polymers
b Chemical and physical degradation; loss of mechanical properties
c... d Formation of cell biomass and carbon dioxide, leading ultimately to complete mineralisation

Fig.6. Time requirements for biodegradable plastics in different environments [19] (with permission, Royal Society of Chemistry, *Polymers and the Environment*, 1999)

Plastics wastes that are disposed of in sewage systems or in rivers are required to disintegrate and biodegrade rapidly - generally within weeks - to ensure that there is no accumulation of plastics debris to cause damage to man, animals or fish. By contrast, plastics that are deliberately treated in industrial compost in order to recover fertiliser should fragment to particulates that biodegrade in a similar manner to natures lignocellulosic wastes and which should be indistinguishable from normal fertiliser biomass during the composting process. However, in order to avoid the rapid formation of "greenhouse" $CO_2$ and retain the benefit of carbonaceous biomass in the soil, bioassimilation should ideally take place over a year or more. Plastics terminating in or on land either deliberately as part of their function, as in the case of polyethylene agricultural films, or accidentally, as in the case of packaging litter, are required to fragment and peroxidises rapidly in sunlight. However, as is the case of nature's lignocellulosic waste, provided the modified residues can be shown to be ultimately bioassimilable either by mass loss to cell biomass or by mineralisation in laboratory tests, the time scale is not critically important provided that, like Nature's slow biodegrading wastes, they do not accumulate in the soil [19,20]. Finally, although landfill should in principle be the last resort for biodegradable materials, much household waste is still sent to landfill even in the developed countries and will eventually biodegrade. This process can be accelerated by the use of plastics packaging that disintegrates and by the use of landfill covers that fragment after burial [20]. The rate of ultimate biodegradation is of minor significance. It is important to recognise that domestic and industrial plastics packaging can terminate in more than one and in some cases all of the above disposal options. From this it follows that not only must they be capable of being manufactured in conventional processing equipment, but they should also be recoverable through composting. An additional rerquirement is that the waste polymers should be mechnically reprocessable without loss of mechanical properties. If they intentionally or inadvertantly end up as litter they should be photo-biodegradable. In principle, the oxo-biodegradable polyolefins are capable of meeting any of these end-of-lfe requirements.

# 6 The Development of International Standards for Biodegradable Polymers

The principle of a "level playing field for business" is outlined in the "Green Report" of the Attorneys General of the USA [47]. The same principle is also enshrined in the EU Packaging and Packaging Waste Directive [48], whose primary objective is stated to be as follows.

> ".....to harmonise national measures....to ensure the functioning of the internal market and to avoid obstacles to trade and distortion of competition within the community"

The "Green Report" also requires that Standards should be based on the best scientific evidence available. However, it has been seen above that the development of biodegradable polymers has been beset by misinterpretation and misrepresentation of the way in which Nature deals with its waste products. In particular, the importance of abiotic processes has not been given sufficient emphasis in the process of bioassimilation [49]. Consequently existing international standards for biodegradable polymers are not based on scientific evidence since they ignore completely the role of abioic chemistry [50-53]. The earliest test methods for the biodegradation of plastics were based on biological oxygen demand (BOD) developed originally to evaluate the environmental persistence of synthetic detergents that were the cause of pollution in inland waterways during the 1950s. More recent standard test methods for water quality have been comprehensively reviewed together with their subsequent application to the biodegradation of plastics [50]. The main critical parameters proposed by the International Standards Organisation (ISO) are either oxygen absorption or carbon dioxide evolution in the presence of microorganisms. The latter procedure was taken over directly by the European Standards Organisation (CEN) in EN 13432 [53]. However, low molar mass chemicals are quite different from plastics. The first evidence of physical deterioration of plastics is due to abiotic chemical processes, which commence at an early stage during exposure to the environment. Consequently the measurement of mechanical properties is an important indicator of degradation, although not necessarily of biodegradation. It is argued with some justification that these changes alone do not guarantee that the residues are eventually harmless to the environment and that complete mineralisation of biodegradable plastics by biometric means is the simplest and most convenient way of measuring biodegradation rate. Unfortunately "simple and convenient" in this case masks the important environmental requirement that carbon is retained in the "land carbon sink" as long as possible, rather than being rejected to the environment as $CO_2$.

In principle, kinetic measurement of abiotic peroxidation process at various temperatures allows the complete oxidation time to be extrapolated to ambient temperatures. Such kinetics have been measured for aged or weathered hydrocarbon polymers at the end of the induction period. In practice this must be the point at which biodegradation begins and this in turn is determined by the post treatment they receive during or after their service life – for example during composting for packaging plastics or during outdoor weathering for agricultural products. The durability of degradable plastics before the end of their useful life is of great importance to manufacturers of packaging and agricultural products such as mulching films [2,20,22]. Plastics technologists use standard forced air oven ageing or UV weatherometer tests to predict this in practice [54-61] and considerable amount of work has been done to establish good correlations between these standard tests and exposure during service [62]. Provided the same standard procedures are used to simulate the effects of the environment on degradable plastics in the pre-biodegradation (user) stage, they have no direct relevance to the rate of bioasimilation. Consequently, the concerns of the Standards Organisations begin only when the product is discarded as litter in the environment.

While hydro-biodegradable polymers may in principle begin to biodegrade in a biometric test before they have ever been used as packaging, this is not normally possible with oxo-biodegradable hydrocarbon polymers. Polyolefins do not begin to biodegrade until chemical modification of the polymer has produced a hydrophilic polymer surface during ageing that supports microbial colonisation [26,29]. In the case of thin films, this inevitably also involves a loss of mechanical properties, such as elongation at break ($E_b$) and it is normally assumed that polymer fragmentation occurs under mild physical pressure at 90% loss of $E_b$, The transformation of the polymer surface from a hydrophobic to a hydrophilic environment thus enables the microorganisms to utilise the polymer oxidation product as a sole source of carbon [26,29].

## 6.1 Standards for the Composting of Packaging Plastics

The initially developed Standard for composting of packaging plastics is typfied by CEN's EN 13432 "Packaging – Requirements for packaging recoverable through composting and biodegradation – Test scheme and evaluation criteria for final acceptance of packaging" [53] is elaborated in EN 14045 [63] and EN 14046 [64]. In these standards, compostability is assessed by the following criteria, all of which must be satisfied.

(1) Identification of packaging constituents, dry solid content, ignition residues, hazardous metal residues.
(2) Biodegradability: 90% of the total theoretical $CO_2$ evolution in compost or simulated compost in 6 months.
(3) Disintegration: Not more than 10% shall fail to pass through a 2mm sieve.
(4) Compost quality: No negative effects on density, total dry solids, volatile solids, salt content, pH, total nitrogen, ammonium nitrogen, phosphorus, magnesium and potassium eco-toxicity effects on 2 crop plants.
(5) Recognisability: "must be recognisable as compostable or biodegradable by the end user by appropriate means"

These criteria were pragmatically designed to cause the minimal interference with the commercial operations of the industry [65] and have little to do with protection of the environment. For example criterion 2 which is mandatory is less concerned with recovery of fertiliser from waste, as required by the Directive than with the easy disposal of carbon dioxide in the environment. Criterion 3 governing the choice of particle size for "disintegration" is entirely arbitrary since it is not based on scientific evidence. Indeed, scientific considerations played very little part in arriving at the above criteria [49,65].

The above weakness of EN 13432 was recognised by the EU Environment Directorate [66], in the following critique.

- ISO 14851 (0xygen consumption) and ISO 148 (Sturm test) do not simulate composting conditions.
- What is really needed is to know what is the fate of materials under composting conditions and what happens once it is released to the soil.
- If the packaging material does not completely biodegrade during the composting process, it should be demonstrated that it eventually degrades in the soil"

As currently formulated, EN 13432 discriminates against manufacturers and users of oxo-biodegradable polyolefins since certification of biodegradable plastics as compostable at present requires compliance with this published Standard. This clearly violates the EU Directive, quoted in Section 5.1, since EN 13432 and EN 14046 require that the unaged plastics components must mineralise within six months at 58 ±2°C. As observed above, oxo-biodegradable plastics as normally manufactured do not undergo mineralisation to any significant extent under these conditions. In fact this requirement has little to do with composting, since it uses as a reference standard crystalline cellulose that is 68% mineralised under the chosen conditions in 38 days [53]. Pure cellulose is rarely found in Nature's wastes. In woody materials and straw it chemically bound to lignin as lignocellulose, which reduces its biodegradation rate [6]. A practical consequence of this is that polymers that mineralise at the same rate as pure cellulose make only a minor contribution to the nutritional value of compost. Instead of contributing to the "land carbon sinks" [67], the carbon is converted to "polluting" $CO_2$ during the composting process. It is therefore not available for slow release to growing plants in the same way as natural lignocellulosic materials. The European Waste Framework Directive, however, requires that organic waste is "recovered" so that the product can be "spread on land resulting in benefit to agriculture and ecological improvement". Consequently, a much better reference standard, based on the behaviour of Nature's waste polymers, would be lignocellulose (e.g. straw).

It is now accepted that a new Standard is required for oxo-biodegradable polymers that while fulfilling their role as land carbon sinks do not accumulate in the soil or cause toxicity to plants and animals. The British Standards Institution, BSi PKW/4 "Packaging and the Environment" is in the process of developing a new Standard "Packaging – Determination of the compostability (including biodegradability and eco-toxicity) of packaging materials based on oxo-biodegradable plastics" and ASTM Committee D20.96 has already issued D 6954 - 04 "Standard Guide for exposing and testing plastics that degrade by a combination of oxidation and biodegradation". Both incorporate a controlled pre-ageing processes before the mineralisation test. The protocol shown in Fig. 7 outlines the BSi "twin-track" route to mineralisation.

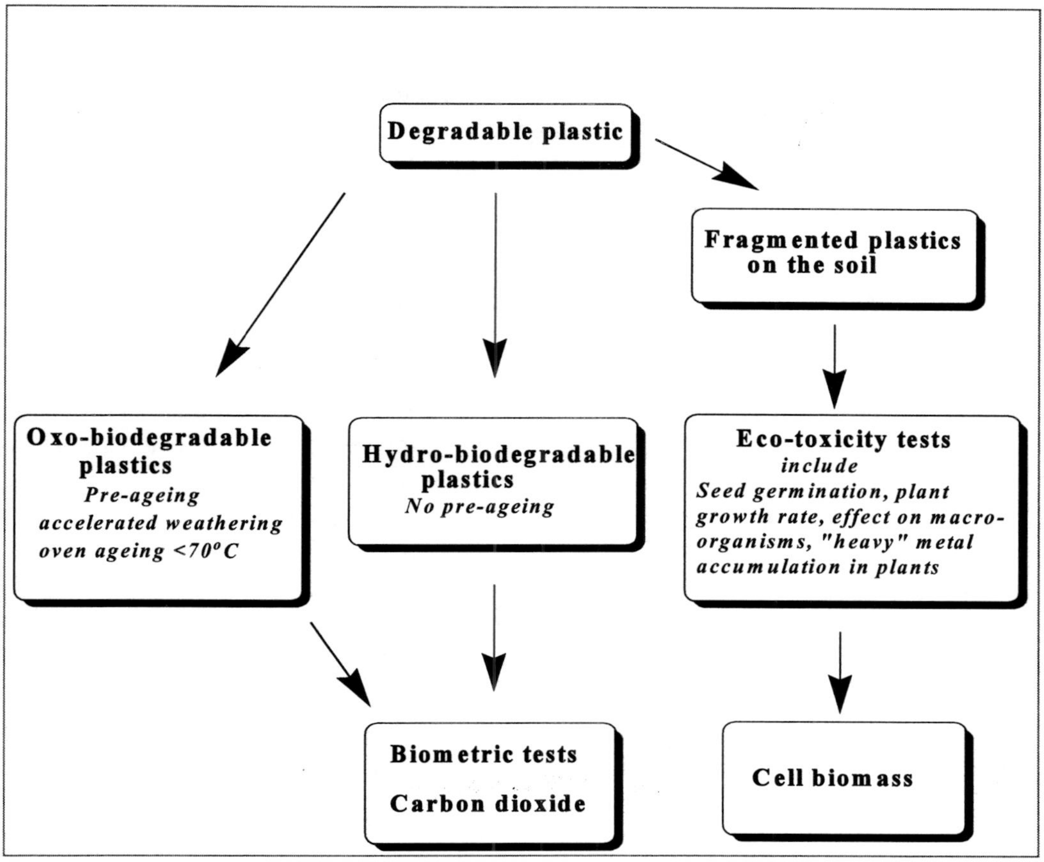


Fig. 7. "Twin-track" testing protocol for biodegradable packaging plastics

Key features of the test method are as follows.

(1) The plastic is subjected to a laboratory ageing procedure to the point at which it is oxidised to fragments. For plastics ending up in compost, this involves heat ageing to the temperatures experienced during industrial composting processes greater than 60°C and normally not exceeding 75°C.
(2) Appropriate mineralisation procedures in a laboratory composting test with adequate aeration using matured compost or soil as inocula to provide assurance of ultimate biodegradability [34,49,67].
(3) Soil toxicity tests on polymer fragments to ensure that any degradation products released from the plastic have no long-term deleterious effects on plants or on animals that may imbibe them.

## 6.2 "Heavy Metals" and Essential Trace Elements

Regarding requirement (3) above, concern has been expressed by environmentalists about the effect of transition metal compounds (often inappropriately called "heavy metals") present in oxo-biodegradable plastics when spread on land. Extensive research has been carried out on potential eco-toxicity effects of particulate and extensively degraded plastics when mixed with soil. Such effects include seed germination and plant growth rates, compared with the same soil without degraded plastics [43,68], the effects on macroorganisms (worms, daphnia, etc.) in the soil [68] and on the accumulation of transition metal ions in the stems, leaves and fruit of plants during the growing season [39,69]. So far with the present range of degradable plastics used in agriculture, which incorporate fractions of a percent of transition metal ions, have no negative effects in any of the above tests.

The commonly used transition metal compounds in commercial oxo-biodegradable plastics are manganese, iron, cobalt and nickel. None of these have been shown to be toxic and until recently have not been in national lists of dangerous substances. All the above transition metal ions, which are required in human nutrition, are absorbed from foodstuffs and water. They are therefore considered to be "essential" minerals [70] required in oxygen transport systems. The non-toxicities of iron, which is present in haemoglobin, catalase and peroxidases and of manganese, required for manganese peroxidase, have not been questioned.

An Expert Group on Vitamins and Minerals of the UK Food Standards Agency has carried out a risk assessment [70] on trace elements and the following is a summary of their findings on the risk to humans of cobalt and nickel.

High concentrations of cobalt are found in fish (0.01 mg/kg), nuts (0.09 mg/kg), green leafy vegetables (0.009 mg/kg) and fresh cereals (0.01 mg/kg). Most of the cobalt ingested is inorganic. Fresh water concentrations of Co range from 0.001 to 0.01 mg/L. The mean population intake of Co is 0.012 mg/day. Cobalt is also included in some multi-constituent licensed medicines, at a maximum daily dose of 0.25 mg. Although cobalt is an essential trace element, Co deficiency has not been reported in humans (presumably because of its widespread availability from food and water). Gastrointestinal absorption of cobalt depends on the dose. Very low doses are almost completely absorbed, whereas larger doses are less well absorbed. Most excess cobalt is excreted in urine. The only toxicity data for cobalt reported in the literature was in 1960, when heavy beer drinkers suffered cardiomyopathy as a result of the use by the brewing industry of cobalt chloride as a "foam stabiliser" at 1.0-1.5 ppm. Ethanol and cobalt have a synergistic effect in reducing blood flow causing damage to the heart. Massive doses of cobalt salts (30 mg/day), evaluated as a treatment for anaemias led to skin rashes and hot flushes. Prolonged use of cobalt "therapy" led to depression in iodine uptake.

Nickel is present in a number of enzymes in plants and microorganisms and in humans it influences iron absorption and metabolism. It is found in a variety of foods as ionic Ni, particularly in pulses and oats (0.18 mg/kg in miscellaneous cereals), and in nuts (1.77 mg/kg). Lower levels are found in water. Total intake of nickel by humans from all sources is up to 0.26 mg/day and no potential high intake groups have been identified. The average intake from food and drinking water is 0.16 mg/day. Nickel is excreted in urine and in sweat. Acute nickel exposure is associated with nausea, vomiting abdominal discomfort and diarrhoea. The lowest reported oral dose associated with acute effects of nickel in humans was 1.2 mg in a 60 kg adult. Chronic inhalation of nickel and its compounds is associated with lung cancer in humans and in animals but orally administered nickel was found not to be carcinogenic. It was the exposure of humans to nickel during mining that led to the believe that nickel is carcinogenic however it is imbibed but administration of nickel compounds orally has shown that the main effects in humans is in skin sensitisation but only over 5.6 mg.

From the above, it is now understood how and why the common transition metals are obtained by humans as essential nutrients. It will also be useful when discussing "dangerous substances" in the environment to see how they are absorbed into the food chain from the soil. In fact, the amounts of transition metal ions available to plants from common soils is very much greater than could be produced from degradable plastics in the soil and are much higher than can be absorbed by plants [71]. Particular attention has been paid to cobalt and nickel for the reasons discussed above. Volcanic soils contain very high concentrations of cobalt oxide (up to 100 ppm) and nickel oxide (up to 750 ppm). Sandstone and limestone contain 90 ppm and 10-20 ppm of nickel respectively [25]. However, the amount of nickel taken up by the plant appears to have little to do with its concentration in the soil. Table 6 shows the effect on plant uptake of nickel sulphate applied to the soil to simulate the deposition of nickel from degradable polyethylene mulching films up to 180 years of application to the same soil [39].

Table 6. The accumulation of nickel in melons (ppm, measured by atomic absorption) grown in soils containing increasing amounts of nickel sulphate [39].

| | **Control** | **60 years*** | **120 years*** | **180 years*** |
|---|---|---|---|---|
| **Leaves** | **17.3** | **15.2** | **13.5** | **13.7** |
| **Stems** | **5.0** | **4.5** | **5.2** | **5.0** |
| **Flesh** | **2.7** | **2.0** | **3.0** | **3.2** |
| **Skin** | **3.0** | **3.5** | **3.2** | **3.0** |

***The soil was sprayed with $NiSO_4$ to give nickel concentrations in the topsoil equivalent to the accumulation from SG mulching films used for the number of years indicated.**

It is clear that the accumulation of nickel in various parts of the plant remains constant within experimental limits, whatever the concentration of nickel in the soil. Furthermore, It can be calculated that in the 'worst case scenario', it would take 500 years to increase the nickel content of soil using typical nickel contents of degradable polyethylene mulching films by 1 ppm [25].

## 6.3 Mineralisation Test Procedures

Polyolefins that have been thermally oxidised or exposed to light as described above have been shown to generate $CO_2$ to over 60% of theoretical in 18 months to two years in the presence of mature compost or when inoculated with soil over a longer period of time [34,67]. The laboratory test procedure for mineralisation of plastics at ambient temperature described in EN 14046 [64] is a simple plastic box containing a low-carbon soil inoculated with mature (2-4 months) compost. The plastic is introduced as produced and the maximum temperature permitted in is 58±2°C. The BSi proposal is similar to this except that there is a more efficient distribution of air within the compost (see Fig. 8). The ageing procedure that precedes the mineralisation test must in this case be matched to the environmental exposure conditions.

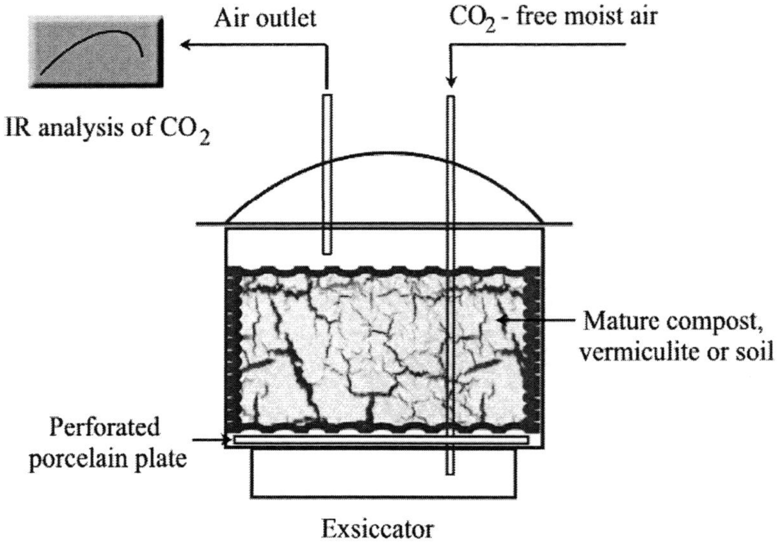


Fig. 8 Biometer for the oxo-bidegradation of polyethylene (Reproduced with kind permission of I.Jakubowicz, unpublished work)

# 7 Degradable Plastics in Soil

As already indicated, degradable polyolefins that terminate on the surface of soil as litter have been used in agriculture for many years. No accumulation of these materials has been observed and it will be evident from the evidence already cited that they do indeed biodegrade in fertile soils and compost.

It is now accepted by CEN (TC 249/WG9) that, in the case of oxo-biodegradable polymers, biometric tests should be preceded by weathering tests (UV exposure in a weatherometer) before mineralisation. This together with the eco-tests, shown in simplified form in Fig. 7, provides a basis for ensuring that not only is the carbon of the plastic ultimately converted to useful fertiliser but also that no non-biodegradable toxic products are produced in the process.

Chiellini et al [71] have shown that thermally aged oxo-biodegradable plastics mineralise in forest soils to over 60% in 18 months and CEN TC 249/WG9 has concluded that 2 years at temperatures up to 28°C in the bio-active stage of degradation to achieve 60% mineralisation is acceptable for polymers that require an ageing period before biodegradation commences. The evidence from studies of Jakubowicz and Chiellini et al. have shown that biodegradable polyethylenes evolve $CO_2$ in an auto-accelerating mode, which continues progressively after this time. However, in the event that some polymers may not reach 60% mineralisation after 2 years at 28°C, it is proposed to allow biometric mineralisation at temperatures up to 50°C. This is fully justified in view of the fact that the surface of the soil may reach this temperature than this during the summer season [2] so that the rate of peroxidation may be quadrupled.

The rate of abiotic peroxidation is known to be markedly accelerated under a plastic film due to the synergistic influence of heat and light. This process is described schematically in Fig. 9. Since abiotic peroxidation and bio-oxidation occur together during the bioassimilation of oxo-biodegradable plastics, it is of crucial importance that the mineralisation chamber is perfused with air during the whole of the mineralisation procedure. This is also important to minimise the potential production of methane due to anerobic conditions. Fig. 8 shows the introduction of air at the bottom of perforated plate, which ensures that the air is evenly distributed through the biologically active soil [32].

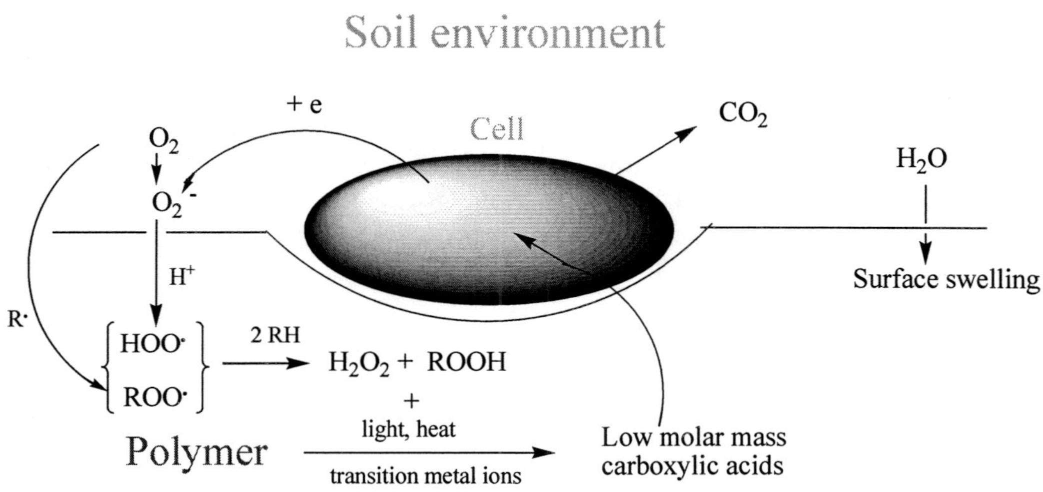


Fig. 9. Synergism between peroxidation and bio-oxidation in the bioassimilation of oxo-biodegradable polymers [19]

Several commercial weatherometers provide similar temperatures and provided the UV shorter wavelength cut-off is not less than that experienced in sunshine (~ 290 nm) [62], these are realistic practical conditions for plastics that end up on the soil particularly in the "sun-belt" countries, where they are wisely used. International standards describe the use of typical commercial laboratory techniques. Since ageing and weathering procedures have been standardised for the benefit of manufacturers and users of plastics, it is not considered that the details of the weathering procedure should be part of the standard itself, provided that full reference is made to the standard method chosen. ASTM D 5510-94 [60] describes a convenient heat-ageing oven for degradable plastics.

## 8 Sustainability of Biodegradable Polyolefins

The term 'sustainable' is often assumed to be synonymous with 'renewable'. The concept that the source of the carbon structure of polymer molecules is the only criterion of sustainability is simplistic and is not supported by life-cycle assessment (LCA) of degradable plastics products [72]. Some typical statements expressing the views of bioplastics manufacturers and hence legislators are as follows.

***"Carbon dioxide emissions from the degradation of polyolefins will contribute to global warming"***

This statement is ignores the energies involved in the manufacture and disposal of both fossil-based and bio-based polymers. In general, more fossil fuel is used and correspondingly more $CO_2$ is produced in the agricultural production, extraction of the feed stocks and synthesis of bioplastics than in the corresponding processes for fossil-based polymers [1,72,73]. During bioassimilation after use, oxo-biodegradable plastics contribute to the land carbon sink much more effectively than hydro-biodegradable plastics to the benefit of growing plants and reduction in greenhouse gases.

***"Life-cycle assessment does not result in a definitive answer as to which is the preferred option".***

It is true that LCA has not yet been able to demonstrate the ecological superiority of bio-based plastics, since most published studies consider only the energy utilised in the manufacture of products and do not include the energy used or produced in ultimate recovery. For example incineration produces the same amount of energy when hydrocarbon polymers are incinerated, whereas most biobased polymers produce much less. Unless or until the complete life-cycle energy has been determined, this statement must remain a hypothesis [72-74]. However, there is already sufficient information in the technical literature to conclude that further LCA studies will have not have a significant influence on whether intermediates for the plastic industry will in the future be more sustainable based on renewable resources or upon fossil fuels [73,74].

***"The National Non-food Crops Strategy...provides a unique opportunity to address wide-ranging issues, as diverse as climate change, environmental degradation, rural development and agricultural diversification"***

"Non-food crops" is a misleading term in the present context since bioplastics are currently manufactured from food crops (polysaccharides, edible oils, etc.), not from non-food crops. Consequently, the manufacture of bioplastics must be in competition with the production of food [72,73]. No serious attempts have so far been made by to estimate the respective usage of agricultural land currently used in food production that would be required to produce the equivalent tonnage of bioplastics to satisfy the commodity thermoplastics markets.

***"The (UK) Government supports the development of biodegradable polymers derived from renewable resources"***

Most polymers can be made from "renewable resources" at a cost. For example polyethylene, the major commodity polymer used in packaging, was one of the earliest biopolymers. For some years ethylene was manufactured from sugar by fermentation to alcohol, followed by dehydration to ethene and, if and when the economics of alcohol manufacture justify its use as a biofuel, the polyolefins will again become biopolymers. So far as resource depletion is concerned, if fossil carbon resources were used only for the manufacture of polymers and not for energy production, the former would last for approximately 300 years based on present estimates. By that time, polyolefins will in any case probably be manufactured from renewable resources anyway for purely economic reasons. At present energy utilisation and hence economics are unfavorable for the manufacture of polyolefins from agricultural products, just as they are for most bioplastics.

***"Oxo-degradable plastics are designed to degrade entirely to carbon dioxide and water with no biomass residue."***

This statement is incorrect. All carbon-based degradable polymers, including rubbers, lignin and the polyolefins are converted to biomass in addition to carbon dioxide and water under biotic conditions [6,19,75]. In practice, oxo-biodegradable polymers, after loss of mechanical properties and converted to oxidised fragments, become attached to or buried in soil, where they support the growth of cell biomass. It has recently been shown that, when fully biodegraded, hydrocarbon polymers actually produce more biomass than hydro-biodegradable polymers (e.g. cellulose). Furthermore, as already noted oxo-biodegradable polymers contribute more effectively to the land carbon sink than hydro-biodegradable polymers (e.g. poly(lactic acid) or starch-based polymers). They release low molar mass biodegradable oxidation products to the growing environment after the fragmented plastic has been returned to the soil. They thus have a similar soil conditioning effect to natural lignocellulosic materials (e.g. leaves and twigs).

***"A product, which self-destructs, does not fit with recycling options"***

Oxo-biodegradable plastics do not self-destruct. They are designed to give a controllable service life and, when used in short-term applications such as clean bottles and containers, they can be readily reprocessed (recycled) with recovered commodity polyolefins after use. Alternatively they may be incinerated with energy recovery equivalent to that of the oil from which they were manufactured [1]. In the case of food packaging films and waste bags, they are preferably composted after use. When the energy expended in the collection and cleansing of domestic waste plastics is added to that of the reprocessing operation (about one third of the energy originally used in the manufacture of the commodity plastics), it is questionable whether reprocessing (mechanical recycling) of contaminated domestic packaging can be justified so long as energy is based on fossil fuels [1,46,72].

It must be concluded then that polymers made from renewable natural resources are not necessarily more eco-efficient than synthetic hydrocarbon polymers. Polymers manufactured from polysaccharides utilise more non-renewable fossil fuels and are more polluting during manufacture than petro-based polymers. Furthermore, sustainable polymers also have to be industrially acceptable and although ultimate biodegradability in the natural environment is important, polymer-based products are required to biodegrade in a controlled way. There are some applications of biodegradable plastics in which inherent resistence to water is an essential technological requirement; notably the use of polyolefins in mulching films, silage wrap and baler twines. It is uneconomic to collect these materials after use and because of contamination they cannot be usefully recycled. Indeed, the main purpose of using biodegradable mulching films and binder twines is that, by their congtrolled biodegradability, they do not need to be recovered from the fields, with considerable cost benefits to the farmer [25].

# 9 Conclusions

It has become evident in recent years that there is a place for both bioplastics and synthetic plastics in packaging. Bio-based hydro-biodegradable plastics have an obvious advantage in products that are likely to end up in watercourses or in sewage systems due to the requirement that they must be substantially biodegraded during the waste treatment process. However oxo-biodegradability plastics are much more useful in applications that require assured and reproducible durability to random attack by microorganisms. The rate of hydro-biodegradation of natural polymers can be retarded by chemical modification. However, this process also retards biodegradation after discard, so that the balance between usefulness and biodegradability is rather precarious [46] Oxo-biodegradable plastics by contrast degrade under the influence of specific aspects of the outdoor environment - notably by light, heat, and mechanical action, whereas hydro-biodegradable plastics based on cellulose or starch are relatively resistant to these influences.

The bioassimilation of synthetic carbon-chain polymers has much in common with that of their natural analogues (notably natural rubber, resins and lignin). In all cases, nature uses abiotic oxidation chemistry together with biotic chemistry, very often together. The degradation products formed by oxo-biodegradation are of benefit to the agricultural environment as biomass and ultimately in the form of humus. Carbon is retained in the land carbon sink during oxo-biodegradation in a form accessible to growing plants, and is not therefore eliminated to the environment as carbon dioxide immediately as is the case with hydro-biodegradable polymers (e.g. pure cellulose, starch and aliphatic polyesters). The time-scale for complete oxo-bioassimilation of the synthetic polyolefins is very similar to that for their natural analogues such as *cis*(polyisoprene) and lignocellulose, the structural material of plants.

Time control of biodegradation of the synthetic carbon-chain polymers is achieved by antioxidants that behave similarly to naturally occurring antioxidants present in lignin and tannin [6]. The abiotic processes that lead to the peroxidation of the hydrocarbon polymers involve prooxidant transition metal ions that are analogous to the oxygenase and peroxidase enzymes used by nature in the oxo-biodegradation of rubber and lignin. Consequently the oxidation products (notably the carboxylic acids) are rapidly absorbed a nutrient by biological organisms.

# References

[1] Scott G (1999) "Management of Polymer Wastes", *Polymers and the Environment*, Royal Society of Chemistry Paperbacks, Chapter 4.

[2] Gilead D and Scott G (1982) 'Time Controlled Stabilisation of Polymers' in Scott G, ed., *Developments in Polymer Stabilisation*-**5,** Appl. Sci. Publ., Barking, Chapter 4.

[3] Scott G (1999) 'Polymers in modern life', *Polymers and the Environment*, Royal Society of Chemistry Paperbacks, Chapter 2.

[4] Scott G (1965), *Atmospheric Oxidation and Antioxidants*, Elsevier, pp. 379-407.

[5] Scott G (1993) in *Atmospheric Oxidation and Antioxidants*, 2nd edition, Vol. I, Elsevier, Chapter 1.

[6] Scott G (2002) 'Degradation and Stabilisation of Carbon-Chain Polymers' in Scott G, ed., *Degradable Polymers: Principles and Application,*2nd edition, Kluwer Acad. Pub., Amsterdam, Chapter 3.
[7] Scott G (1999) 'Antioxidant Control of Polymer Biodegradation' in *Degradability,Renewability and Recycling; 5th International Scientific Workshop on Biodegradable Plastics and Polymers*, Macromolecular Symposia, Eds.,Albertsson A-C, Chiellini E, Feijen J, Scott G and.Vert M, Wiley-VCH, Weinheim, 1999, 113-125.
[8] Grassie N and Scott G (1985) 'Antioxidants and Stabilisers', Cambridge University Press, Chapter 5.
[9] Karlsson S, Hakkarainen M and Albertsson A-C (1997) 'Dicarboxylic acids and ketones formed in degradable polyethylenes by zip- depolymerisation through a cyclic transition state', *Macromolecules*, **30**, 7721-7728.
[10] Li S and Vert M (2002) 'Biodegradation of aliphatic polyesters' in Scott G *Degradable Polymers: Principles and Applications*, 2nd Edition, (Kluwer Acad. Pub.), Chapter 5.
[11] Scott G (1997) 'Chain-breaking and Preventive antioxidants', *Antioxidants in science, technology, medicine and nutrition,* Albion Chemical Science Series, Chapters 3,4.
[12] Scott G (1965) *Atmospheric Oxidation and Antioxidants,* Elsevier, Chapters 4, 5, 8-10.
[13] *Atmospheric Oxidation and Antioxidants,* 2nd edition, Vol. II, ed. G. Scott, Elsevier, Chapters 1-9.
[14] Scott G (1999) 'Evironmental Stability of Polymers', *Polymers and the Environment,* Royal Society of Chemistry Paperbacks, Chapter 3
[15] Al-Malaika s and Scott G (1983) "Thermal stabilisation of Polyolefins" in *Degradation and Stabilisation of Polyolefins*, ed. N.S. Allen, Chapter 5.
[16] Al-Malaika s and Scott G (1983) "Photostabilisation of Polyolefins" in *Degradation and Stabilisation of Polyolefins*, ed. N.S. Allen, Chapter 7
[17] Scott G (1983) "Peroxidolytic Antioxidants: Sulphur Antioxidants and Autosynergistic Stanilisers based on Alkyl and Aryl sulphides" in *Developments in Polymer Stabilisation*, ed. G.Scott, Appl. Sci. Publ., Chapter 2.
[18] Scott G (1983) "Peroxidoytic Antioxidants: Metal Complexes containing Sulphur Ligands" in *Developments in Polymer Stabilisation*-**6**, ed. G.Scott, Appl. Sci. Publ., Chapter 3.
[19] Scott G (1999) "Biodegradable Polymers", *Polymers and the Environment*, Royal Society of Chemistry, Chapter 5.
[20] Scott G and Wiles D M (2002) 'Degradable hydrocarbon polymers in waste and litter control' in *Degradable Polymers: Principles and Applications*, 2nd Edition, Scott. G, ed., Kluwer Acad. Pub., Chapter 13.
[21] Scott G (1974) "New Polymer Compositions", *UK Patent Specification* **1**,356,107; *US Patent* **4**,121,025 (1978).
[22] Gilead D and Scott G (1984) "Controllably Degradable Polymer Compositions", *US Patent* **4**,461,853.
[23] Scott G and Gilead D (1990) "Controllably and Swiftly Degradable Polymer Compositions", *US Patent* **4**,939,194.
[24] Amin M U and Scott G (1974) *Europ. Polym. J.,* **10**, 1019-1028.

[25] Gilead D (1995) 'Photodegradable Plastics in Agriculture' in Scott G and Gilead D, eds., *Degradable Polymers: Principles and Application,* 1st edition, Chapman & Hall (Kluwer Acad. Pub.), Chapter 10.

[26] Bonhomme S, Cuer A., Delort A-M, Lemaire J, Sancelme M and.Scott G, (2003) 'Environmental biodegradation of polyethylene', *Polym. Deg. Stab.,* **81**, 441-452.

[27] Scott G (1993) "Antioxidants: Chain-breaking Mechanisms" in *Atmospheric Oxidation and Antioxidants*, 2nd Edition, Vol. I ed., ed. G.

[28] Scott, Elsevier, Chapter ? 28. Scott G (1984) "Stable Radicals as Catalytic Antioxidants in Polymers" in *Developments in Polymer Stabilisation*-**7**, Elsevier Appl. Sci. Publ., Chapter 2.

[29] Arnaud R, Dabin P, Lemaire J, Al-Malaika S,.Chohan S, Coker M, Scott G,.Fauve A and Maarooufi A (1994) 'Photooxidation and Biodegradation of Commercial Photodegradable Polyethylenes' *Polym. Deg. Stab.*, **46**, 211-224

[30] Harlan G and Kmiec C (1995) 'Ethylene-carbon monoxide copolymers' in Scott G and Gilead D, eds., *Degradable Polymers: Principles and Application,* 1st edition, Chapman & Hall (Kluwer), Chapter 8.

[31] Griffin G J L (1994) 'Particulate starch-based products' in Griffin G J L *Chemistry and Technology of Biodegradable Polymers*, Ed. Griffin J G L, Blackie Academic and Professional, London.

[32] Jakubowicz I (2003) 'Evaluation of biodegradable polyethylene (PE), *Polym. Deg. Stab.*, **80**, 39-43.

[33] Pandey J K and Singh R P (2001) 'UV-irradiated biodegradability of ethylene-propylene copolymers, LDPE and I-PP in composting and culture environments', *Biomacromolecules*, **2**, 880-885.

[34] Sadrmohaghegh C, Scott G and Setudeh E (1985) 'Recycling of mixed plastics', *Polym. Plast. Tech. Eng.,* **24**, 149-185.

[35] Ikram A, Alias O and Napi D (2000) 'Biodegradability of NR gloves in soil', *J. Rubb. Res.,* **3** 104-114.

[36] Ikram A, Alias O, Bahri A R S, Fauzi M S and Napi D (2001), 'Effects of added mitrogen and phosphorus on the biodegradation of NR gloves in soil', *J. Rubb. Res.,* **4**, 102-117.

[37] Heisey R M and Papadatos S (1995) 'Isolation of microorganisms able to metabolise purified natural rubber' *App. Environ. Microbiol*, **61**, 3092- 3097.

[38] Lee B, Poletto A L, Fratzke A and Bailey T B (1991), *Am. Soc. Microbiol.,* **57**, 678-685.

[39] Fabbri A (1995) 'The role of Degradable Polymers in Agricultural Systems'in Scott G and Gilead D, eds., *Degradable Polymers: Principles and Application,* 1st edition, Chapman & Hall (Kluwer Acad. Pub.), Chapter 11.

[40] Yang S-R (2000) Personal communication, Tainan Agricultural Improvement Station, Taiwan.

[41] Unpublished Report (1999) Yorkshire Dales National Park.

[42] Kiawai F, Shibata M, Yokoyama S, Maeda S, Tada K and Hayashi S (1999) "Biodegradability of Scott-Gilead Photodegradable Polyethylene and Polyethylene wax by microoganisms" in *Degradability,Renewability and Recycling; 5th International Scientific Workshop on Biodegradable Plastics and Polymers*, Macromolecular

Symposia, Eds.,Albertsson A-C, Chiellini E, Feijen J, Scott G and.Vert M, Wiley-VCH, Weinheim, 1999, 78-84.

[43] Yang S-R and Wu C-h (1999) "Degradable plastic films for agricultural applications in Taiwan" in *Degradability, Renewability and Recycling, 5$^{th}$ International Scientific Workshop on biodegradable Plastics and Polymers, Macromolecular Symposia* **144**, Eds., A-C. Albertsson, E. Chiellini, J. Feijen, G. Scott and M. Vert., Wiley-VCH, 101-112.

[44] Eggins H O W, Mills J, Holt A and Scott G (1971) 'Biodeterioration and biodegradation of synthetic polymers' in Sykes G and Skinner F A, *Microbial Aspects of Pollution*, Academic Press, London and New York, pp 267-277.

[45] Sadun A G, Webster, T F and Commoner B (1990) *Breaking down the degradable plastics scam,* for Greenpeace, Washington D.C.

[46] Scott G and Wiles D M (2001) 'Programmed-life plastics from polyolefins: A new look at Sustainability' *Biomacromolecules*, **2**, 615-622.

[47] *Green Report'* (1990) Report of a task force set up by the Attorneys General of the USA to investigate 'Green marketing'

[48] European Union (1994) 'Packaging and Packaging Waste Directive'

[49] Scott G (2002) 'Science and Standards' in Chiellini E and Solaro R, eds., *Biodegradable Polymers and Plastics*, Kluwer Academic/Plenum Pub., 3- 32.

[50] Műller R-J (2003) 'Biodegradability of polymers: Regulations and methods for testing' in Steinbűchel A, *Biopolymers* Vol. 10, Wiley – VCH, pp. 365- 392.

[51] ISO (1999) 'Determination of the ultimate aerobic biodegradability of plastic materials in an aqueous medium – Method by measuring the oxygen demand in a closed respirometer' ISO 14851.

[52] ISO (1999) (International Standards Organisation) 'Determination of the ultimate aerobic biodegradability of plastic materials in an aqueous medium - Method of analysis of evolved carbon dioxide' ISO 14852.

[53] CEN EN 13432 (2000) 'Packaging – Requirements for packaging recoverable through composting and biodegradation – Test scheme and evaluation criteria for the final acceptance of packaging..

[54] ISO 4892-1 Plastics - *Methods of exposure to light sources – Part 1*: General guidance.

[55] ISO 4892-2 Plastics – *Methods of exposure to laboratory light sources – Part 2*: Xenon-arc sources.

[56] ISO 4892-3 Plastics – *Methods of exposure to laboratory light sources – Part 3*: Fluorescent UV lamps.

[57] ISO 4892-4 Plastics – *Methods of exposure to laboratory light sources – Part 4*: Carbon arc sources.

[58] ASTM D 5071- 99 – *Standard practice for operating xenon- arc type exposure apparatus with water for exposure of photodegradable plastics*

[59] ASTM D 5208-01 – *Standard practice for operating fluorescent Ultraviolet (UV) and condensation apparatus for exposure of photodegradable plastics.*

[60] ASTM D 5510-94 – *Standard practice for heat aging of oxidatively degradable plastics.*

[61] ASTM D5272-92 – *Standard practice for outdoor exposure of photo- degradable plastics.*

[62] Davis A and Sims D, (1983) *Weathering of Polymers*, App. Sci. Pub., Barking, Chapters 3,4,8..

[63] CEN EN 14045 (2003) "Packaging – Evaluation of the disintegration of packaging materials in practical oriented tests under defined composting conditions",

[64] CEN EN 14046 (2003) 'Packaging – Evaluation of the ultimate aerobic biodegradability and disintegration of packaging materials under controlled composting conditions – Method by analysis of released carbon dioxide.

[65] Innocenti F D (2002) 'Biodegradability and compostability" in Chiellini E and Solaro R, eds., *BiodegradablePolymers and Plastics*, Kluwer Academic/Plenum Pub., Chapter 2, 33-45.

[66] European Commission Directorate E, Industry and the Environment (2000) 'Comment on prEN 13432; Organic recovery'

[67] "The role of land carbon sinks in mitigating global climate change", Royal Society Policy Document 10/01, July 2001 (see www.royalsoc.acc.uk).

[68] Wiles D M, Tung J-F, Cermak B E, Hare C W J. and Gho J G (1990) in P*roceedings of the Biodegradable Plastics 2000 Conference*, Frankfurt, June 6 & 7.

[69] Cassalicchio G, Bertoluza A and Fabbri A (1990) *Plasticulture*, 86, 21-28.

[70] Food Standards Agency Expert Group on Vitamins and Minerals (2003), *Risk Assessment*

[71] Chiellini, E, Corti A and. Swift G (2003)'Biodegradation of thermally oxidised, fragmented low-density polyethylenes', *Polym. Deg. Stab.*, **81**, 341-351.

[72] Scott G (2002) 'Why biodegradable polymers' in *Degradable Polymers: Principles and Application,* 2nd Edition, ed. G. Scott, Kluwer Acad. Pub., Amsterdam Chapter 1.

[73] Guillet J (2002) 'Plastics and the environment' in *Degradable Polymers: Principles and Application,*2nd edition, in G. Scott, Editor, Kluwer Acad. Pub., Chapter 12.

[74] Patel M (2003) 'Do Biopolymers fulfil our expectations concerning environmental benefits' in *Biodegradable Polymers and Plastics*, Eds. Chiellini E and.Solero R, Kluwer Acad. Pub., Chapter 7.

[75] Linos A and Steinbűchel A. (1998) 'Microbial degradation of natural and synthetic rubbers by novel bacteria belonging to the genus *Gordon*', *Kauchuk Gummi Kunstsoffe,* **51**, 496-499.

# INDEX

## A



## B

## D



## E

## F



## G



## H

## I



## J



## K

## L



## M



## N

## O



## P

## Q



## R

## W



## X



## Z